Springer Texts in Business and Economics

More information about this series at
http://www.springer.com/series/10099

Martin Kolmar

Principles of Microeconomics

An Integrative Approach

 Springer

Martin Kolmar
Institute for Business Ethics
University of St. Gallen
St. Gallen, Switzerland

ISSN 2192-4333   ISSN 2192-4341 (electronic)
Springer Texts in Business and Economics
ISBN 978-3-319-57588-9   ISBN 978-3-319-57589-6 (eBook)
DOI 10.1007/978-3-319-57589-6

Library of Congress Control Number: 2017951848

Printed on acid-free paper

This Springer imprint is published by Springer Nature
The registered company is Springer International Publishing AG
The registered company address is: Gewerbestrasse 11, 6330 Cham, Switzerland

To my daughter,
Carlotta

Acknowledgements

One may wonder why I think that it makes sense to add yet another introductory textbook to the overfilled shelf of well-established books on microeconomics. There are three reasons that motivated me to do so.

First, a lot of textbooks in economics want to make one believe that the theories presented are more or less context-free and objective. This is a wrong and dangerous belief. First of all, all theories are embedded in an intellectual milieu from which they borrow and on which they build. No man is an island, and no scientific theory is either. The tendency to shun any contextualisations of the theories comes at the risk of blindness towards the implicit assumptions, value judgements and epistemes on which the theory depends. This makes economics prone to being misused for ideological purposes. Economic literacy does not only mean that one is able to understand the rules and patterns of modern economies, but also that one understands how economic theories relate to other social sciences and the culture from which they emerge. This textbook is an attempt to contextualize modern economics in the hope that students will get a better overview of its strengths and weaknesses. It puts also a specific emphasis on case studies that range widely from the functioning of coffee markets, the logic of overfishing, to price discrimination in the digital age. This approach makes this book also potentially interesting for students who study economics as a minor and who want to understand how economic theories relate to other social sciences and how they can be used to better understand markets as well as phenomena like climate change, among many others.

To make it easier to identify the most important contextualizations in this book, I work with a series of icons that one will find in the margins of this book. $\mathcal{L}$ indicates a legal, $\mathcal{B}$ a business, and Φ a philosophical (broadly speaking) context. Furthermore, one will find the most important definitions and technical terms highlighted with a ✍-sign in the margins of the book.

Second, textbooks that give an introduction to economics have become a million-dollar business over the last decades, with thousands of universities and colleges teaching the same basic principles worldwide. The globalization of this market has led to a commodification of textbooks in the attempt to sell as many copies as possible. As a result, the lion's share of the market is served by textbooks that are very elementary and only scratch the surface of most theories. This strategy makes them commercially successful, because of the appeal to the mass market but, at the same time, denies the students a deeper and more sophisticated understanding of the strengths and weaknesses of the theories. One could argue that such an in-depth understanding of theories is not necessary in an introductory class in economics, because there will be plenty of intermediate and advanced courses that will fill these gaps later on. The reality is, however, that a significant fraction of students gets all of its knowledge from the introductory course. It is never too early to educate independent and critical minds.

Third, most textbooks that I am aware of are not tailored to the needs of a business school where students study economics, business administration and maybe law. Economics is about the functioning of institutions and most institutions have a legal backbone. Bringing this fact to the foreground creates synergies between law and economics. By the same token, economic theory allows one to identify the key parameters that a firm must know in order to be successful in the markets in which they compete. Examining the common ground between management and economics allows one to better understand the implications of different market contexts and industries for managers and it shows one how closely economics and business administration can and should be linked. Economics, law and business administration are really three perspectives on the same phenomenon: the logic of social interaction.

This book took shape over many years during which I have been teaching "Principles of Economics" and "Microeconomics" to undergraduate students. I would like to thank all of my former students for their patience and for their countless discussions that all contributed, in their own ways, to this book. Special thanks are due to my present and former PhD students and research assistants Philipp Denter, Magnus Hoffmann, Hendrik Rommeswinkel and Dana Sisak, all of whom had a major influence on the content and the didactics of this book. This is also true for Thomas Beschorner, Friedrich Breyer, Claudia Fichtner, Jürg Furrer, Michael Heumann, Normann Lorenz, Ingo Pies, Alfonso Sousa-Poza and Andreas Wagener, who gave me me detailed feedback on earlier versions of the book and helped me with valuable suggestions. I would also like to thank Maya G. Davies, Corinne Knöpfel, Leopold Lerach, Jan Riss and Jan Serwart, who supported me in finishing this book and who did a great job in making it more student friendly and accessible. It is definitely not their fault if you find yourself struggling with some of the material.

St. Gallen, May 2017   Martin Kolmar

Contents

Part I Introduction

1 First Principles 3
  1.1 What Is Economics About? 3
  1.2 Some Methodological Remarks 10
    1.2.1 True and Reasonable Theories 12
    1.2.2 Theories and Models 13
    1.2.3 The Virtue of Thriftiness 14
    1.2.4 Do Assumptions Matter? 15
    1.2.5 An Example 16
    1.2.6 Critical Rationalism 17
    1.2.7 Positive and Normative Theories 18
    1.2.8 Schools of Economic Thought 20
  References 23

2 Gains from Trade 25
  2.1 Introduction 25
  2.2 An Example 28
  2.3 How General is the Theory of Comparative Advantage? 32
  2.4 Comparative Advantage and the Organization of Economic Activity 38
  References 41

Part II A Primer in Markets and Institutions

3 Introduction 45
  3.1 General Remarks 45
  3.2 Taxonomy of Markets 49
  References 53

4   Supply and Demand Under Perfect Competition   55
   4.1   Introduction   55
   4.2   Determinants of Supply and Demand   56
   4.3   Equilibrium   69
   4.4   Equilibrium Analysis   73
   References   82

5   Normative Economics   83
   5.1   Introduction   83
   5.2   Normative Properties of Competitive Markets   87
   5.3   Is One's Willingness to Pay One's Willingness to Pay?   94
   References   97

6   Externalities and the Limits of Markets   99
   6.1   Introduction   99
   6.2   Transaction Costs 102
      6.2.1   An Example 103
      6.2.2   Analysis of Externalities on Markets 107
      6.2.3   The Bigger Picture 114
   6.3   Four Boundary Cases 129
   References 140

Part III   Foundations of Demand and Supply

7   Decisions and Consumer Behavior 145
   7.1   Basic Concepts 145
      7.1.1   Choice Sets and Preferences 146
      7.1.2   Indifference Curves 150
      7.1.3   Utility Functions 154
   7.2   Demand on Competitive Markets 158
      7.2.1   Graphical Solution 160
      7.2.2   Analytical Solution 163
      7.2.3   Three Examples 168
      7.2.4   Comparative Statics and the Structure of Market Demand . 176
      7.2.5   Changes in Income 177
      7.2.6   Changes in Price 178
   References 181

8   Costs 183
   8.1   What Are Costs, and why Are They Important? 183
   8.2   A Systematic Treatment of Costs 188
   References 197

Part IV   Firm Behavior and Industrial Organization

9   A Second Look at Firm Behavior Under Perfect Competition 201
  9.1   Introduction 201
  9.2   The Relationship Between Production Technology and Market
         Structure 204
  9.3   The Short Versus the Long Run 209
  9.4   Firm and Market Supply 216
  References 220

10   Firm Behavior in Monopolistic Markets 221
  10.1   Introduction 221
  10.2   Conditions for the Existence of a Monopoly 222
  10.3   Profit Maximization in Monopolistic Markets 225
  10.4   Monopoly Without Price Discrimination 226
  10.5   Price Discrimination 232
         10.5.1 First-Degree Price Discrimination 232
         10.5.2 Second-Degree Price Discrimination 238
         10.5.3 Third-Degree Price Discrimination 245
  10.6   Monopolistic Competition 249
  References 254

11   Principles of Game Theory 255
  11.1   Introduction 255
  11.2   What Is a Game? 256
  11.3   Elements of Game Theory 257
  11.4   Normal-Form Games 259
         11.4.1 Multiple Equilibria 266
         11.4.2 Collectively and Individually Rational Behavior 269
         11.4.3 Simple Games as Structural Metaphors 271
  11.5   Extensive-Form Games 273
  11.6   Summary 276
  References 279

12   Firm Behavior in Oligopolistic Markets 281
  12.1   Introduction 281
  12.2   Cournot Duopoly Model 286
  12.3   The Linear Cournot Model with n Firms 290
  12.4   The Bertrand Duopoly Model 291
  12.5   Conclusion and Extensions 294
  References 300

Part V   Appendix

13   A Case Study 303
   13.1 The Grounding of Swissair as a Case Study for the Use of
        Economic Theory 303
   13.2 Some Facts About the Aviation Industry in Europe 305
   13.3 Applying Economic Theory to Understand the Aviation Industry . 306
        13.3.1 Costs 307
        13.3.2 The Linear Cournot Model with n Firms 308
        13.3.3 Extension: Network Choice, Acquisitions, and Strategic
               Alliances 313
   13.4 How About Swissair? 316
   13.5 Concluding Remarks 318

14   Mathematical Appendix 319
   14.1 Functions with Several Explanatory Variables 320
   14.2 Solution to Systems of Equations 325
   14.3 Elasticities 326

Index 331

First Principles

1

This chapter covers . . .

- enough philosophy of science to be able to have a qualified opinion about the status and scientific role of economic theories.
- the basic paradigm of economics: to understand the functioning of societies as an adaption to the underlying principle of scarcity.
- why economics considers itself a methodology, and not a field of application.
- why the opposite of positive is no longer negative but normative.
- the relevance of opportunity costs.
- how to think and make decisions like an economist.

1.1 What Is Economics About?

> Economics is the science which studies human behavior as a relationship between ends and scarce means which have alternative uses. (Lionel Robbins)

If one looks at economics departments all over the world, one may be surprised to see what economists are doing. Of course, economists deal with "the economy," but modern economics is extremely diverse and covers a wide range of fields, which few laymen would intuitively associate with economics. Here is a list of examples: economists deal with the "big old" questions about the sources of growth and business cycles, poverty and the effects of unemployment, or the effects of monetary policy on the economy. More generally, they want to find out how markets allocate goods and resources and how markets have to be regulated in order to make sure that they function properly. An important field of research is the economic role of the government: the ways it can levy taxes and provide services. However, economists also deal with problems related to political institutions, like the effects of different voting systems, the causes and consequences of political and military conflicts, or the relationship between different levels of government. They are involved in evolutionary biology, the design of products on financial markets, auctions, and internet

© Springer International Publishing AG 2017
M. Kolmar, *Principles of Microeconomics*, Springer Texts in Business and Economics,
DOI 10.1007/978-3-319-57589-6_1

market platforms; they work with lawyers to understand the consequences of legal rules and cooperate with philosophers.

The reason for this diversity of fields stems from the evolution of the modern definition of the science of economics. Economics is not the science that studies "the economy:" it is not defined by an object of study. Instead, it defines itself by a particular perspective from which it tries to make sense of the social world: scarcity. Paul A. Samuelson (1948), one of the most influential economists of the 20th century, provided what is still the most concise definition of economics: "Economics is the study of how men and society choose, with or without the use of money, to employ scarce productive resources which could have alternative uses, to produce various commodities over time and distribute them for consumption, now and in the future among various people and groups of society." This definition may not be as elegant as the one by Lionel Robbins, but it has the advantage of larger concreteness: economists try to understand how resources are used to alleviate scarcity. Economics is therefore a scientific method: economists start with the premise that it is possible to understand the logic of individual behavior and collective action as a result of scarcity. This is why the above list of examples covers such a broad array of fields. Whenever one has the conjecture that scarcity plays a role in the functioning of a situation, economists can be brought on board.

But what is scarcity? *Scarcity* refers to situations where the wants exceed the means. In economics, the wants are usually restricted to human wants, and means includes the resources and goods that contribute to fulfilling these wants. The reference to wants implies that scarcity has its origin in human physiology as well as psychology. The human metabolism requires a certain intake of energy in order to function and, if food intake falls below a certain threshold, human beings cannot develop and will eventually become sick and die. These physiological wants can be called objective, and their fulfilment is indispensable for life. However, a lot of wants are not of this type. Fast cars, big houses, and fancy clothes are not necessary for healthy survival, but are merely pleasant. These wants can be called subjective. Economics is the study of how individuals and societies manage goods and resources, which can be objectively as well as subjectively scarce.

Digression 1. Increasing Means or Increasing Autonomy?
Economics has no monopoly on scarcity as a starting point for the scientific endeavor. Philosophies like Buddhism start from a similar premise, although phrased in a different terminology. The first two of the so-called "Four Noble Truths" state that (1) *dhukka* exists and (2) that it arises from one's attachment to desires. Dhukka is often translated as suffering, but this blurs its meaning. It refers to misaligned desires and needs or, in other words, scarcity.

It is interesting to see, however, that the impulse that resulted from this same premise points in opposite directions. Most "Western" economists try to find out how scarcity can be alleviated by *increasing means* (through technological progress, growth, etc.). The intuitive reaction to the phenomenon of scarcity points *outwards*: increasing the means to fulfill the given wants. This impulse is even reflected in the idea of individual freedom that is, by and large, conceptualized in the Western tradition as political freedom: as the absence of external compulsion.

On the contrary, the reaction to scarcity in Buddhism points *inwards*: overcoming the wants to make them match the means. To see this, consider the two other noble truths: (3) suffering ceases when attachment to desire ceases and (4) freedom from suffering is possible by practicing the Eightfold Path. Freedom, according to this view, is interior freedom: autonomy from the "dictatorship" of desires. One sees the same starting point, but two completely different conclusions.

Scarcity immediately leads to one of the most powerful tools of economics: the concept of *opportunity cost*. If one makes decisions under the conditions of scarcity, then going one way necessarily implies that one cannot go another way. On the other hand, the other way looks interesting, as well, so deciding to go this way incurs a cost, in this sense. To be a little more specific, assume that one has to choose an alternative a from a set of admissible alternatives A and assume further that one can rank the admissible alternatives according to the joy and fulfillment that one is expecting to experience when one chooses them. If a_1 is the best and a_2 is the second-best alternative, according to this measure, then the opportunity costs of choosing a_1 is the joy and fulfillment that one would have expected to enjoy from alternative a_2.

This sounds rather abstract, but it need not be. The concept of opportunity cost allows one to better understand how one makes decisions and how one should make decisions (this distinction will be discussed in more depth later). If one goes to the movies, one cannot go to a restaurant; if one spends one's money on a new car, one cannot afford an expensive trip to Japan; if one studies Economics, one cannot, at the same time, study Physics; and so on. In order to make the right decisions, one should be aware of the value one attaches to the other alternatives that one cannot realize. The value one connects to the next-best alternative forgone is the opportunity cost of one's choice.

Digression 2. Generosity for Nerds: Opportunity Costs and Donations
The concept of opportunity costs is helpful when considering the consequences of any kind of behavior. To illustrate this point, I would like to focus on a recent trend called *effective altruism*, sometimes also called "generosity for nerds." Effective altruism seeks to maximize the good from one's charitable donations. Here is an example that illustrates the problem. Assume that one graduates and wants to make a great difference in the world by devoting one's career to doing something good. A lot of students with this type of mo-

tivation consider careers at Oxfam, or some other charity. However, this may not be the smartest idea. Assume that one would earn CHF 50,000 with a job at a charity and could be replaced by some other graduate student, who does an equally good job. Now, assume that one considers a career at a major bank, where one would earn CHF 120,000 instead and then gives CHF 70,000 away to charity. This decision creates CHF 70,000 that can be used for doing good. In fact, it finances the position at Oxfam and still leaves CHF 20,000 for other charitable purposes. If the person replacing one at Oxfam does not have this career option, it is better if one works for the bank, even if it seems to contradict one's intentions of devoting one's life to doing good. (But please, make sure to actually donate the money!)

The importance of this example is not the career advice that it provides, but the principle that can be elicited from it. Consider a simple version of this problem in which one wants to donate a certain amount of money and wants to make sure that it does as much good as possible. Effective altruism makes the point that one should think in terms of opportunity costs when one makes one's decision: what are the alternative uses for one's money and how much good could be done with the different uses? One should then spend one's next Swiss Franc in a way that would maximize the additional good that the money can create.

This idea of donations may look like economics on steroids but, in fact, it is an important regulative idea to alleviate suffering. There is a lot of evidence, for example, that donations are highly irrational. Disaster relief following earthquakes and tsunamis is a good example. These events are horrible and create a lot of human suffering. However, media attention often creates "superstar effects," where people want to help and thereby crowd out other needs. In the end, earthquake-relief programs end up with more money than they can usefully spend to alleviate the suffering from the earthquake. To illustrate, if everyone spends a fixed amount of money on charitable projects, then one additional Swiss Franc for earthquake relief reduces the money that is available for less prominent (but equally urgent) projects. Some charities are aware of this problem and want to use part of the earmarked donations for other projects, but they are often criticized for doing so, because the people want to make sure that their money is spent "in the right way." On the other hand, what is the right way to spend their money? If saving an additional life in the earthquake region is expected to cost CHF 50,000 and it will likely cost CHF 10,000, if the money is spent on malaria prevention in some low-key project, then it may make sense to apply economic principles to save as many lives as possible. Thinking in terms of opportunity cost allows for a more rational allocation of scarce resources from a utilitarian perspective.

When economists study social phenomena, they usually distinguish between three different levels of analysis:

- **Individual level:** The individual level focuses on the question of how individual people behave under conditions of scarcity. A typical question is, for example, how a person spends his or her income on consumption. Will she go to the movies or to dinner? Will she spend her income on clothes or travel? Decision-theory is the field of research that develops theories about individual behavior.
- **Interaction level:** Typically, individual behavior does not take place in isolation. If A decides to go to the movies tonight, and B would like to meet A, then B must go to the movies, as well. Equilibrium models of trade and market behavior or game theory are examples of fields of research that investigate how human beings interact with each other.
- **Aggregate level:** Phenomena that are studied at the aggregate level are, for example, inflation, growth, or unemployment. They are a result of individual decisions and the rules that govern individual interactions, but an analysis of a certain phenomenon at the aggregate level usually abstracts from a lot of the details of individual decision-making and interactions in order to still be able to see the forest for the trees.

I have already mentioned that individual human beings ultimately cause all social phenomena at the aggregate or interaction levels. *Methodological individualism* is a scientific position that requires that all social phenomena be explained with reference to individual behavior. According to this view, it is not sufficient to assume that abstract laws exist, which explain, for example, growth and inflation: these laws must be derived from the behavior of individuals in a society. Methodological individualism is a widely recognized position among economists, according to which ultimately all phenomena, which are studied on the aggregate level, need to be traced back to patterns of behavior on the individual and interaction levels.

Digression 3. Homo Oeconomicus
Economics is infamous for a character that populates most of its tales: the *homo oeconomicus*. Any theory that explains social phenomena as a result of individual behavior needs a *decision theory* that allows for the making of predictions. The term *homo oeconomicus* summarizes a number of assumptions about the way individuals make the decisions that are used in mainstream economics to make predictions about behavior.

Different economists use the term differently, but there is a broad consensus that the minimum requirements are as follows: first, economists usually do not use the concept to explain the motivations that drive behavior in an exclusively descriptive way. This approach goes back to Vilfredo Pareto, John Hicks, Roy Allen, and Paul Samuelson, who eliminated psychological concepts from economics and based economic theory on principles of rational choice. The idea is that all one can observe are individual choices, but not the mental processes that motivate or cause behavior.

All one has to know is that people make decisions in a structured way that allows one to infer a so-called *preference relation* from the observed behavior of the individuals. This is the *revealed-preference approach* in economics, which makes the point that, if behavior follows certain consistency assumptions, then the individual behaves *as if* she maximizes her preferences or her utility function. Note that the formulation says "as if," which implies that the theory does not claim that individuals *have* preferences or utility functions somewhere in their heads. Pareto justified this approach in a letter from 1897: "It is an empirical fact that the natural sciences have progressed only when they have taken secondary principles as their point of departure, instead of trying to discover the essence of things. [...] Pure political economy has therefore a great interest in relying as little as possible on the domain of psychology."

The consistency assumptions, which guarantee that an individual behaves as if she maximizes preferences, are as follows: she can rank the alternatives from which she can choose according to some relation representing her preferences (if I have the choice between broccoli and potato chips, then I prefer broccoli to potato chips). This ranking is unique and stable over a sufficiently long period of time. Furthermore, the ranking is complete (I can rank any two alternatives) and transitive (if I prefer broccoli to potato chips and potato chips to ice cream, then I also prefer broccoli to ice cream). Such preferences are called an *ordering*. Last but not least, it is assumed that individuals always choose the best alternative that is available to them (*maximization*). The maximization of a preference ordering is the core of the *rational-choice paradigm*, which is integral to the concept of homo oeconomicus.

Please note that this view on rationality is purely instrumental: it refers to the consistency of a ranking and the relationship between ranking and behavior. (If preferences are inferred from behavior, then there is conceptionally no gap between behavior and preferences; the latter are a workaround to systemize choices.) It can be discussed whether completeness and transitivity capture the idea of rationality and if individuals always choose an alternative that is best for them, but both assumptions are considered indispensable for rational decision-making. The concept has been further refined to be able to cover choice situations under conditions of risk and uncertainty.

It is a widespread misunderstanding that homo oeconomicus is conceptualized as a selfish actor. Given that the above concept wants to eliminate any deliberations about motives for action from the theory, it cannot, in its purest form, say anything about selfishness, altruism, or fairness concerns, because these concepts refer to the individual's motives for taking action. Admittedly, a lot of authors added assumptions about the structure of a preference ordering that can be interpreted as selfishness to the theory, but it should be noted that selfishness is not an integral part of what economists conceptualize as rationality.

A detailed discussion of the concept of homo oeconomicus is beyond the scope of an introductory chapter, but some remarks are important. The different aspects of the concept have been subject to critique. Psychologists and behavioral economists have shown that preferences need not be transitive and that individuals do not consistently choose alternatives that are best for them (a statement that cannot even be made within the revealed-preference paradigm). Furthermore, people do not act selfishly in a number of situations. Despite these empirical anomalies, the concept is popular in economics. From a methodological point of view, it is important as a regulative idea that helps one to better understand the structure of limited rationality and non-selfish behavior, even if everyone agrees that real people often deviate from the ideal of rational decision-making. Boundedly rational behavior follows patterns and it is easier to explore these patterns with reference to the standard of full rationality. In addition, as will become clear throughout this chapter, good theories do not rely on "realistic" assumptions in a naïve understanding of the word. The predictive power of a theory that, for example, explains the behavior of prices in markets can be high, even if the underlying assumptions abstract from a lot of factors that may be important in reality.

Claiming scarcity as the exclusive paradigm of economics would be imperialistic and wrong without further deliberation. The definitions of economics by Robbins (1932) and Samuelson are vague about the exact role that human beings play in economic theory. Methodological individualism is what, for the most part, distinguishes mainstream economics from other sciences that are also built on the idea of scarcity.

Evolutionary biology is a good example. Evolution in populations is the result of three basic principles: (1) There are traits that are heritable, (2) there is variability in them, and (3) some traits are more adaptive than others, which implies that the organisms pass more copies of their genes on to the next generation. The crucial point is that a mismatch between means and ends must be underlying the "mechanics" of evolution, because otherwise traits could reproduce indefinitely. Therefore, to get the theory of evolution off the ground, one has to start from scarcity. What distinguishes evolutionary biology from economics is not the underlying paradigm, but the smallest unit of study, genes versus human beings, and it would be preposterous to declare evolutionary biology as part of economics (even though there is a subfield of economics called evolutionary economics that applies the above three principles to study economic phenomena and to lay the foundations for human behavior). Evolutionary biologists distinguish between *ultimate* and *proximate* causes. According to this view, the human brain, with its desires and preferences and, more generally, a human being is a proximate cause of behavior, shaped by forces of evolution. It is an organism that makes its genes more or less well-adapted to its environment. The laws of genetic evolution are, therefore, the ultimate causes of human behavior. Economists recognize that human brains are ultimately shaped by evolution, but nevertheless take the individual human being as the ultimate cause of behavior and as the starting point of their scientific endeavor. As shown in the next subchapter, this shortcut is neither right nor wrong, good nor bad: it merely simplifies the analysis.

Another common distinction is made between *microeconomics* and *macroeconomics*. Microeconomists study individual decision-making at the individual and interaction levels, whereas macroeconomists study economic phenomena at the aggregate level. The distinction is, however, not as sharp as it may seem. Traditional macroeconomics postulated regularities for aggregates like national savings, which were then used to predict the consequences of, for example, an increase in economic growth. One could find in the data that about 30% of national income Y goes into savings S, which would yield a savings function $S(Y) = 0.3 \cdot Y$. Combined with other regularities of this kind, such a function can be used, for example, for economic forecasting. The problem with this approach is, however, that it remains unexplained why, on average, 30% of national income goes into savings. It is not a theory that fulfills the requirements of methodological individualism. In the end, it is the single individual that makes savings decisions, so the implications of such aggregate models of the economy are more reliable if the behavior of aggregates is linked to individuals. This observation led to the so-called *micro-foundation* of macroeconomics, i.e. the attempt to relate aggregate phenomena, like unemployment or growth, to individual decision-making. Mainstream macroeconomics is micro-founded in this sense, so it is more appropriate to distinguish between micro and macro, as they are colloquially called, by their fields of application. Macroeconomists usually study phenomena like growth, unemployment, business cycles or monetary policy, whereas microeconomists focus on the functioning of markets and other institutions or the role and design of incentives in economic decision-making, among other things.

1.2 Some Methodological Remarks

> Everything should be made as simple as possible, but not simpler. (Attributed to Albert Einstein)

There are cookbooks and scientific theories. In a cookbook, one learns that it takes a hot pan, eggs, flour, milk, baking soda and a pinch of salt to make pancakes. If one follows the recipe, one ends up with a tasty meal, but one does not really understand why. A scientific theory tells one how heat changes the molecular structure of proteins present in egg white, how baking soda reacts with acids and how gluten builds elastic networks. This knowledge may not inform one about how to make a pancake, but it can tell one a lot about the deeper reasons why the recipe works. Moreover, one can use this information to develop new, innovative recipes. Both cookbooks and scientific theories complement each other: understanding the physical, chemical, and biological mechanisms underlying the transformation of in-gredients into meals helps one improve recipes, and the evolved recipes are a source of inspiration for scientific discoveries.

Economics comes in the form of both cookbooks and scientific theories. A stock-broker may just "follow his gut" about profitable picks. He has no explicit theories about the functioning of capital markets in the back of his mind, which inform him about the future development of stocks. Like an experienced cook, he just "feels" or "sees" which stocks will be profitable. Scientific reasoning would require try-ing to understand the mechanisms that make one stock successful and the other a failure. Alternatively, take the manager of a firm as an example. When she sets up the organization of the firm and the compensation packages for the employees, she might follow custom and her intuition. The scientific approach to organization and compensation would be to develop theories about the consequences or organiza-tional designs and the incentive effects of different ways to compensate employees. These theories might not be directly applicable to a specific problem but, over time, they feed into the "culture" of a society and shape the intuitions of decision-makers. John Maynard Keynes made this point quite poignantly: "The ideas of economists and political philosophers, both when they are right and when they are wrong are more powerful than is commonly understood. Indeed, the world is ruled by little else. Practical men, who believe themselves to be quite exempt from any intellec-tual influences, are usually slaves of some defunct economist."

Economics, as a social science, develops such theories. It is not the primary pur-pose of these theories to inform decision-makers about the consequences of their decisions, but rather to give one a better understanding of the logic of economic interactions by developing scientific theories. Deepening one's understanding of the functioning of, for example, labor or financial markets will, in the end, allow decision-makers to make better decisions, but this is only a byproduct of the en-deavor.

The following subchapters give a short introduction to some philosophical issues that are important for understanding economics as a *social science*. It is very dif-ficult to find the right point in time to discuss these fundamental issues because, if one discusses them before one starts studying economics, then one discusses issues that are still very abstract; one gives answers to questions that the students would not even have asked. If one covers the material at the end of a course, then stu-dents will have difficulty understanding what is going on in economic theories and it is very likely that they will be led astray in their ideas about their relevance and potential. The third option is also suboptimal: integrating the philosophical debate into the presentation of the theories. If one sprinkles economic theories with little bits of the philosophy of science here and there, it is very likely that one will miss the forest for the trees. (The fourth option would, of course, be to skip the material completely and to rely on the methodological and philosophical intuitions of the students.) This is why I have decided to cover some ground at the beginning of this book, with the idea that one need not understand every little detail after the first reading. It is good enough if one gets the basic concept in the beginning and then returns to this subchapter later on when one has a better understanding of economics. All the bits and pieces will fall into place eventually if one perseveres.

1.2.1 True and Reasonable Theories

✍ I used the term *scientific theory* in the last subchapter. For the purposes of this text, a *theory* is defined as a relatively broad conceptual approach that makes reasonable conjectures about causal relationships in the world.

When is a conjecture reasonable and why does one find the word "reasonable" instead of "true" in the above statement? Given the limitations of one's sense-organs and one's mind, it is impossible to say that a conjecture is true in the sense that it is in total accordance with reality. A nice way to briefly grasp the epistemic problems that come with a naïve idea of truth is a short elaboration of the so-called Münchhausen Trilemma. The basic problem is that scientific reasoning requires that one is prepared to provide proof for any of one's statements. However, such a proof can only be given by means of another statement, which must also be provable. The Münchhausen trilemma makes the point that one has the choice between exactly three unsatisfactory options to deal with this situation:

- **Infinite regress:** Each proof requires a further proof, *ad infinitum*. This process will, of course, never end, such that one never "breaks through" to the truth. It is, understandably, impossible to give an example for an infinite regress.
- **Circularity:** The statement and the proof support each other, maybe in a complex chain of arguments. An example is a flawed interpretation of the theory of evolution that defines the species that fits best in an environment as the one that survives, and then one argues that the species one observes must fit best into its environment.
- **Dogmatism:** One finally reaches a stage in the process of statement and proof, where the underlying assumptions have no further justification. A wonderful example of dogmatism is the second sentence of the US Declaration of Independence, even if it is not a scientific theory: "We hold these truths to be self-evident, that all men are created equal, that they are endowed by their Creator with certain unalienable Rights, that among these are Life, Liberty and the pursuit of Happiness."

For all practical intents and purposes, only dogmatism is an option. That means that truth cannot be achieved by a process of scientific reasoning, but necessarily relies on intuition that must be nurtured by other sources. Dogmatism, for the same reason, also implies that every scientific theory must start from *value judgments* about the basic, self-evident principles. Coming back to the discussion about theories, calling a conjecture reasonable bites the trilemma-bullet by requiring the much more moderate standard of being consensual. Ludwig Wittgenstein (1972, *94 and 110*) expressed this beautifully: "But I did not get my picture of the world by satisfying myself of its correctness; nor do I have it because I am satisfied of its correctness. No: it is the inherited background against which I distinguish between true and false. [...] As if giving grounds did not come to an end sometime. But the end is not an ungrounded presupposition: it is an ungrounded way of acting." In order to reach a consensus among experts, one must at least reach an agreement of the different dogmas that (perhaps subconsciously) taint one's own perspective.

Digression 4. Transcending Reason   Φ

The necessity of a dogmatic starting point of any scientific research project points towards the limits of language in expressing reality. Some spiritual traditions even claim that some truths can and must be assessed by means other than scientific reasoning – for example an act of revelation – and that reason is not a means of perceiving the truth but an obstacle on the way. This thought is most clearly expressed in Zen Buddhism, where the practice of meditation leads to a state of pure consciousness in which one sees the world "as it really is." In order to get closer to this state, students are expected to work on *kōans* which, from the point of view of a Western understanding, are unanswerable questions or meaningless statements. Working with them helps one to transcend reason, which is seen as an obstacle on the way to an enlightened state of being. The idea of enlightenment in Buddhism differs from the use of the word in the West, where it refers to the independent use of reason to gain insight into the true nature of this world.

The first dogmatic principles of a theory closely resemble what Thomas Kuhn called a *paradigm*. Paradigms are sets of practices that define a scientific discipline at any particular period of time. Paradigms come in at different levels of abstraction. The most general level of abstraction within the Western paradigm of Enlightenment is the belief that the independent use of *reason* allows one to gain insight into the true nature of this world (see the above digression). Neoclassical economics, for example, has the ideas of rational choice and methodological individualism as important parts of its paradigm.

1.2.2 Theories and Models

In mainstream economics, at least, theories have *models* as their "logical backbones." A model is a collection of assumptions and hypotheses that are linked by the rules of logic and mathematics. A model makes several assumptions about an aspect of reality and derives hypotheses from these assumptions in a logically consistent way. To understand the difference between theories and models, look at the following example.

Assume one wants to develop a theory about the functioning of the price mechanism on markets. In order to do so, one thinks about, for example, the way individuals sell and buy their stuff and how these buying and selling decisions explain the formation of prices. This structured way of thinking is one of the models underlying one's theory.

The crucial function of a model, in the context of theory-formation, is to make sure that the key causal mechanisms underlying a theory are made explicit and logically consistent. Look at the following model to understand why:

Model 1
Assumption 1: All human beings are in the streets.
Assumption 2: Peter is a human being.
Hypothesis: Peter is sitting at my home.

✍   "Model 1" is a model because it has a set of *assumptions* and a *hypothesis*, but the hypothesis does not follow logically from the assumptions. In this case, the model is logically inconsistent, even though the hypothesis might be correct empirically (Peter is sitting right next to me). The point is that the assumptions cannot explain my observation, which makes the model useless for any theory. A consistent model is therefore a *necessary* condition for a good theory.

However, is it also *sufficient*? Here is an example of a very simple model:

Model 2
Assumption 1: All the dead are looking over one's shoulder.
Assumption 2: Karl Marx is dead.
Hypothesis: Karl Marx is looking over one's shoulder.

The above set of assumptions and hypotheses fulfills all the requirements of a good model. All the assumptions are spelled out explicitly and the hypothesis follows from the assumptions in a logically consistent way. Would the model be a good ingredient of a *theory of the dead*? It is difficult to imagine that it would get the approval of many experts. Logical consistency is therefore of obvious importance for scientific reasoning, but it is not enough. In order to evaluate the "soundness" of a theory, one needs additional, "softer" criteria like adequacy, simplicity or plausibility.

1.2.3   The Virtue of Thriftiness

An important criterion for good models is simplicity, frugality or thriftiness. The idea is often referred to as Ockham's razor (named for an English Franciscan Friar in the 14th century), which states that, among competing models, the one with the fewest assumptions should be selected. However, this concept is much older. Aristotle (2004), in his *Posterior Analytics*, stated that, "we may assume the superiority ceteris paribus [all things being equal] of the demonstration which derives from fewer postulates or hypotheses." Ockham's razor is widely accepted among economists. Robert Solow (1997, p. 43) summarizes the self-image of the profession in a very concise way: "Today, if you ask a mainstream economist a question about almost any aspect of economic life, the response will be: suppose we model that situation and see what happens. [...] A model is a deliberately simplified representation of a much more complicated situation. [...] The idea is to focus on one or two causal or conditioning factors, exclude everything else, and hope to understand how just these aspects of reality work and interact."

Ockham's razor necessarily implies that the assumptions of a model should not be realistic in the naïve sense that the assumptions shall fit reality. Scientific theory-building necessarily reduces complexity to make a situation comprehensible for the human mind. Joan Robinson (1962) found a nice expression for the problems implied by models built on "realistic" assumptions: "[a] model which took account of all the variegation of reality would be of no more use than a map at the scale of one to one." However, the epistemic problem goes even deeper, as illustrated by the novel *Tristram Shandy* by Laurence Sterne (2003). The book is the autobiography of the protagonist, which is so detailed that it takes the author one year to write down a single day of his life. From this perspective, the map is even more detailed than the territory and the level of detail one considers adequate must be based on a subjective value-judgment.

Maps have to simplify in order to be useful. On the other hand, is there a "right" way to simplify? The answer to this question must also be "no," because it depends on what one wants to do with the map. If one is driving a car, contour lines are not essential and may easily distract attention from other, more important information. However, if one is planning to hike in the mountains, contour lines are crucial. Therefore, a good simplification depends on its purpose.

1.2.4   Do Assumptions Matter?

If assumptions shall not be realistic, then maybe one can conclude that assumptions do not matter at all. This position has, in fact, been put forward by Milton Friedman (1953, p. 14), one of the most influential economists of his time. He proposed that "Truly important and significant hypotheses will be found to have 'assumptions' that are wildly inaccurate descriptive representations of reality, and, in general, the more significant the theory, the more unrealistic the assumptions (in this sense)."

There is some debate as to whether Friedman adheres to the extreme position that assumptions do not matter at all (called *instrumentalism*) or not but, for the sake of argument, consider this position and see where it leads. According to an instrumentalist's view, one should judge a theory according to the validity and usefulness of the *hypotheses*, whereas the assumptions are irrelevant. Does this position make sense? Look at the following model.

Model 3
Assumption: Seatbelts reduce the likelihood of fatal accidents.
Hypothesis: Seatbelts reduce the likelihood of fatal accidents.

Model 3 looks like a pretty nonsensical waste of time and is an example of a circular argument, but why does one find it intuitively unconvincing? The hypothesis can be empirically tested and it has been confirmed by the data. Therefore, according to an instrumentalist's view, a theory that is built on this model passes the test of usefulness. The idea that assumptions are completely irrelevant is, of course, flawed because it prevents one from learning anything about the causal mechanisms that *drive* the hypotheses, if one cannot rule out the trivial model where hypotheses and assumptions coincide. Even if one's mind can never grasp the true causal mechanism, and thus one has to be satisfied with crude narratives and heuristics, declaring the assumptions irrelevant leaves one with only cookbooks.

Instrumentalism is an extreme position and there are reasons to assume that Friedman's own position is more balanced. He argues that the role of a positive science "is the development of a 'theory' or 'hypothesis' that yields valid and meaningful (i.e. not truistic) predictions about phenomena not yet observed." It can be argued that the term "truistic" refers to models of the above type that are only uninteresting tautologies. In the end, scientific theory-building has a subjective component, because the balance between, on the one hand, meaningful simplifications of the assumptions and of the supposed causal mechanisms and, on the other hand, the explanatory power of the hypotheses cannot be precisely nailed down. It is the art and craft of experienced scientists to see if a theory is "in balance" in this sense.

1.2.5 An Example

To illustrate the role assumptions play in models, I will introduce the concept of the *production-possibility frontier* that will reappear in Chap. 2. In a modern, complex economy, there are millions of people, who all go to work, consume goods, enjoy their friends and families, and so on. This maze of interactions would be impossible to analyze without simplifying assumptions that make it comprehensible to the scientist's mind. One question, which is relevant to economists, is about the trade-offs the economy faces when it produces goods and services. Goods and services are produced by all kinds of resources, using tools and skills. The production-possibility frontier abstracts from all these complexities. In the simplest case, one makes the assumption that the economy can produce exactly two goods, 1 and 2, whose quantities are drawn along the axes of Fig. 1.1.

The quantity of good 1 is drawn along the abscissa (horizontal axis) and the quantity of good 2 along the ordinate (vertical axis). The downward-sloping graph in the figure is the production-possibility frontier for goods 1 and 2. The graph shows the various combinations of the two goods that can be produced in the economy, in a given period of time. This illustration of production possibilities relies on drastically simplifying assumptions, but it has two great advantages: it is easy to grasp and it allows for analyzing some of the basic trade-offs a society faces. The graph must be downward-sloping, because scarcity implies that an increase in the production of one good must have opportunity costs: an increase in the produc-

Fig. 1.1 An example for a
production-possibility frontier

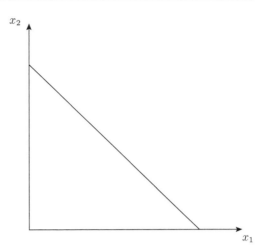

tion of one necessitates a reduction in the production of the other good. The slope
of the function is a measure of the opportunity costs, because it measures by how
much the production of one good must be reduced, if one produces an additional
(small) unit of the other good. It remains to be shown how useful this tool actually
is. The purpose of this simple model is to illustrate what the role of simplifying
assumptions means.

1.2.6   Critical Rationalism

The workhorse epistemology in economics is *Critical Rationalism*, which has been
advocated by Karl Popper, among others. According to this view, scientific theories can never be verified (for reasons that have been discussed before) but can,
in principle, be falsified by bringing in empirical evidence that is in conflict with
the hypotheses of the theory. Hence, theories should be formulated such that their
hypotheses are falsifiable, which means that it has to be possible to give empirically
accessible conditions under which the theory is considered incompatible with observations. Good theories are those that have a large empirical content, but have not
been falsified so far. In addition, Ockham's razor is also an integral part of Critical
Rationalism.

Critical Rationalism leads to a back-and-forth between theoretical and empirical
reasoning: the (empirically) falsifiable hypotheses of models must be empirically
tested. These tests can either falsify the theory or not. Non-falsified theories are
preliminarily accepted, if no other non-falsified theory exists that explains more
cases with less restrictive assumptions. If the theory is falsified, the insights from
the process of falsification can be used to modify the models.

A weak spot of Critical Rationalism is its unscrutinized belief in empirical falsification. The problem is that one does not have direct proof of facts; that any

empirical observation relies on a theory, as well. A lot of statistical data, for example, is collected within a highly complex system of national accounting and data generated in lab experiments depends on the interpretation of the experiment. Therefore, from an epistemic point of view, falsification is no more or less than the proof of the logical inconsistency of two different theories: a theoretical theory and an empirical theory. Which one will be refuted relies, again, on value judgments or on experts' "gut feelings."

This problem is an example of the much more profound problem known as *underdeterminacy of scientific theory* or the *Duhem-Quine problem*. According to Quine (1951), "[t]he totality of our so-called knowledge or beliefs, from the most casual matters of geography and history to the profoundest laws of atomic physics or even of pure mathematics and logic, is a man-made fabric which impinges on experience only along the edges. Or, to change the figure, total science is like a field of force whose boundary conditions are experience. A conflict with experience at the periphery occasions readjustments in the interior of the field. But the total field is so underdetermined by its boundary conditions, experience, that there is much latitude of choice as to what statements to reevaluate in the light of any single contrary experience. No particular experiences are linked with any particular statements in the interior of the field, except indirectly through considerations of equilibrium affecting the field as a whole."

Some philosophers of science, like Imre Lakatos (1976), draw the conclusion that underdetermination makes scientific "progress" largely a function of the scientist's talent, creativity, resolve and resources. Even more radical work, by Thomas Kuhn (1962), suggests that ultimately the social and political interests determine the conclusions one draws from the inconsistencies within or between theories. In order to take this sting out of the program of scientific reasoning, one must conclude that the best one can hope for is to create a level playing field for scientific debate, where the success of an argument is not influenced by money or power, and a consensus among experts is reached by a "non-hierarchical discourse" (Jürgen Habermas (1983)).

1.2.7 Positive and Normative Theories

Theories come in two flavors: *positive* theories are used to explain a phenomenon. With the exception of the underlying "dogmatic" first principles, they contain no value judgments. They generate insights into the causal mechanisms linking causes and effects. Positive economics, therefore, tries to explain how people deal with the phenomenon of scarcity. Statements about "what is" are also called *descriptive*.

Normative theories, on the other hand, make *prescriptions* about what people should do under certain circumstances. They, therefore, rely on a value judgment. Statements like "you should lose some weight" or "Switzerland should reduce its corporate income-tax rate" are normative statements. Whether one considers them relevant or not depends on two things. First, one has to share the basic normative principles underlying the advice ("living longer is better than living shorter", or "Switzerland should maximize its national income"). Second, one has to agree with the positive theories bridging normative principles and normative advice ("overweight people live, on average, shorter than lean people because they have a higher risk of cardiovascular diseases", or "lower corporate-income tax rates attract capital investments, which increase national income"). Amartya Sen (1970) calls the first class of value judgments *basic* and the second *nonbasic*. Basic value judgments only depend on first ethical principles, whereas nonbasic value judgments are an amalgam of first principles and positive theories.

The distinction between basic and nonbasic value judgments is important for political debates, because it is possible to debate about positive statements. The claim that lower tax rates increase investments which, in turn, increase national income, can be empirically tested, so people can, in principle, settle disputes about positive statements (in the sense of consensus among experts in a non-hierarchical discourse). The mainstream view, however, claims that this is not the case for basic value judgments. According to David Hume, there is a qualitative difference between descriptive and prescriptive statements. Prescriptive statements, according to this widely accepted view, are not facts of life that can be discovered by scientific reasoning. Here is the famous passage where Hume (1739) makes his point: "In every system of morality, which I have hitherto met with, I have always remarked, that the author proceeds for some time in the ordinary ways of reasoning, and establishes the being of a God, or makes observations concerning human affairs; when all of a sudden I am surprised to find, that instead of the usual copulations of propositions, is, and is not, I meet with no proposition that is not connected with an ought, or an ought not. This change is imperceptible; but is however, of the last consequence. For as this ought, or ought not, expresses some new relation or affirmation, 'tis necessary that it should be observed and explained; and at the same time that a reason should be given, for what seems altogether inconceivable, how this new relation can be a deduction from others, which are entirely different from it. But as authors do not commonly use this precaution, I shall presume to recommend it to the readers; and am persuaded, that this small attention would subvert all the vulgar systems of morality, and let us see, that the distinction of vice and virtue is not founded merely on the relations of objects, nor is perceived by reason."

Likewise, George Edward Moore (1903) coined the term *naturalistic fallacy* for the category error someone commits when he defines the "good" (in the sense of intrinsically valuable) by certain properties of things. It could, for example, be that pleasant things are "good" things, and that pleasure should therefore be the basis for one's value judgments (a position called hedonism). Others may argue that meaningful things are "good" things, and society should therefore provide the necessary means to promote meaningfulness. The point of the naturalistic fallacy, as Moore describes it, is that, even if one agrees to a certain position, one cannot define the idea of goodness with reference to these properties, because it is no natural property. It is "one of those innumerable objects of thought which are themselves incapable of definition, because they are the ultimate terms by reference to which whatever is capable of definition must be defined." (Principia Ethica, §10)

Scientific reasoning cannot prove the goodness of things or acts; goodness can only reveal itself in, for example, a moment of bliss or awe. The fact that ethical insights lie beyond scientific reasoning is also clearly expressed by Wittgenstein

(1998) in that "there can be no ethical propositions. Propositions cannot express anything higher. [...] It is clear that ethics cannot be expressed. Ethics is transcendental. [...] There are, indeed, things that cannot be put into words. They make themselves manifest."

The *fact-value dichotomy* has not remained undisputed, however. As shown, dogmatism is the only practical solution to the Münchhausen Trilemma. However, dogmatism implies that the distinction between facts and values is not as clear as was envisioned by Hume. Apparently, value-free facts are tainted by value judgements about first principles (Hilary Putnam 2002). The opposite point of view is the position that ethical sentences express propositions, which refer to objective features of the world. It is called *moral realism*.

The fact-value dichotomy is widely accepted by economists, which has important consequences for economics as a social science, because it constrains the role of the economist to that of an expert (hopefully) in descriptive statements. Economists can clarify the effects of changes in the tax system, monetary policies, or labor market policies like minimum wages, to name only a few. However, the economist is no expert in the basic normative question, of what the members of society should want. The division of labor between economists and the general public is, according to this view, that the general public articulates what it wants (in terms of first principles) and that the economist uses her toolbox to figure what to do.

This restricted role of economists sounds pleasantly humble and innocuous but, in fact, it is not necessarily so. First of all, and at a very profound level, even positive theories have a normative core that defines the acceptable practices (remember the problem of dogmatism discussed before). However, from a more practical point of view, the division of labor outlined above does not exist in practice. Usually, average people – and even politicians – have very opaque and conflicting ideas about their normative principles. As a result, they mix conflicting emotions and narratives into an amalgam of ideas that Sen (1970) has called nonbasic value judgments. Economic advisors have to somehow fill the resulting gaps, which necessarily grants them authority in normative matters that they should not have. There is not much that one can do about that, but an awareness of this fact can be very helpful. In addition, it should be a unanimously accepted principle of scientific hygiene that economists, who are asked for advice in a situation of unclear normative principles, disclose and actively communicate the normative premises of their work. Otherwise, one crosses the border from scientist to ideologist.

1.2.8   Schools of Economic Thought

The majority of the theories discussed in this book stem from two different schools of thought: *neoclassical economics* and *new institutional economics*. Despite the fact that neoclassical economics is the mainstream school of thought and taught at most universities around the world, the underlying paradigm is far from uncontroversial. The purpose of this subchapter is, therefore, to give a short overview over these, as well as other schools of thought, to better understand the paradigms and to put them into perspective.

Neoclassical economics is not a monolithic theory with undisputed first princi- ✍ ples. Despite its heterogeneity and versatility, some underlying unifying principles can be identified: (1) methodological and normative individualism, (2) consequentialism (and, more specifically, welfarism, an ethical theory that we will discuss in Chap. 5), (3) rational or rationality-seeking agents, and (4) society as a network of mutual transactions that follow the logic of opportunity costs. These basal axioms are enriched by other, more specific assumptions. Neoclassical economics is especially dominant in microeconomics, but it also developed into macroeconomics where, together with Keynesian economics, it forms the so-called *neoclassical synthesis*. Keynesianism was initially a fundamental critique of some of the implications of neoclassical thinking (like the neutrality of money) but was later integrated into the neoclassical theory (at the price of changing Keynes' initial theory beyond recognition, as some Keynesians would stress).

As the name suggests, neoclassical economics emerged from *classical economics*, which is also called political economy. The main differences between classical and neoclassical economics represented a shift in attention regarding the most relevant economic problems and in the underlying theory of value.

Classical economics originated at a time when capitalism was gradually replacing feudalism and innovations were fueling the Industrial Revolution that was completely changing society. One of the most pressing problems, in such a period of change, was how society could be organized, if every individual seeks his or her own advantage. This is why the idea that free markets have the ability to regulate themselves was of such profound importance, because it expressed the belief that a decentralized society, built on the principles of self-interest, can work. Important proponents of this school of thought were Adam Smith, Jean-Baptiste Say, David Ricardo, Thomas Malthus and John Stuart Mill.

In addition, the classical economists had a focus on economic growth and production. This is why they shared the view that the economic value of a product depends on the costs involved in producing that product. This theory of value is one of the key differences between classical and neoclassical thinking, which replaced this comparatively more objective standard of value with a subjective one that is based on utility. The idea is that economic value does not stem from any objective property of a good (like the amount of labor that went into its production), but from the importance the good has for the achievement of the individuals' goals.

A second distinctive feature of early neoclassical thinking is what is sometimes called the "marginalist revolution" in economics. When individuals make decisions, they think in terms of trade-offs and they think "at the margin," which means that they compare the satisfaction they get from an additional unit of a good with the costs of this additional unit. Market prices then reflect these marginal exchange rates. Thinking "at the margin" allowed neoclassical thinking to resolve puzzles that resulted from objective theories of value, like the fact that water is more important than diamonds, but the price for diamonds is higher than the price for water.

While water has greater total utility, diamonds have greater marginal utility, which is relevant for prices.

The neoclassical approach started to replace classical economics in the 1870s. Important early proponents of neoclassical economics were William Stanley Jevons, Carl Menger, John Bates Clark, and Léon Walras. They were followed by Alfred Marshall, Joan Robinson, John Richard Hicks, George Stigler, Kenneth Arrow, Paul Anthony Samuelson and Milton Friedman, to name only a few.

Institutional economics emphasises that economic transactions are embedded in a complex network of culture, norms and institutions. The functioning of, for example, markets, according to this view, cannot be understood in isolation, which brings this school of thought into sharp contrast with neoclassical economics and it is considered a heterodox school of thought. However, an important variant of this school evolved in the second half of the 20th century: *new institutional economics*. It has its roots in two articles by Ronald Coase. They made it clear that transaction costs are at the heart of an institutional analysis, which allows for the characterization of the relative merits of markets, firms or the government in achieving the normative goals of society. New institutional economics is critical of neoclassical economics, because of its focus on markets and rationality, but uses and builds on elements of this theory. Therefore, it does not contrast with, but instead complements neoclassical thinking. Important figures in this field are Armen Alchian, Harold Demsetz, Douglass North, Elinor Ostrom and Oliver Williamson, among others.

There are, of course, other schools of economic thought, and it would be beyond the scope of this book to do justice to them all and to show their relationships with neoclassical and new institutional economics. Some influential schools of the past have declined in influence, including the *historical school of economics* (for example, Gustav von Schmoller, Etienne Laspeyres, Werner Sombart), *Marxian economics* (for example, Karl Marx, Antonio Gramsci), *Austrian economics* (for example, Carl Menger, Eugen von Böhm-Bawerk, Ludwig von Mises and Friedrich von Hayek, some of whom are better classified as neoclassical economists), and *institutional economics* (for example, Thorstein Veblen, John R. Commons, and John Kenneth Galbraith), and are now often considered heterodox approaches. However, there are also more recent developments like *feminist economics* (represented by, for example, Marylin Waring Marianne Ferber or Joyce Jacobson) and *ecological economics* (represented, for example, by Herman Edward Daly or Nicholas Georgescu-Roegen). Up until now, however, these schools mostly criticize specific aspects of mainstream economics without developing independent schools.

It is unclear how other, more recent developments, like evolutionary economics, behavioral economics and neuroeconomics will relate to the neoclassical paradigm. As demonstrated by the neoclassical synthesis, which integrated a specific interpretation of Keynesian with neoclassical thinking, neoclassical economics proved to be extremely versatile in the past, adapting to changes and adopting approaches that started as a critique of the mainstream.

References

Aristotle (2004). *Posterior Analytics*. Whitefish: Kessinger Publishing.

Franklin, B., Adams, J., Sherman, R., Livingston, R., & Jefferson, T. (2015). *United States Declaration of Independence*. CreateSpace Independent.

Friedman, M. (1953). *Essays in Positive Economics*. University of Chicago Press.

Habermas, J. (1983). *Moralbewusstsein und kommunikatives Handeln*. Frankfurt a. M.: Suhrkamp.

Hume, D. (1739)[2004]. *A Treatise of Human Nature*. Dover Publications.

Kuhn, T. (1962). *The Structure of Scientific Revolution*. University of Chicago Press.

Lakatos, I. (1976). *Proofs and Refutations*. Cambridge University Press.

Moore, G. E. (1903). *Principia Ethica*. Cambridge University Press.

Pareto, V. (1897). The New Theories of Economics. *Journal of Political Economy*, 5(4), 485–502.

Putnam, H. (2002). *The Collapse of the Fact/Value Dichotomy and Other Essays*. Harvard University Press.

Quine, W. V. O. (1951). Main Trends in Recent Philosophy: Two Dogmas of Empiricism. *The Philosophical Review*, 60(1), 20–43.

Robbins, L. (1932). *An Essay on the Nature and Significance of Economic Science*. Auburn: Ludwig von Mises Institute.

Robinson, J. (1962). *Economic Philosophy*. Aldine Transaction.

Samuelson, P. A. (1948). *Economics: An Introductory Analysis*. McGraw-Hill.

Sen, A. (1970). *Collective Choice and Social Welfare*. San Francisco: Holden Day.

Solow, R. (1997). Is there a Core of Usable Macroeconomics We Should All Believe In? *American Economic Review*, 87(2), 230–232.

Sterne, L. (2003). *The Life and Opinions of Tristram Shandy*. Penguin Classics.

Wittgenstein, L. (1972). *On Certainty*. Harper and Row.

Wittgenstein, L. (1998). *Tractatus Logico-Philosophicus*. Dover Publications.

Further Reading

Albert, H. (2014). *Treatise on Critical Reason*. Princeton University Press.

Hausman, D. (Ed.). (1994). *The Philosophy of Economics: An Anthology*. Cambridge University Press.

Kincaid, H., & Ross, D. (Eds.). (2009). *The Oxford Handbook of Philosophy of Economics*. Oxford University Press.

Reiss, J. (2013). *Philosophy of Economics a Contemporary Introduction*. Routledge.

Simon, H. A. (1957). *Models of Man: Social and Rational*. John Wiley.

Gains from Trade

2

This chapter covers ...

- the application of the concept of opportunity costs toward understanding why the process of specialization and trade is potentially beneficial.
- why the principle of comparative advantage is crucial to an understanding of why societies organize economic activities and to the economic role of institutions.
- why institutions matter and what an economic theory of institutions has to explain in order to provide a better understanding of societal phenomena like growth, unemployment, globalization, or anthropogenic climate change.

2.1 Introduction

What is prudence in the conduct of every private family, can scarce be folly in that of a great kingdom. If a foreign country can supply us with a commodity cheaper than we ourselves can make it, better buy it of them with some part of the produce of our own industry, employed in a way in which we have some advantage. The general industry of the country, being always in proportion to the capital which employs it, will not thereby be diminished [...] but only left to find out the way in which it can be employed with the greatest advantage. (Adam Smith, The Wealth of Nations, Book IV: 2)

Under a system of perfectly free commerce, each country naturally devotes its capital and labour to such employments as are most beneficial to each. This pursuit of individual advantage is admirably connected with the universal good of the whole. [...] It is this principle which determines that wine shall be made in France and Portugal, that corn shall be grown in America and Poland, and that hardware and other goods shall be manufactured in England. (David Ricardo (2004))

Economics has always been an underdog among the sciences, with the aspiration to play in the same league as the natural sciences but, at the same time, lacking the general theories and insights that characterize, for example, modern physics.

© Springer International Publishing AG 2017
M. Kolmar, *Principles of Microeconomics*, Springer Texts in Business and Economics,
DOI 10.1007/978-3-319-57589-6_2

This fact is nicely expressed by an exchange between Paul Samuelson, one of the most influential economists of the 20th century and Nobel price winner, and the mathematician Stanislaw Ulam, who once challenged Samuelson to "name me one proposition in all of the social sciences which is both true and non-trivial," obviously expecting the question to be left unanswered. Apparently it took Samuelson several years before he found one: the theory of comparative advantage. "That it is logically true need not be argued before a mathematician; that it is not trivial is attested by the thousands of important and intelligent men who have never been able to grasp the doctrine for themselves or to believe it after it was explained to them" (Samuelson 1969). It should, therefore, come as no surprise that this theory is still at the heart of economics and it is very useful for understanding how societies cope with the problem of scarcity. Perhaps surprisingly, after all these years, the theory of comparative advantage still gives rise to misunderstandings and ideology-tainted controversies. It is the purpose of this chapter to illustrate the basic insight of this theory, its implications not only for economics but also for business administration and law, as well as the potential fallacies one is prone to when applying the theory.

Ricardo developed the theory of comparative advantage to explain why it is mutually beneficial for nation states to allow for international trade. His famous example is trade between England and Portugal. This focus on international trade was obvious in the political atmosphere of the days when powerful political forces in England opposed free trade, because they feared that they could not compete with Portugal. At the same time, it planted the seed for potential misunderstandings and misinterpretations, as will be discussed below. This is why I approach the theory from a different angle where individuals rather than states contemplate exchanging goods and services.

A fundamental aspect of societies, which must cope with scarcity, is that the acts of individuals are interdependent. If person A eats a sandwich, then B cannot eat the same sandwich; if B wears a red sweater, then A has to look at it, and so on. This interdependency can explain the phenomenon of exchange and specialization. The starting point of this endeavor is a situation where all individuals in a society abstain from trading goods and services, i.e. an individual must produce everything she consumes. This situation is called an *autarky*.

The theory of comparative advantage is developed here by means of an intuitive example: assume there are two individuals, Ann (A) and Bill (B), and two commodities, pears (P) and tomatoes (T). Ann and Bill are initially in a situation of autarky.

- **Case 1:** Each individual can produce exactly one and wants to consume only this good. This is the trivial benchmark case when autarky is, in fact, a perfectly adequate way to organize economic activities. Interactions between Ann and Bill cannot reduce scarcity because neither person has goods or services to offer that the other person wants. The only challenge imposed by scarcity is self-management: how should Ann and Bill organize their days such that they can consume enough pears and tomatoes? If this situation were an adequate and exhaustive description of reality, the economic journey would end before it really got started, because there would be no need to think about how people organize interactions and develop institutions to alleviate scarcity. Human beings would live happily from the yields of their own gardens. Fortunately for social scientists, this is not what the world is about.

- **Case 2:** Each individual can produce exactly one good, but wants to consume both. This case gives a first explanation for *trade* between the individuals. Ann can grow only tomatoes in her garden and Bill only pears, but they prefer to have both for lunch and dinner. In this case, it is obvious that it makes sense for them to establish a "trade agreement" that specifies how many pears and tomatoes shall be exchanged between A and B. If this scenario were to adequately describe the world, one should observe trade.

- **Case 3:** Each individual can produce and wants to consume both goods. However, one individual is better at producing tomatoes (Ann) and the other at producing pears (Bill). In this situation, each individual has an *absolute advantage* in the production of exactly one good. (One can interpret case 2 as a special case of case 3 where the absolute advantage of each individual is pushed to its extreme.) In this situation, Ann and Bill do not necessarily rely on each other, if they want to have tomatoes together with pears for lunch, but cooperation can make life easier. With absolute advantages in the production of goods, it makes sense if individuals *specialize and trade*. The total number of pears and tomatoes that is available, if Ann specializes in the production of tomatoes and Bill in the production of pears, increases. However, specialization and trade are two sides of the same coin. Assume that the production of pears and tomatoes in an autarky reflects Ann's and Bill's "taste" for both goods (economists call these tastes the individuals' *preferences*) such that they are the most preferred bundles they can produce, given their production possibilities. Deviating from this production plan and specializing, according to the absolute advantage, makes them worse off without trade. Given that specialization increases total production, it is always possible to guarantee both individuals their autarky consumption and still leave a surplus.

- **Case 4:** Each individual can produce both goods and wants to consume both. One individual is better at producing both goods, tomatoes and pears. This is the critical case, which required Ricardo's ingenuity to understand that both, Ann and Bill, can be better off by specialization and trade. The intuition as to why this is the case is built around the idea of scarcity itself. Assume A has an absolute advantage in the production of pears and tomatoes. With unlimited resources, she could easily outperform B. However, resources are not unlimited. Assume that both individuals spend all of their work time producing either P or T. In this case, the only way for A to produce more T is by reducing the production of P, because she has to devote more of her scarce time to the production of T, which leaves less for P. Assume that the rate of transformation is 2 to 1, i.e. a reduction of one unit of P increases the production of T by two units. Given that her autarky production was optimal for her needs, such a change is not advisable. At this point, B can make his appearance. Assume that A and B team up and arrange that B compensates the loss of P by producing less T. Remember that his rate of transformation is 1 to 1. To compensate for the increase of P by one unit, he therefore has to reduce the production of T by one unit. However, if one does the math, one can see how the apparent magic of comparative advance works: the change in the total quantity of pears is -1 (by A) $+1$ (by B) $= 0$, so the total amount of pears remains unchanged. How about tomatoes? The change in total the quantity of tomatoes is $+2$ (by A) -1 (by B) $= +1$: total production goes up, because Bill has a *comparative advantage* in the production of pears. Even though Ann is twice, or even a hundred times, more productive than Bill in absolute terms, she is still subject to resource (time) constraints. Hence, as long as Ann and Bill differ in their rates of transformation of the two goods at the point where the resource (time) constraints become binding (their opportunity costs), there is room for mutually beneficial specialization and trade.

To illustrate the point, compare Jonathan Ive, the chef designer of Apple Inc., with me. I am doing OK as an economics teacher, but I am a pretty bad designer. Jonathan Ive is a brilliant designer and, I assume, he would also be a very good economics teacher. Thus, he is better than me in both respects: he has an absolute advantage in designing computers and in teaching economics. I can sympathize with the position that a student would like to be educated by a very good teacher, not only a decent one, but does it make sense from an overarching perspective? If Jonathan Ive were to teach economics, he would not be able to use the time to develop new designs for the next generation of whatever kind of gadget is in Apple's pipeline. Thus, one basically has the choice between two alternatives: learning economics from the best teacher available, but using computers that are, well, interestingly designed, or learning economics from an OK teacher but with well-designed computers. Alternatively, consider another perspective. It is very likely that I would be unemployed as a designer. If I replace Jonathan Ive as an economics teacher, then I am better off, because I make at least some money. He would, however, be profoundly better off because, despite the fact that he loses the money he would make as professor, he makes much more money as Apple Inc.'s head of the design group.

Looking at it from this perspective, the theory of comparative advantage is very comforting for people like me: you think that you are nothing special? Do not worry, even if there are people out there who outperform you in every possible aspect of life, they have only a limited amount of time. That is your chance.

2.2 An Example

The next step is to develop a more precise understanding of the theory of comparative advantage by specifying the production possibilities of Ann and Bill. Assume that both have a total of 100 hours that they can spend on the production of either

Table 2.1 Productivities and production possibilities for Ann and Bill

	Time for 1 kilo of tomatoes	Time for 1 kilo of pears
A	1 hour	1 hour
B	2 hours	4 hours

	Maximum quantity of tomatoes	Maximum quantity of pears
A	100 kilos	100 kilos
B	50 kilos	25 kilos

pears or tomatoes. Table 2.1 gives an overview of the productivities of the two and the implied maximum production levels. The table shows that A is, in fact, more productive than B: she needs half the time to produce a kilo of tomatoes and one quarter of the time to produce a kilo of pears. Assume that A and B can divide their time freely between the production of both goods and that productivities are constant. With this assumption, one can analyze the example by means of the concept of the production-possibility frontier that was introduced in Chap. 1.

Figure 2.1 shows the production-possibility frontiers for A and B. x_T^A, x_P^A, x_T^B, x_P^B denote the quantities of T and P for A and B. The functional forms can be defined as:

$$x_T^A = 100 - x_P^A, \quad x_T^B = 50 - 2 \cdot x_P^B.$$

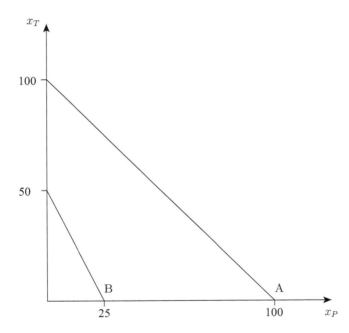

Fig. 2.1 Ann's and Bill's Production-Possibility Frontiers

Table 2.2 Opportunity costs for Ann and Bill		OC tomatoes (in terms of pears)	OC pears (in terms of tomatoes)
	A	1	1
	B	0.5	2

The total production of pears (in kilos) is drawn along the abscissa and the total production of tomatoes (in kilos) is drawn along the ordinate. The *absolute* advantage of A in the production of both goods is reflected in the fact that her production-possibility frontier lies to the north-east of B's. Note, however, that the *slopes* of both frontiers differ, which will be crucial for the identification of a *comparative* advantage.

In order to get there, one starts with the determination of the opportunity costs of production (OC). The opportunity cost of an additional kilo of, for example, tomatoes is the reduction in the production of pears, which is necessary due to the reallocation of time from pear to tomato production. It is equal to the slope of the production-possibility frontier (in absolute terms). Alternatively, one can determine the opportunity cost of pears in terms of tomatoes, which is equal to the inverse slope of the production-possibility frontier (in absolute terms). Table 2.2 summarizes these costs.

A comparison of the opportunity costs allows one to identify the comparative advantage of Ann and Bill. Figure 2.2 shows the increase in the production of

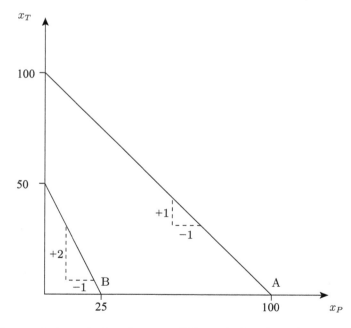

Fig. 2.2 Increase in the production of tomatoes, if the production of pears is reduced by one kilo

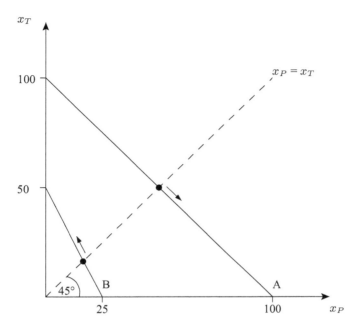

Fig. 2.3 Specialization according to Ann's and Bill's comparative advantage

tomatoes, if Ann and Bill are willing to reduce the production of pears by one kilo. It is one kilo for Ann and half a kilo for Bill. This observation shows that it is *relatively* easier for Bill to increase the production of tomatoes: an additional kilo costs him half a kilo of pears, whereas Ann would have to reduce the production by one kilo. Hence, Bill has a *comparative advantage* in the production of tomatoes.

The concept of comparative advantage is *relational*, because the fact that Bill has a comparative advantage in the production of tomatoes *necessarily* implies that Ann has a comparative advantage in the production of pears: if she reduces the production of tomatoes by one kilo, she gets an additional kilo of pears, whereas Bill only gets 500 grams.

After identifying the comparative advantages of Ann and Bill, one can set up an example that illustrates how specialization, according to comparative advantage, can increase total production. In order to do so, assume that both individuals want to consume pears and tomatoes in equal quantities. The 45°-line in Fig. 2.3 denotes the set of most preferred combinations of pears and tomatoes (consumption bundles).

The consumption levels in autarky are given at the intersections of the 45°-line and the production-possibility frontiers. Analytically, they are the solution to the two systems of equations $x_T^A = 100 - x_P^A \wedge x_T^A = x_P^A$, and $x_T^B = 50 - 2 \cdot x_P^B \wedge x_T^B = x_P^B$. The solutions are $x_T^A = x_P^A = 50$ and $x_T^B = x_P^B = 50/3$. The important question is if both Ann and Bill can be made better off, if they specialize according to their comparative advantages and exchange goods. Table 2.3 shows

Table 2.3 The effects of specialization, according to comparative advantages

	Change in tomatoes	Change in pears
A	$-3/4$ kg	$+3/4$ kg
B	$+1$ kg	$-1/2$ kg
$A+B$	$+1/4$ kg	$+1/4$ kg

how individual and total production changes, if A produces more pears and B more tomatoes. As predicted by the theory of comparative advantage, specialization can increase the total production of both tomatoes and pears. It is completely irrelevant that B is less productive than A in everything he can do. The only thing that matters is that they differ in their opportunity costs.

The increase in production is called the *material gains from trade*, which one distinguishes from the *subjective gains from trade* or, simply, *gains from trade*. What is the difference? The material gains from trade measure the increase in total production. The production of material goods is, however, only a means and not the end of economic activities. What ultimately counts is what material goods can do for people, how they contribute to their well-being, and the term 'gains from trade' refers to a measure of this increase in subjective well-being.

At this point, one has to ask two different questions.

- First, it is important to understand how comprehensive the result is. Is it an artifact of the above model or does it hold under general conditions?
- Second, if it is a general result, then one has to ask what it implies for an economy. Are gains from trade exploited automatically or must societies organize economic activity in a specific way to ensure that the potential gains from trade will in fact be exploited?

The next two subchapters are dedicated to the discussion of these two topics.

2.3 How General is the Theory of Comparative Advantage?

A peculiar feature of the above model is the linearity of the production-possibility frontier. In this case, comparative advantage is a well-defined *global* concept and the results are completely general: with the exception of the limiting case of equal opportunity costs, there is always a way to increase production by specialization. There may, however, also be cases where production possibilities are more accurately described by a strictly concave (outward-bending) frontier. In this case, comparative advantage is no longer a global, but rather a local concept and it depends on the autarky points along the production-possibility frontier. Production-possibility frontiers are concave, if productivity is decreasing in production. Figure 2.4 illustrates such a situation and the possibility for the reversal of comparative advantages.

Points X and Y represent two possible autarky situations. The slopes of the frontiers are a measure for *local* opportunity costs. As one can see, A has a comparative advantage in the production of T in X and a comparative advantage in the

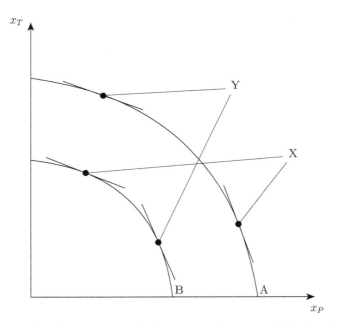

Fig. 2.4 Comparative advantage with a strictly concave production-possibility frontier

production of P in Y. Beyond that, however, there is no difference from the model with constant opportunity costs: if opportunity costs in autarky differ, there is room for mutual improvement by specialization and trade.

If the frontier can be concave, it could also be convex (inward-bending). Production-possibility frontiers are convex, if productivity is increasing in production. Figure 2.5 illustrates this case.

Assume that one starts in a situation of autarky X, where neither A nor B have a local comparative advantage. Even in this case, it makes sense to specialize and trade, because specialization allows them to increase productivity. In this case, specialization creates a comparative advantage that was nonexistent in autarky. This is shown with point Y in Fig. 2.5 in which it is assumed that A and B completely specialize.

The theory of comparative advantage, therefore, seems to be robust with respect to the laws governing production. In this sense it is, in fact, completely general. If this were the end of the story, however, there should be no resentment towards the process of globalization, which seems to be shared by many people. It can often be read that globalization does not only create winners but also losers and that it is, therefore, a lie to claim that everyone is better off. If, in the words of Adam Smith, "the extent of this division [of labor] must always be limited by the extent [...] of the market," and if *division of labor* is a good thing, the process of global market integration must be a good thing, as well. The only explanation as to why skeptics are skeptics is that they do not get it.

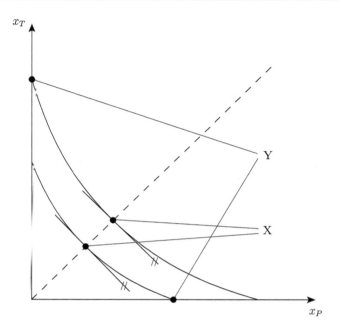

Fig. 2.5 Comparative advantage with a strictly convex production-possibility frontier

However, this may not be the case. I claimed, at the beginning of this chapter, that there is an important difference in whether the theory of comparative advantage is applied to individuals or states and that one of the reasons for the confusion can be traced back to Ricardo's decision to apply it to the England-Portugal case. However, what is the difference?

The key to finding an answer to this question is the realization that the sequence of market integration may matter, and this phenomenon cannot be understood in a two-person example. To illustrate this, I will, therefore, add Charles (C) to the picture and assume that he can produce tomatoes and pears, as well. One distinguishes between two scenarios.

1. Simultaneous integration In this scenario, assume that all the individuals start negotiating about specialization and trade simultaneously, starting from autarky. In this case, the theory of comparative advantage readily applies in the modified situation: A, B, and C will specialize according to their comparative advantages, and the surplus of production can be distributed in a way that makes at least one individual better off, without making any other individual worse off, since each individual can at least ensure himself the autarky level of consumption. Therefore, with simultaneous integration, the benchmark is autarky consumption and, given this benchmark, there can be no loser from integration.

2. Sequential integration In this scenario, assume that A and B have already negotiated a trade agreement when C enters the picture. Thus, the benchmark for comparisons for A and B is no longer the autarky consumption, but the consumption with 'partial' integration. Compared to this situation, including C into the trade agreement need no longer be mutually beneficial. It could, for example, be that the availability of C as a potential trading partner for A motivates A to drop B as a trading partner and to replace him with C. In this case, B falls back to autarky, which worsens his position. This is, in a nutshell, exactly the situation a lot of manufacturing industries in Europe were facing when China entered the world markets. Its rise as the "workbench of the world" induced deep structural changes in Europe, which lost the better part of its manufacturing jobs to China. The crucial fact one has to understand, in order to square this fact with the theory of comparative advantage, is that the theory can be correct and yet, at the same time, market integration creates losers. It is a problem of the adequate benchmark that is used for welfare comparisons.

The fact that a process of sequential integration can create losers on the way to a fully globalized world economy also gives a hint as to why Ricardo's example is potentially misleading. Ricardo's original example had cloth and wine as goods and England and Portugal as countries, with Portugal having an absolute advantage in the production of both. Market integration between countries is a special form of sequential integration in the above sense: the starting point is not autarky, but a situation where the English and the Portuguese have partially integrated markets. Starting from this benchmark, any further integration of markets can produce losers.

The decisive difference between Portugal and Ann and England and Bill is that countries are no unitary actors. Saying that "England" can profit from trading with "Portugal" is not the same as saying that Bill can profit from trading with Ann. If, in the process of specialization, the necessary restructuring of the two economies leads to a loss of jobs in the winemaking industry in England, it is hardly comforting for the now unemployed to learn that the total size of the cake for the English has increased. On the contrary, knowing that the total size of the cake is larger, but one's own piece is smaller, may even foster social tensions. The restructuring processes in the economies create winners and losers and, even if the winners could, in principle, compensate the losers because goods become more abundant, this is rarely done in practice. The problem does not exist in the Ann and Bill example because it is a single individual that has to reorganize her or his day. Treating countries like individuals blurs the underlying distributional conflicts that necessarily exist when new players enter a market. Forcing one to think all the way back to the level of the individual is one of the strengths of methodological individualism.

There are two additional aspects of the theory of comparative advantage that should, at least briefly, be discussed: the vulnerability towards exploitation and the phenomenon of alienation.

- **Exploitation:** An economy that opens up for international trade will undergo restructurings when the industries adapt, according to the comparative advantages of the economy, which is usually not in the short-run interests of the workforce of the declining industries. An often-heard argument in the preceding political debates is that relinquishing autarky makes a country more vulnerable, because of the increasing dependence on exports and imports. Switzerland, for example, sees the security of vital goods like food as sufficiently important to give it the rank of constitutional law. Article 102 of the federal constitution specifies two principles. "(1) The Federation ensures the supply of the country with vital goods and services for the case of power-politics or martial threats, or of severe shortages which the economy cannot counteract by itself. It takes precautionary measures. (2) The Federation may, if necessary, deviate from the principle of economic freedom." By and large, agriculture is not a sector where Switzerland has a comparative advantage by international standards, which implies a tension between economic freedom, market integration, and food security. These enacted policies to protect domestic agriculture lead to higher domestic food prices and subsidies of the agrarian sector.

It is not easy to understand whether ensuring food security in times of political crises is a welcome narrative for the farmers to support protectionism or not. What can be said, however, is that vulnerability due to specialization cuts both ways, because different economies become more dependent on each other. In order to get the right perspective on the problem, one therefore has to distinguish between two scenarios. The first can be called the *ceteris paribus* scenario: *if* a crisis occurs, it is important that one does not have to give in to the unjustified demands of an aggressor. Food security is, potentially, a way to achieve this goal. The second can be called the *General equilibrium* scenario, which tries to understand the effect of specialization on the *likelihood* of political crises. Here, the basic idea is that mutual interdependencies make crises less likely, because international interdependencies also increase the risks for potential aggressors.

However, there are conditions under which countries are vulnerable towards exploitation. Such a scenario exists, if a relatively small country is highly dependent on a relatively large, powerful country, and the restructuring of the local economy is not easily reversible. Economists have coined the term *hold-up problem* for a situation where specialization creates power asymmetries that can be exploited by the relatively more powerful partner. It is, ultimately, an empirical question whether the pacifying effects of specialization or the potential exploitability dominate in a specific context.

Digression 5. "Taming the Passions:" How Early Theorists of Capitalism Looked at Trade and Competition

Early theorists of capitalism, like Montesquieu, James Steuart and Adam Smith, had a complex understanding of the interplay between individuals and society. Albert Hirschmann (1977), for example, pointed out that it was a widely shared conjecture among these philosophers that a major merit of an economic system, based on specialization and trade, is its ability to "tame" the passions of men: "Money making [was seen] as an 'innocent' pastime and outlet for men's energies, as an institution that diverts men from the antagonistic competition for power to the somewhat ridiculous and distasteful, but essentially harmless accumulation of wealth." This view on markets (as institutions) is fundamentally different from the later view, held by 20th century mainstream economics, which has almost exclusively focused on the ability of competitive markets to achieve efficiency. It was reanimated in the 20th century by John Maynard Keynes (1936, p. 374), who argued that "[...] dangerous human proclivities can be canalized into comparatively harmless channels by the existence of opportunities for money-making and private wealth, which, if they cannot be satisfied in this way, may find their outlet in cruelty, the reckless pursuit of personal power and authority, and other forms of self-aggrandisement. It is better that a man should tyrannise over his bank balance than over his fellow-citizens; and whilst the former is sometimes denounced as being but a means to the latter, sometimes at least it is an alternative."

Profit-seeking behavior and competition defined a vision of a better society where the darker passions of human beings are kept under control by the pursuit of profit. With free trade, according to this view, one need not see an enemy in a stranger, but instead a potential trading partner. Free trade and competition present a form of moral education that brings about relatively harmless bourgeois virtues and that suppresses the darker aspects of human nature. Competition, within the context of free markets, has an explicitly *moral* quality, because the alternatives are so much worse. This view found its expression in Milton Friedman's famous example of discrimination in a competitive economy (Friedman 1962, Ch. 7): "It is a striking historical fact that the development of capitalism has been accompanied by a major reduction in the extent to which particular religious, racial, or social groups have operated under special handicaps in respect of their economic activities; have, as the saying goes, been discriminated against. The substitution of contract arrangements for status arrangements was the first step toward the freeing of the serfs in the Middle Ages. The preservation of Jews through the Middle Ages was possible because of the existence of a market sector in which they could operate and maintain themselves despite official persecution."

- **Alienation:** Specialization implies the division of labor. As soon as one gives   Φ
  up autarky, one devotes one's professional life to a specialized task. The limit of this process is, according to Adam Smith, only defined by the number of potential trading partners: "As it is the power of exchanging that gives occasion to the division of labour, so the extent of this division must always be limited by the extent of that power, or, in other words, by the extent of the market." (Adam Smith 1776, p. 21). Thus, the larger the number of potential trading partners is, the more specialized the individual activities can be. The only additional lim-

its come from coordination and transportation costs, which are inevitable when production is partitioned into specialized tasks.

This view focuses exclusively on the material side: the production of "stuff," material goods and services. However, this ignores the psychological consequences for the people. Karl Marx (1844) was, perhaps, the most prominent thinker, who stressed the implications of the division of labor on the ability of human beings to experience a happy, autonomous, and meaningful life. He described the concept of *Entfremdung*, or *alienation*, as the costs of living in a socially stratified and specialized society where each individual is just a little cog in a big wheel, without autonomy of his time and the products he contributes to, and his managers merely value him as a factor of production.

However, the idea of alienation goes back, at least, to Adam Smith. He expresses the idea very poignantly in *The Wealth of Nations*: "In the progress of the division of labour, the employment of the far greater part of those who live by labour [...] comes to be confined to a few very simple operations; frequently to one or two. But the understandings of the greater part of men are necessarily formed by their ordinary employments. The man whose whole life is spent in performing a few simple operations, of which the effects too are, perhaps, always the same, or very nearly the same, has no occasion to exert his understanding, or to exercise his invention in finding out expedients for removing difficulties which never occur. He naturally loses, therefore, the habit of such exertion, and generally becomes as stupid and ignorant as it is possible for human creature to become."

He follows up on these ideas in his *Lectures*: "Where the division of labour is brought to perfection, every man has only a simple operation to perform; to this his whole attention is confined, and few ideas pass in his mind but what have an immediate connection with it. [...] These are the disadvantages of a commercial spirit. The minds of men are contracted and rendered incapable of elevation. Education is despised, or at least neglected, and heroic spirit is almost utterly extinguished. To remedy these defects would be an object worthy of serious attention."

In order to be able to understand if specialization leads to alienation, in this sense, one has to understand the meaning work has for human beings, which is to a certain extent culture specific. Thus, the concept is inherently a cultural and psychological one. What one can say, without digging deeper into this matter, is that the production of goods and services is a means for some underlying end: call it happiness, meaningful life, or whatever. A focus on the materialistic side of production can, therefore, be too narrow to understand the implications of the division of labor on individual wellbeing.

2.4 Comparative Advantage and the Organization of Economic Activity

Growing up, I slowly had this process of realising that all the things around me that people had told me were just the natural way things were [...] weren't natural at all. They were things that could be changed. (Aaron Swartz (2010))

When will we realise that the fact that we can become accustomed to anything [...] makes it necessary to examine carefully everything we have become accustomed to? (George Bernard Shaw (1930))

I committed a major scientific crime in the last subchapter, because I was very sloppy with respect to the terminology I used. I repeatedly referred to markets, although the theory of comparative advantage is formulated as a *technological* property, without any reference to specific institutions like markets. However, Adam Smith (1776)[1991] and David Ricardo (2004) are brothers in arms, which makes this crime, hopefully, forgivable. Examples of the effects of specialization and trade are easier to find in market contexts and the political tensions and debates are also results of market integration. For David Ricardo, it was reasonable to think of comparative advantages in market contexts, because he was thinking of opening up the English economy to competition. What he found, however, was much more general than a property of markets.

- Consider the theory of comparative advantage within the organizational context of a firm. The divisional structure of the firm (like accounting, marketing, strategy, production, ...) reflects a specific form of division of labor. Employees specialize as engineers, workers, or accountants with the expectation that the whole is bigger than the sum of its parts. The within-firm exchange of goods, services, and resources is, in general, not organized like a market, but follows hierarchical rules that may, but need not, simulate market mechanisms (as, for example, in the case of profit centers that exchange goods and resources on the basis of firm-internal transfer prices).
- Alternatively, looking at organizations like public research institutes or universities, scientists are extremely specialized, but their exchange of ideas is not governed by market forces. Instead, to a large extent, this takes place in conferences and in research seminars where scientists give away their ideas "for free." Competition for public research funds more closely resembles a contest where the relatively best or most promising proposals get the funding.

The examples show that one can have a division of labor and an exchange of goods, services, and resources without markets. A more fruitful perspective on the relationship between the theory of comparative advantage and markets is to ask whether markets are a good *means* to enable specialization and trade. Theory states that people can alleviate scarcity by a process of specialization and exchange. The next question has to then be: do they have to organize economic activities in a specific way to make sure that the potential becomes actual gains from trade? This is the question about *institutions*, which is at the heart of economics.

According to Douglas North (1991), "[i]nstitutions are the humanly devised constraints that structure political, economic and social interaction. They consist of both informal constraints (sanctions, taboos, customs, traditions, and codes of conduct), and formal rules (constitutions, laws, property rights). Throughout history, human beings devised institutions to create order and reduce uncertainty in exchange, either consciously or by cultural evolution. Together with the standard constraints of economics they define the choice set and therefore determine transaction and production costs and hence the profitability and feasibility of engaging in economic activity. They evolve incrementally, connecting the past with the present and the future; history in consequence is largely a story of institutional evolution in which the historical performance of economies can only be understood as a part of a sequential story. Institutions provide the incentive structure of an economy; as that structure evolves, it shapes the direction of economic change towards growth, stagnation, or decline."

Economics, as the study of how men and society choose to employ scarce productive resources, is therefore primarily the study of institutions. A market is one example for an institution, a firm is a second, and centralized government planning is a third. The study of the properties of different institutions will help develop an understanding of how they function (positive science), to what extent they alleviate scarcity and how they distribute goods, services, and resources (positive as well as normative science).

From a philosophical point of view, institutions are very peculiar. Searle (2010) reconstructs the ontology of institutions as a specific class of speech acts called "declarations." Humans possess "the capacity to impose functions on objects and people where the objects and the people cannot perform the functions solely in virtue of their physical structure. The performance of the function requires that there be a collectively recognized status that the person or object has, and it is only in virtue of that status that the person or object can perform the function in question." Institutions come into being by the repeated applications of specific linguistic representations (declarations), and they cease to exist as soon as people no longer collectively recognize their status. They are, at the same time, epistemologically objective and ontologically subjective: there can be no doubt that an institution exists within a given convention, but the convention itself is, to a certain extent, arbitrary.

Take the convention or institution "Switzerland" as an example. There is a mutually recognized consensus that Switzerland exists, as a legal entity and as an institution; thus, it makes no sense for a single human being to deny its existence and act on its territory according to, for example, Russian law. In this sense, the institution "Switzerland" objectively exists. However, as soon as the people in Switzerland (and the rest of the world) deny that Switzerland exists, it actually ceases to exist, which makes it ontologically subjective. Its existence relies on a convention. This is different, for example, from the Matterhorn, which continues to exist even if seven billion people deny its existence; it is ontologically objective. The hybrid nature of institutions distinguishes them from most phenomena studied in the natural sciences: they are products of "shared fantasies." Property rights, as an integral requirement for markets, do not exist independently of human beings: their existence relies on a mutual consensus. The same is true for money, the state, firms, and so on. The most fundamental declarative speech act is language itself: there is nothing inherent in the ontological object "chair" that requires one to call it a chair: it could just as well be called a "Stuhl." The partition of phenomena, according to the rules of a language, has far-reaching implications for one's perception of reality.

Digression 6. What is Ontology and Epistemology?
In philosophy, ontology is the study of "what there is," of the nature of being and reality. It studies problems concerning the entities that do exist and their properties. Examples of ontological questions include the following: *What is existence? What is the nature of existence? What principles govern the properties of matter?*

Epistemology is the study of knowledge and justified belief. Questions that it addresses may include: *What are the necessary and sufficient conditions of knowledge? How does one separate true ideas from false ideas? How does one know what is true?*

In other words, ontology is about what is true and epistemology is about the methods of figuring out those truths.

The specific ontology of institutions makes economics special among the sciences. In the words of Rosenberg and Curtain (2013), "[u]nlike the physical world, the domain of economics includes a wide range of social "constructions" – institutions like markets and objects like currency and stock shares – that even when idealized don't behave uniformly. They are made up of unrecognized but artificial conventions that people persistently change and even destroy in ways that no social scientist can really anticipate. We can exploit gravity, but we can't change it or destroy it. No one can say the same for the socially constructed causes and effects of our choices that economics deals with." This potential fluidity of institutions makes them inherently difficult to study. Here is an example: money relies on the social convention that people are willing to accept it as a medium of exchange, because it cannot directly be consumed. As soon as this convention starts to unravel, money loses its value; it is ontologically subjective. In order to have a reliable theory of money, economists cannot simply assume its existence; they have to identify the individual and group processes that determine its emergence and sustainability.

To summarize, the theory of comparative advantage can explain why people organize economic activities. They are organized by means of institutions, which is why economics is the study of such institutions. The major part of this book will be devoted to the analysis of markets as – together with democracy – the most prominent institution in bourgeois societies. However, it should be clear by now that markets are only one way to organize economic activities, among many others.

References

Friedman, M. (1962). *Capitalism and Freedom*. Chicago: University of Chicago Press.
Hirschmann, A. (1977). *The Passion and the Interest*. Princeton University Press.
Keynes, J. M. (1936). *The General Theory of Employment, Interest and Money*. Palgrave Macmillan.
Marx, K. (1988). *The Economic and Philosophic Manuscripts of 1844*. Prometheus Books.

North, D. C. (1991). Institutions. *Journal of Economic Perspectives*, 5(1), 97–112.

Ricardo, D. (2004). *The Principles of Political Economy and Taxation*. Dover Publications.

Rosenberg, A., & Curtain, T. (2013). What is Economics Good For? *The New York Times*.

Samuelson, P. A. (1969). The Way of an Economist. In P.A. Samuelson (Ed.), *International Economic Relations: Proceedings of the Third Congress of the International Economic Association* (pp. 1–11). London: MacMillan.

Searle, J. R. (2010). *Making the Social World: The Structure of Human Civilization*. Oxford University Press.

Shaw, G. B. (1930). *A Treatise on Parents and Children*. In Misalliance. London: Constable.

Smith, A. (1776)[1991]. *An Inquiry into the Nature and Causes of the Wealth of Nations*. Everyman's Library.

Swartz, A. (2010). Interview with Ruairí McKiernan. http://www.ruairimckiernan.com/articles1/previous/3. Accessed 7 July 2017.

Further Reading

Deardorff, A. (1980). The General Validity of the Law of Comparative Advantage. *Journal of Political Economy*, 88(5), 941–957.

Deardorff, A. V. (2005). How Robust is Comparative Advantage? *Review of International Economics*, 13(5), 1004–1016.

Dixit, A., & Norman, V. (1980). *Theory of International Trade: A Dual, General Equilibrium Approach* (pp. 93–126). Cambridge University Press.

Dornbusch, R., Fischer, S., & Samuelson, P. (1977). Comparative Advantage, Trade and Payments in a Ricardian Model with a Continuum of Goods. *American Economic Review*, 67, 823–839.

Ehrenzeller, B., Schindler, B., Schweizer, R. J., & Vallender, K. A. (2014). *Die Schweizerische Bunderverfassung*. Dike.

Findlay, R. (1987). Comparative Advantage. *The New Palgrave: A Dictionary of Economics*. V.

Krugman, P., & Obstfeld. M. (1988). *International Economics: Theory and Policy*. Prentice Hall.

Part II
A Primer in Markets and Institutions

Introduction

3

This chapter covers ...

- the importance of property rights.
- the different forms of markets.
- the importance of "money" as a barter good.
- basic terminology that is useful in analyzing markets.
- where economics and law meet.

3.1 General Remarks

The chapter on the theory of comparative advantage has revealed that the problem of scarcity can be alleviated, if individuals are willing to specialize according to their comparative advantages and then find a way to allocate goods and services that is mutually beneficial. I have further argued that this process cannot be expected to unfold without an adequate institutional "frame" within which specialization and exchange can take place. A *market* is one such institution; it is the most important institution that fosters specialization and exchange and is the foundation on which modern capitalist societies are built. Informally speaking, a market is a framework that allows potential buyers and sellers to exchange goods, services, and information.

In order to make these transactions possible, a market relies on *private property rights* and *contract law*. Property rights define individual spheres of control over objects and they allow individuals to determine in which ways these objects shall be used and thus create a distinction between 'mine' and 'yours.' Without such a distinction, it would be impossible to establish markets and trade, because it would be unclear who has the right to control and use these objects. Property rights can be absolute, giving the owner of an object the freedom to use it in any way she wants, but in most societies there are socially agreed upon restrictions on the use of one's

© Springer International Publishing AG 2017
M. Kolmar, *Principles of Microeconomics*, Springer Texts in Business and Economics,
DOI 10.1007/978-3-319-57589-6_3

property. Restrictions may occur, if some uses impede on the well-being of others or are in conflict with moral values.

An important example of objects for which many countries have constrained the rights of the owner is the ownership of land, which is called real estate or immovable property. Land development, types of uses and the architecture of buildings are subject to constraints and regulations, and some countries limit individual rights even further by preventing them from using real estate in the way most preferred by the owner (for example by construing the right to abandon one's buildings). Therefore, it is more adequate to think of property rights as those user rights that society leaves to the formal owner. The technical term for these rights is *residual control rights*.

Φ

Digression 7. Property Rights Enforcement
It is vital to distinguish between the mutual recognition and the enforcement of property rights. People are used to thinking of property-rights enforcement as a centralized activity delegated to "the state." An important proponent of this view was Max Weber (1988), who observed that the modern state has monopolized the legitimate use of force. According to this point of view, the state provides for public enforcement and, with a few exceptions like self-help, limits private enforcement of property rights. This has not always been the case. The private enforcement of rights has been of considerable importance historically, for example in late medieval Europe. The development of the code of conduct called "Lex Mercatoria," in the 11th and 12th century, is seen as one of the key factors for the economic success of Europe, which arched over into the Renaissance. This helped to overcome the limited possibilities of centralized law enforcement in a politically fragmented Europe. According to Berman (1983), "[t]his legal system's rules were privately produced, privately adjudicated, and privately enforced." The system became effective exactly because medieval Europe was plagued by a maze of fragmented states, whose rulers more closely resembled self-interested elites. In certain respects, the situation in medieval Europe looks similar to the situation of the globalized economy of today, where multinational firms are confronted with nation-states that lack a centralized agency, which enforces contracts.

𝓛

The fact that markets rely on property rights implies that every transaction on a market has a "physical" and a "legal" side. The physical side of a transaction is the exchange of goods, services, or information (I will henceforth speak of goods and services, implicitly assuming that information can be interpreted as a specific kind of service) whereas, from a legal perspective, a transaction is an exchange of rights. In order to be able to exchange rights, it is necessary to specify the conditions under which such a transaction is binding. An exchange of rights is specified in a *contract* and the rules that apply to the establishment of such contracts are specified in a society's *contract law*.

Digression 8. Self Ownership Φ

An often bypassed constituent element of private property is *self-ownership*, which is an important virtue and achievement of modern bourgeois society. Self-ownership excludes serfdom and slavery and is a necessary prerequisite for ownership rights over objects in the outside world. It is also important for the establishment of transactions of services like, for example, the time and expertise a person offers on labor markets. Usually, a labor contract specifies the duties of the employer as well as of the employee. Self-ownership makes these contracts possible and, at the same time, defines limits to contractual freedom, because it, for example, prohibits a person from voluntarily selling herself into slavery.

The very brief discussion of the institutional prerequisites for a market economy £
– private property as residual control rights plus contract law – reveals that there is a close relationship between the legal and economic aspects of the study of markets. The civil law of a society implicitly defines the extent to which markets can develop and what they can achieve, while the economic analysis of the functioning of markets can inform the legal scholar about the likely consequences of legal rules. The importance of the interaction between a legal and an economic perspective is reflected in the fact that a whole field of analysis called "Law and Economics" has emerged, which is devoted to the analysis of the relationship between legal rules, individual behavior, and social outcomes.

Assume that a society has established a system of private property rights, which assigns residual rights of control over objects to individuals and contract laws, which specify the conditions under which the ownership of rights can be transferred. The individuals can now start to exchange these rights, given the rules specified in contract law. The rights-based approach to markets is straightforward but, at the same time, may be a little too abstract to define a good or service as any (bundle of) right(s) an individual may be interested in buying or selling. These rights can be anything from the right to eat an apple to the right to acquire a share in a company twelve months from now, if the share price is above a certain threshold.

Basically, there are two ways to establish trade. In a *barter economy*, goods and ✍
services are exchanged for other goods and services, like two apples for a loaf of bread. Most modern societies, however, rely on an abstract medium of exchange: money. At this point, it is not necessary to explicitly distinguish between economies that barter and economies that use money as a medium of exchange, but the following digression discusses the "nature" of money:

Digression 9. Money

One of humanity's major achievements has been the invention of an abstract medium of exchange for facilitating the exchange of goods or trade. This medium of exchange is called money. Money is traditionally regarded as having three functions: it acts as a medium of exchance, a unit of accounting and as a means of storing value.

Given that most people grew up in societies where money is almost as pervasive as the air we breathe, it is easy to oversee these three really peculiar aspects of money. First, compared to a barter economy where transactions can only take place if the supply and demand of two individuals perfectly align (which is called the "double coincidence of want"), the use of money dramatically facilitates this exchange, because it no longer depends on this coincidence.

Second, given that money has no intrinsic value and merely represents an abstract promise to be convertible into directly useful goods and services in the future, it is a convention in the sense of Searle, see Chap. 2. Thus, its invention relies on abstract thinking and trust (it most likely evolved from debt certificates) and the historic development of money shows people's increasing ability of thinking in abstract ways about the use and nature of money. The step from gold and silver coins (used by the Lydian's around 500-600 BC) to paper money (from the 7th century AD in China and the 13th century AD in Europe), and then from Banknotes backed by Gold (Bretton-Woods System) to unbacked money, and finally to a perfectly abstract unit of exchange in the digital age, show an increasingly abstract way of thinking.

Third, in opposition to directly useful goods and services, the value of money results from a social convention. Money has value only insofar as people are willing to accept it as a medium of exchange. This explains why the value of money, and of currencies, is inherently fragile, because the value of banknotes and coins (and, even worse, of purely abstract forms of money) drops to almost zero (which is an extreme form of inflation) as soon as people lose faith in its future value and start rejecting it as a medium of exchange, despite the fact that everyone would be better off, if money was accepted.

Assume that an exchange rate between goods and services, or a monetary price, exists. In the case of money, a person who is willing to give away (some of) her residual rights of control in exchange for the given price is called a *seller* of these rights (and the associated goods and services), while a person who is willing to acquire (some) residual rights of control from another person, in exchange for the given price, is called a *buyer*. The example of a barter economy, where one good is necessarily exchanged for another, makes it clear that a person is necessarily a buyer and a seller at the same time, because she has to give up apples for potatoes or *vice versa*. This reciprocity of supply and demand carries over to monetary economies, if one reminds oneself that money is an abstract promise to acquire goods and services in the future and, therefore, a bundle of rights. Thus, buying apples for money means that one person acquires control rights over apples (buyer) and the other gives up control rights over future consumption (seller). Hence, one should bear in mind that any transaction in a market is necessarily complemented by a transaction on some other market.

3.2   Taxonomy of Markets

The remainder of this book will take the existence of property rights and contract law as given and develop a taxonomy of different markets. Table 3.1 gives an overview of the most important market structures. It is common to distinguish supply and demand according to the number of buyers and sellers on a market. It is also customary to distinguish between one buyer or seller, a few buyers or sellers, and many buyers or sellers. This taxonomy defines nine prototypical market structures, each one with its own distinctive, functional logic. First of all, one should focus on the three market structures that will be analyzed in greater detail in the following chapters: polypoly, oligopoly, and monopoly.

A polypoly has many buyers and sellers of a homogenous good or service. Goods or services from different suppliers are called *homogenous*, if the potential buyers are not willing or able to distinguish between them and, therefore, consider them as perfectly interchangeable. The term "many" has a specific meaning in this context, as well. It refers to a situation where each buyer or seller considers her influence in the market so negligible that she does not have any influence on the market price. The buyers and sellers are therefore *price takers*, and the market is also called *perfectly competitive*. A market with perfect competition is the workhorse model for a lot of problems analyzed by economists, ranging from the determination of market prices, to the effects of taxes and to the determinants of international trade. In addition, this market is relatively easy to analyze, which is why this analysis of market economies starts with this case. Examples for markets that approximate perfect competition are:

- Some agrarian resources, like wheat, approximate perfect competition, because an international commodities market exists for these approximately homogenous resources, which implies a large number of producers (farmers) and buyers.
- The stock exchange is, in principle, also a good example for a competitive market, but one has to be cautious, because of institutional investors who can, generally, influence prices.

However, for reasons that will become apparent later on, not many markets can be adequately described as polypolistic. The reason why economics textbooks still focus on a market structure that is apparently unrealistic or not very common is because its simplicity allows one to understand fundamental properties of market transactions. Furthermore, it also acts as a reference point for more complicated markets. More realistic markets, like monopolies or oligopolies, are more complex to analyze but, fortunately, the additional complexity is relatively easy to digest

Table 3.1 Taxonomy of market structures

Sellers	Buyers		
	One	Few	Many
One	Bilateral monopoly	Restricted monopoly	**Monopoly**
Few	Restricted monopsony	Bilateral oligopoly	**Oligopoly**
Many	Monopsony	Oligopsony	**Polypoly**

because it is, in a sense, additive: the functioning of the most basal monopolistic market can be analyzed using the understanding derived from competitive markets, plus additional layers of complexity.

These additional layers of complexity exist because the seller on a monopolistic market understands that she is the only seller of a specific good or service, which gives her a certain power to influence prices. Hence, the assumption of price-taking behavior is no longer adequate and one has to understand how this additional factor influences supply and demand. The first known mention of a monopoly goes all the way back to Aristotle who, in his "Politics," describes the market for olive presses as a monopoly. More recently, De Beers had a monopoly in raw diamonds before countries like Russia, Canada, and Australia emerged as alternative distributers of diamonds. Public utilities that maintain infrastructures like electricity, water, sewage, etc. usually also have regional monopolies.

On the same note, the functioning of the most basic oligopolistic market can be analyzed using the understanding derived from monopolistic markets, plus yet another layer of complexity. With only a few suppliers, each of them has, in principle, some control over prices, but they have to take their competitors' likely behavior into consideration. Such strategic considerations are not necessary in monopolistic markets, because there are no competitors. They are also obsolete in perfectly competitive markets, because no supplier is able to influence the market. This is no longer the case with a limited number of competitors, because the optimal behavior of one supplier, in general, depends on the behavior of her competitors. This situation is defined as *strategically interdependent* and an analysis of markets with strategically interdependent decisions will be the capstone of this introductory textbook. Formally, an oligopoly is a market where a limited number of suppliers sell homogenous goods. Here are some examples:

- The grocery market in Switzerland is dominated by Coop and Migros.
- The market for wireless telephone services in Switzerland is dominated by Swisscom, Sunrise, and Salt.
- The worldwide accountancy market is dominated by PriceWaterhouseCoopers, KPMG, Deloitte Touche Tohmatsu, and Ernst & Young.
- The worldwide aircraft market is dominated by Boeing and Airbus.

Bearing the increasing complexity of different market structures in mind, the structure of the following chapters is straightforward: they start with a relatively simple market structure that is easy to understand, continue with a market structure that better describes a lot of markets that one is confronted with on an almost daily basis and that is of moderate complexity, and finish with the most complex market structure.

The taxonomy in Table 3.1 encompasses not only three, but nine market structures. Even if I only explicitly cover the above-mentioned three in this introductory textbook, I will discuss the other structures' peculiarities briefly.

Monopsonistic and oligopsonistic markets are mirror images of monopolistic and oligopolistic markets and the main insights that can be derived from the latter can also be applied to the former.

A bilateral monopoly, however, confronts one with a totally different situation, because both sides of the market possess some market power, which derives from the fact that the trading partner cannot fall back on some other identical alternative, if trade does not take place. Such a situation arises because manufacturers and suppliers often customize their production processes and products to the needs of their trading partners, with the result that the manufacturer cannot sell the tailored products at the same price to other trading partners and the trading partner has difficulties finding adequate substitutes on the market. Here are some additional examples:

- Collective bargaining agreements (CBAs) between labor unions and (especially large) companies or employers' associations,
- highly specialized scientists and their employers (e.g. pharmaceutical companies and their lead scientists; both would have difficulties finding adequate alternatives, at least in the short run),
- governments and some of their defense contractors (an extreme example is the market for nuclear-powered aircraft carriers, where the US government is the only buyer and Huntington Ingalls Industries is the only seller);
- marriage (think about it: dissolving a partnership is costly so, even if one thinks that one has found an even better match, one may decide not to dissolve it).

Analytically, the challenge lies in understanding the factors that influence the success of the resulting bilateral negotiations and the distribution of the potential gains from trade between the buyer and seller. The field of research that analyses these questions is called *bargaining theory*.

The three remaining market structures, bilateral oligopoly, restricted monopoly and restricted monopsony, are far less studied. The basic challenge in understanding the functioning of these markets, and the corresponding optimal strategies, is how varying degrees of competitiveness influence the bargaining power of a single buyer or seller. An example is the retail industry: historically, the supply side was concentrated and the demand side was rather competitive in the corresponding markets, but demand-side concentration greatly increased over the last forty years or so. The grocery sector in Germany is a case at hand and the trend towards concentration was reinforced by the formation of buyer groups. In 2014, the five largest buyer groups had a market share of more than 80%.

Another commonly analyzed market form, which does not appear in Table 3.1, is called *monopolistic competition*. The model of monopolistic competition blends elements from the monopoly with elements from the polypoly model. Basically, it is assumed that firms behave like monopolists and that firms can produce similar, but not identical, products. A good example is the market for sports-utility vehicles (SUVs), where each major car company sells its own variant of SUV. All of them are similar, but a lot of customers have their favorite brands. The model of monopolistic competition is very useful, if one is interested in determining the number of competitors that a market can sustain.

The above line of argumentation assumes that the number of buyers or sellers has an important influence on the functioning of a market and one will get a deeper understanding of this conjecture throughout the following chapters. However, there are two questions that should come to mind at the present stage. First, it is unclear what determines the market structure. Is it possible to organize markets for arbitrary goods and services at will, or are there underlying explanatory factors that determine whether a specific good or service is traded on a perfectly competitive or on a monopolistic market, or on a completely different one? Second, how many are "few" and how many are "many?" If the dividing line between few and many is important for the functioning of a market, it would be helpful if one could attach a number to this question.

An exhaustive answer to these questions is beyond the scope of an introductory textbook, but the following chapters will shed a little light on the subject. Regarding question one, economists usually distinguish between markets and industries. An industry is a sector of the economy that produces a specific type of good; it is better characterized by the technological way of production that summarizes the physical, biological, and chemical laws that convert the resources needed for production (inputs) into products (outputs). This relationship between inputs and outputs is also called the *technology of production*. As Chap. 8 illustrates, industries differ with respect to the laws linking inputs with outputs and these laws have an important influence on the possible market structures. Furthermore, the perception of goods and services by the "buyers" (customers) has a direct impact on the market structure. If they can, or are willing to, distinguish between, for example, red wine and white wine, all producers of red wine are in the same market for "red wine." If the customers, however, distinguish between different types of grapes, region of origin, producer, or even characteristics of the vineyard, the market for red wine explodes into a plethora of differentiated markets, where even small local producers may have the market power to influence prices. I will discuss this phenomenon further in Chap. 10. Last, but not least, the legal framework determines market structures. Most countries have, for example, a competition law, the purpose of which is to guarantee a minimum degree of competitiveness on each market, thereby excluding monopolies. However, the opposite can be true as well: patent law, for example, grants the patent holder a temporarily restricted monopoly for those products that can be developed from his patent. In summary, market structure is not completely arbitrary, but it depends, in a complex way, on the technology of production, the perception of goods and services by customers, and the legal framework.

With respect to the second question, the answer is even more difficult. Remember that the dividing line between "few" and "many" is the perception of one side of the market that the price is *de-facto* given. There are industries for which two sellers or two buyers are "many" (there is an example in Chap. 12), and other industries, where a much larger number of competitors is necessary to more closely approach price-taking behavior. Experiments for so-called Cournot markets have shown that the magic number seems to be between two and four.

With these prerequisites, it is now time to analyze the functioning of the first type of market: a market with perfect competition.

References

Berman, H. (1983). *The Formation of the Western Legal Tradition*. Harvard University Press.
Weber, M. (1988). Politik als Beruf. In M. Weber, *Gesammelte Politische Schriften* (pp. 505–560). Tübingen: Mohr Siebeck.

Further Reading
Roberts, J. (1987). Perfectly and Imperfectly Competitive Markets. *The New Palgrave: A Dictionary of Economics*. V. 3.
Stigler J. G. (1987). Competition. *The New Palgrave: A Dictionary of Economics*. V. 3.

Supply and Demand Under Perfect Competition

4

This chapter covers ...

- the functioning of competitive markets as one way of organizing economic activities.
- how supply and demand can be determined, how they can be used to predict market behavior and the effect changes have in the economic environment on economic outcomes.
- how to apply the theory of competitive markets in order to better understand the economy.

4.1 Introduction

From the time of Adam Smith's Wealth of Nations in 1776, one recurrent theme of economic analysis has been the remarkable degree of coherence among the vast numbers of individual and seemingly separate decisions about the buying and selling of commodities. In everyday, normal experience, there is something of a balance between the amounts of goods and services that some individuals want to supply and the amounts that other, different individuals want to sell. Would-be buyers ordinarily count correctly on being able to carry out their intentions, and would-be sellers do not ordinarily find themselves producing great amounts of goods that they cannot sell. This experience of balance is indeed so widespread that it raises no intellectual disquiet among laymen; they take it so much for granted that they are not disposed to understand the mechanism by which it occurs. (Kenneth Arrow (1974))

This chapter will start with a basic initial analysis on the functioning of competitive markets. One may remember, from the last chapter, that everyone acts as a buyer and a seller simultaneously. I buy groceries and other consumer goods in the local stores or on the Internet and, in turn, sell my time and expertise to my employer. In order to have a lean terminology, it is therefore necessary to interpret buyers and sellers as artificial roles, both of which one adopts, depending on the good or service one is considering. For the purposes of this book, the terms seller, firm, and company as well as buyer, consumer, and customer will be used interchangeably,

© Springer International Publishing AG 2017

M. Kolmar, *Principles of Microeconomics*, Springer Texts in Business and Economics,
DOI 10.1007/978-3-319-57589-6_4

keeping in mind that not only firms sell and that not only (end-)consumers buy goods and services.

The two most fundamental concepts in the analysis of markets are supply and demand. These measure the quantity of a given good or service, which customers are willing to buy and firms are willing to sell at a given market price, as well as other explanatory factors like income, expectations about the future, prices of resources, and so on. In order to be able to study the functioning of perfectly competitive markets, one has to develop theories that explain the market's demand-side and its supply-side, as well as the interaction between the two sides.

4.2 Determinants of Supply and Demand

In order to be able to create these theories, one has to distinguish between the demand of a single individual (one's demand of apricots) and market demand (the demand of all individuals, who buy apricots in Switzerland). On that note, one also has to distinguish between the supply of a single firm (the supply of apricots in one's local grocery store) and market supply (the supply of all firms selling apricots in Switzerland).

Given that this chapter is a primer in competitive markets, it will motivate supply and demand heuristically by means of plausibility considerations. A full-sized microeconomic theory of markets replaces these plausibility considerations by a decision-theoretic foundation, which traditionally assumes that individuals can rank alternatives according to their preferences and determine demand and supply by choosing the most highly ranked alternative available. This kind of decision-theoretic foundation of supply and demand is, from a scientific point of view, preferable. However, it comes at the cost of added complexity, so it makes sense to skip it during our first passage through the logic of competitive markets. We will develop the underlying decision theory in Chap. 7.

✐ *Demand* One can start this endeavor with the analysis a single customer's demand for a given good. Assume that there are n goods in total among which a customer, j, can choose. n is a natural number larger than 1, and the goods are numbered $1, 2, \ldots, n$.

- It is reasonable to assume that the quantity of the good i (kilos of apricots), x_i^j, demanded by customer j most likely depends on the price of the good, p_i (CHF per kilo), as well as on the prices of other goods $p_1, p_2, \ldots, p_{i-1}, p_{i+1}, \ldots, p_n$ (for example, the price for a kilo of pears, as well as the price for a kilo of bread).
- In addition to prices, other factors will also likely influence demand. A prominent candidate is the customer's income or wealth, b^j (for budget, which is the amount of money the customer can spend on purchases).
- On a more profound level, the demand is also influenced by the customer's tastes, which are called the customer's preferences. Different customers may have different preferences and they, of course, have an influence on the demand

for goods. A vegetarian will spend nothing on meat, an outdoorsy person will spend part of her budget on hiking boots and other outdoor equipment, etc.

- The last factor, which is likely to influence the customer's demand for some good i, is her expectations for the future. They include expectations about general economic development, life expectancy, career perspectives, and so forth. For example, expectations about the impact of climate change may influence a person's consumption pattern, leading to more environment-friendly choices. Alternatively, the expectations about the future innovative drive of different companies may influence her investment strategy.

This list is not meant to be exhaustive, but it summarizes some of the key explanatory factors for individual demand. Hence, this is a formulation of the first causal hypothesis that will play an important role in the model of the competitive markets that are going to be developed: prices, budget, preferences, and expectations are the *explanatory* or *exogenous* variables, whereas individual demand is the *explained* or *endogenous* variable. Neglecting preferences and expectations for the rest of this chapter, one can write this causal hypothesis of the demand of customer j for good i in the form of a mathematical function $x_i^j(p_1, \ldots, p_n, b^j)$. This complicated piece of notation has the following interpretation: the quantity of the good that is demanded by the customer, x_i^j, is explained by all the prices, $p_1, \ldots, p_n$, and income, b^j. A function of several variables is, therefore, a straightforward extension of a function with one variable, which a college student should know from high-school. The premise of competitive markets, in which all buyers are too small to influence prices, is reflected in the fact that prices are explanatory variables and, therefore, independent of the behavior of the customer.

Table 4.1 shows the demand of customer A, Ann, for apricots based on different prices of apricots. Ann would buy two kilos, if the price were CHF 6 per kilo, and her demand would increase to 6 kilos, if the price were to sink to CHF 2 per kilo. Why should the demand go up, if the price goes down? It could, for example, be that a reduction in the price of apricots induces a change in behavior, because Ann likes different types of fruit equally well and, therefore, decides to save some money by going for the, now, relatively cheaper ones, and substituting, for example, cherries with apricots.

Table 4.1 Ann's demand schedule

p_i	x_i^A
0	8
1	7
2	6
3	5
4	4
5	3
6	2
7	1
8	0

Fig. 4.1 Ann's demand
curve

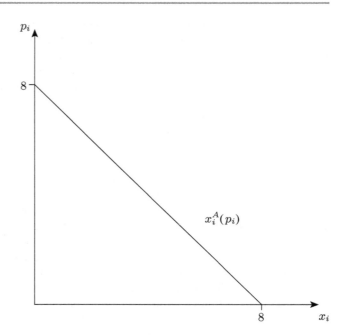

This table is called an *individual demand schedule* and it is a perfectly sound way of representing demand. However, this kind of a table places a heavy load on one's cognitive resources, which is why economists usually analyze demand through the use of a graph.

Figure 4.1 is a graphical representation of the demand schedule in Table 4.1. This figure has the good's price on the ordinate and the quantity demanded on the abscissa. It is easy to verify that the *individual demand function*, $x_i^A(p_i)$, summarizes the same information as the demand schedule.

One will have noticed by now that economists have an apparently odd convention regarding the axes of the diagram. Most students learned in school that the explanatory variable is drawn along the horizontal axis, and the explained variable is drawn along the vertical axis. If this is the case, the opposite convention used by economists will likely drive one crazy during the first few weeks. However, economists have a good reason for this deviating convention: from the point of view of a customer, the price is given. However, what one is ultimately interested in is the determination of prices by the interplay of supply and demand. Thus, the determination of supply and demand is only an intermediate step on one's way towards understanding the market. In the end, the "old" convention will hold again.

In order to determine prices, however, one needs an additional intermediate step that brings one from individual to market demand. Table 4.2 shows A's as well as B's (Bill's) demand for apricots. Assuming that A and B are the only customers, one can get from the individual demand schedules to the *market demand schedule*

Table 4.2 Ann's and Bill's demand schedules

p_i	x_i^A	x_i^B	x_i
0	8	10	18
1	7	8	15
2	6	6	12
3	5	4	9
4	4	2	6
5	3	0	3
6	2	0	2
7	1	0	1
8	0	0	0

Fig. 4.2 Ann's and Bill's demand curves and market demand

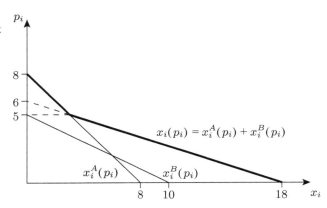

by adding up the individual demands for each price. Market demand is shown in column four.

It is, of course, also possible to analyze market demand by the use of demand functions, as one can see in Fig. 4.2. In this figure, $x_i^A(p_i)$ and $x_i^B(p_i)$ denote Ann's and Bill's demand functions. In order to get from there to the *market demand function* one has to add both demand functions horizontally (i.e. look for the total demand for every possible price). The market demand function is the bold kinked line denoted by $x_i(p_i)$. The kink results from the fact that only Ann is willing to buy apricots, if the price is between CHF 8 and CHF 5.

One can express the same relationship formally. Denote A's and B's demand for good i (apricots, in this example) by $x_i^A(p_1, \ldots, p_n, b^A)$ and $x_i^B(p_1, \ldots, p_n, b^B)$. Then the market demand function for good i can be denoted as:

$$x_i(p_1, \ldots, p_n, b^A, b^B) = x_i^A(p_1, \ldots, p_n, b^A) + x_i^B(p_1, \ldots, p_n, b^B).$$

Realistically, there are more than two customers for a good in the economy, but the same logic applies to the general case. Assume that there are m customers in total, where m is a natural number, and that j is a generic customer with demand function

$x_i^j(p_1, \ldots, p_n, b^j)$. In this case, market demand for good i is:

$$x_i(p_1, \ldots, p_n, b^1, \ldots, b^m) = \sum_{j=1}^{m} x_i^j(p_1, \ldots, p_n, b^j).$$

Now that one has defined individual and market demand, one can introduce some terminology, which will allow one to describe the different causal mechanisms that link explanatory and explained variables that can be formulated for individual, as well as market, demand. One defines them for an individual j. These mechanisms are described with respect to the induced *changes* in demand, which are caused by changes in the explanatory variables. This kind of exercise is called *comparative statics* and lies at the heart of economics as a positive science. The reason why it is so important, is because most of the testable predictions of economic theory are predictions about *changes* in empirically identifiable endogenous variables, caused by changes in empirically identifiable exogenous variables. The *absolute* value of a variable, like demand, is often irrelevant and the models can usually be tailored to meet the empirical patterns. However, *changes* in variables are often more robust and are, therefore, the only means to falsify a theory. The ability to produce falsifiable hypotheses is crucial for Critical Rationalism, which is the mainstream philosophy of science in economics, and is why comparative statics plays such an important role. However, even more than that, most economic-policy questions are also about the effects of changes in taxes, regulations and so on; or on changes in employment, production and so forth.

▶ **Definition 4.1, Ordinary goods** A good, i, is called ordinary, at given prices and budget, if the demand for that good, x_i^j, is decreasing with its price, p_i.

Note that this property is defined as a local one, if it holds for a given combination of explanatory variables. A good can be ordinary at some prices and incomes, and not ordinary at others. The basic idea behind Definition 4.1 is that, if the price increases, the demand decreases; this is the most common type of good. However, this need not be the case and there are a lot of empirical examples showing that the demand for a good can be increasing with its price. Examples for goods that are not ordinary are those that are primarily purchased to signal status, but are otherwise of limited intrinsic value (they have to be expensive to function as a status symbol) or goods whose prices are interpreted as a quality signal by the customers. These goods are called Giffen goods, named after Robert Giffen, who studied this phenomenon.

▶ **Definition 4.2, Giffen goods** A good, i, is called a Giffen good, at given prices and budget, if its demand, x_i^j, increases with its price, p_i.

▶ **Definition 4.3, Normal goods** A good, i, is called normal, at given prices and budgets, if the demand for that good, x_i^j, increases with an increasing budget, b^j.

The name "normal good" is also suggestive, because it seems intuitive that individuals will buy more of a good, if they get richer. However, there are important exceptions from this rule, especially cheaper goods of perceived low-quality that will be substituted by higher-quality goods (as perceived by the customer), if she gets richer. Examples are cheap food that is replaced by high-quality food, or cheap used cars that are replaced by more expensive new ones. These goods are covered by the next definition.

▶ **Definition 4.4, Inferior goods** A good, i, is called inferior, at given prices and budgets, if the demand for that good, x_i^j, decreases with an increasing budget, b^j.

The next definitions describe the relationship *between* different goods:

▶ **Definition 4.5, Substitutes** A good, i, is called a substitute for good k, at given prices and budgets, if the demand for that good, x_i^j, increases with an increase in price, p_k.

An intuitive example for two goods that are substitutes is two different, but similar, types of wine. If the price of, for example, Chianti goes up, the customer substitutes it with Barolo. (Bear in mind, however, that the example may or may not be correct for a given customer: it is ultimately an empirical question whether she is willing to substitute one type of wine for another.) However, a different relationship is also possible:

▶ **Definition 4.6, Complements** A good, i, is called a complement for good k, at given prices and budgets, if the demand for that good, x_i^j, decreases with an increase in price, p_k.

If one has ever wondered why shoes are sold in pairs, the concept of complementarity gives one a clue: for most people, left and right shoes are perfect complements: they always need a pair of them. If they were sold separately, an increase in the price of left shoes would reduce the demand for right shoes and *vice versa*. Another example for two goods that are complementary is printers and toner, because one can only print, if one has both.

Table 4.3 summarizes the comparative-static effects.

With these definitions, one can now analyze the demand-side of the market. This is done graphically and, in order to do, one sticks to the convention that the price of a good is drawn along the ordinate and its quantity along the abscissa. In order

Table 4.3 Overview over the comparative-static effects of an increase in explanatory variables

	Demand goes up	Demand goes down
Increase in income	Normal	Inferior
Increase in own price	Giffen	Ordinary
Increase in other price	Substitute	Complement

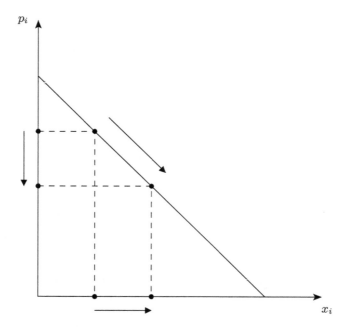

Fig. 4.3 A movement along the demand curve: a change in the explanatory variable, which is displayed in the figure, leads to a movement along the curve

to be able to isolate the effects of a single explanatory variable on an explained variable, economists decompose complex changes in the explanatory variables into simple ones, where the effect of each explanatory variable on the explained variable is analyzed separately (*comparative statics*), and the possible, comparative-static experiments for our model are changes in the price of the good, changes in the price of other goods and changes in income. A change in the price of the good can be analyzed by a movement *along* the demand curve, as illustrated in Fig. 4.3.

Changes in the prices of other goods or income have an influence on the *location* of the demand function. There are two potential effects: a rightward or a leftward shift of the demand function. Both are illustrated in Fig. 4.4.

Please verify that the demand for the good shifts outwards (inwards), if

- the good is normal and income goes up (down),
- the good is inferior and income goes down (up),
- the price of a substitute good goes up (down), or
- the price of a complementary good goes down (up).

Given the comparative-static effects covered by the definitions, this list gives one a comprehensive overview of the possible effects. The art and craft of economics is to identify situations that can be analyzed as price or income changes. An increase in income taxes, for example, decreases the disposable income of the individual.

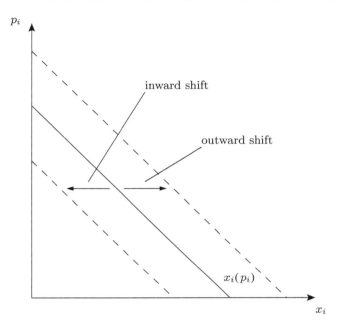

Fig. 4.4 A shift of the demand curve: a change in an explanatory variable, which is not displayed in the figure, leads to a rightward or leftward shift of the curve

The economic effects on the goods markets can, therefore, be analyzed as if the individual's income had decreased. Alternatively, rising immigration to a region might increase the market value of real estate, which implies that the landlords' incomes go up.

Before one can start to see how perfectly competitive markets work, the supply side has to be introduced, as well.

Supply The derivation of the individual and market supply functions follows the same steps as before, assuming that some good, i, is produced by a firm, j, of a total of h firms. What are the likely determinants of firm j's supply decision, y_i^j, (for example kilos of apricots)?

- The price of the good p_i (CHF per kilo of apricots) is likely to have an influence on the quantity supplied.
- Furthermore, the good must be produced by the use of some resources (for the apricot example, land, labor, fertilizer, etc.). Hence, the price of these resources determines the profitability of the firm. These resources are also called *inputs*, whereas the quantity produced is called an *output*. Physical, chemical or biological laws causally link inputs and outputs. The production of goods requires all kinds of inputs, resources and intermediate products. Economists customarily restrict attention to two generic inputs called capital and labor, whose quantities

are denoted by k and l and whose prices (per unit) are denoted by r (which is the interest rate the firm has to pay per unit of capital for borrowing it on the capital market) and w (which is the wage rate the firm has to pay a worker per unit of labor).

- Given that the conversion of inputs into outputs follows the laws of physics, chemistry and biology, the technology of production is also relevant for the supply decisions of the firm. For example, assume that a technological innovation makes labor more productive, thus increasing the output per unit of labor by 20%. In that case, production becomes more profitable, implying that the firm, *ceteris paribus*, is likely to produce more.

- Again, similar to the determination of demand, the supply of some good, i, is likely dependent on the firm's expectations for the future. If, for example, a firm determines the medium- to long-run production capacity of its plants, then it has to form expectations about future output and input prices, exchange rates (if part of the production shall be exported), and so on. The more optimistic the firm's expectations are, the more likely it is to invest in its capacity.

This heuristic allows one to formulate the second causal hypothesis for one's model of competitive markets: output and factor prices, technology of production, and expectations are the explanatory or exogenous variables, whereas firm supply is the explained or endogenous variable. As before, I will neglect all non-price variables for the rest of this chapter, such that one can write this causal hypothesis for the supply of firm j of good i in the form of a mathematical function, $y_i^j(p_i, w, r)$, which reads as: the supply of good i by firm j, y_i^j, is a function of (is explained by) the price of the good, p_i, and the prices of labor and capital, w, r.

Digression 10. What is Capital?
Capital is a key concept in economics and the eponym of the economic system of *capitalism*. It, therefore, deserves some extra attention. The term goes back to the Latin word *caput*, "head," which is also the origin of *cattle*. This is important, because it casts light on two basic properties a resource must have in order to count as capital: the stock of cattle is moveable (which distinguishes it from land) and reproduces. Therefore, capital is any resource that is potentially mobile and bears an interest, if it is not immediately consumed. Adam Smith defined capital as "[t]hat part of a man's stock which he expects to afford him revenue [...]."

The first resources economists had in mind when they used the term capital were livestock, machines and other tools. However, over time, the concept got more abstract, covering other "interest-bearing" phenomena, as well. On a very abstract level, capital consists of resources that enhance a person's power when she uses her time to achieve her goals (Pierre Bourdieu (1983)). This idea is nicely exemplified by the closely related German words "Kapital" (capital) and "Vermögen" (capability, which stems from the latin word *capa-*

bilis, "being able to grasp or hold", but is translated as "assets" in the system of national accounts). One could say that capital is a resource that makes one capable of achieving a goal. (This idea is also reflected by the fact that wealth is counted as an asset, whereas capital is a liability, in the system of national accounts.)

Consequently, contemporary economists distinguish between three or four different types of capital: *physical, human, social and symbolic.*

Physical capital corresponds to the traditional concept, including machines, tools and so on.

Human capital refers to the skills of a human being that make him or her more productive in manipulating physical capital. It is the stock of knowledge that allows an individual to use his or her labor in a productive way.

Social capital refers to the network of friends that allows one to achieve one's goals. It is the stock of social bonds and relationships that helps one succeeding with one's plans and insures one against adverse events. For example, information disseminated in a network of friends may allow one to make better decisions, or one may profit from cooperative and altruistic behavior among friends. This has its roots in the preferential treatment group members can expect from each other.

Symbolic capital is a controversial concept, which is better established in sociology than in economics. It refers to the ability of an individual to achieve her goals because of honor, prestige or recognition and it depends on the cultural norms and language games of a society. The concept allows one to better understand the role of cultural conventions and ideologies within societies and it, therefore, became important in gender studies. Here is an example why: cultural norms and language games impose categories of thought and perception upon individual social agents who, if they accept these categories unscrutinized, perceive the social order as legitimate. If women, for example, do not consider it appropriate to become CEOs of firms, they do not strive for these careers and, thereby, leave them to their male counterparts.

Sometimes, human, social and symbolic capital are difficult to differentiate, and some definitions have social and symbolic capital, as in special cases of human capital.

One can, in principle, analyze individual and market supply using supply schedules that are constructed analogously to the demand schedules introduced in the previous subchapter. Given that a demand or supply schedule is a rather cumbersome instrument for the analysis of markets and, given that this chapter has only introduced demand schedules as an intermediate step to motivate demand curves, one can immediately jump to individual and market supply functions.

Figure 4.5 is a graphical representation of the supply of firm A ("Alpha Limited") of some good i. As before, the figure has the price of the good on the ordinate and the quantity supplied on the abscissa.

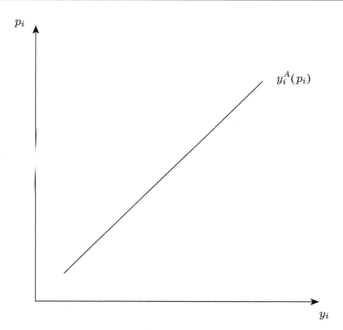

Fig. 4.5 Firm A's supply curve

The graph assumes that supply is increasing in the price of the good, which is a very intuitive assumption, because a higher price – holding all other factors constant – makes the good more profitable and encourages the firm to try to expand production. However, there may be situations where supply is not upward sloping, for example, in the short run, if a capacity constraint is binding. In the apricot case, farming company Alpha limited may not be able to buy or rent additional land to plant and harvest additional apricots, in the short run.

The final step that one has to take is to move from individual to market supply. Figure 4.6 shows A's, as well as B's (Beta Limited's), supply curves for apricots. Assuming that A and B are the only producers, one can get from individual to market supply curves in the following way: $y_i^A(p_i)$ and $y_i^B(p_i)$ denote Alpha's and Beta's supply functions. In order to get from there to the *market-supply function*, one has to add both supply functions horizontally (i.e. look for the total supply for every possible price). The market-supply function is the bold kinked line, denoted by $y_i(p_i)$. The kink results from the fact that only Alpha is willing to sell apricots, if the price is below a certain threshold.

Again, one can express the same relationship formally. Denote A's and B's supplies of good i by $y_i^A(p_i, r, w)$ and $y_i^B(p_i, r, w)$, then the market supply function for good i is:

$$y_i(p_i, r, w) = y_i^A(p_i, r, w) + y_i^B(p_i, r, w).$$

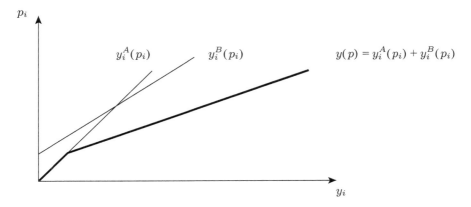

Fig. 4.6 Firm A's and firm B's supply curves and the market supply curve

This formulation implicitly assumes that both firms have access to capital and labor at the same input prices. If both firms hire on different capital and labor markets, they would likely face different interest rates and wages, which would then have to be reflected by firm-specific indices r^j and w^j. This situation is likely to be relevant, if the firms produce in different countries, like apricot farmers who produce in Switzerland and apricot farmers who produce in Italy and export to Switzerland.

If there are more than two firms producing a good, then market supply is the sum of all firms' supplies at a given market price. Thus, with a total of l firms, and j being a generic firm with a supply of $y_i^j(p_i, r, w)$, market supply for good i is given as:

$$y_i(p_i, r, w) = \sum_{j=1}^{l} y_i^j(p_i, r, w).$$

How do changes in input and output prices affect supply? In order to answer this question one follows the same steps as before and graphically analyzes this question in a figure where the price of a good is drawn along the ordinate and its quantity along the abscissa. A change in the price of the good can be analyzed by a movement *along* the supply curve (see Fig. 4.7).

Changes in input prices have an influence on the *location* of the supply curve and, again, it can either shift leftward or rightward, as illustrated in Fig. 4.8.

What are plausible conjectures about the effect of changing input prices? If production gets cheaper (more expensive), it is plausible to assume that firms will increase (reduce) production. If this is the case, one can summarize the potential effects as follows: the supply of the good shifts outwards, if wages or interest rates go down, and *vice versa*. As before, the tricky thing is to identify real-life situations that can adequately be described as changes in input prices.

The fact that this textbook has thus far included only output and input prices as explanatory variables in the formal definition of supply functions is a matter of con-

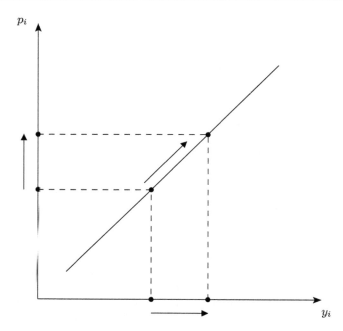

Fig. 4.7 A movement along the supply curve: a change in the explanatory variable, which is displayed in the figure, leads to a movement along the curve

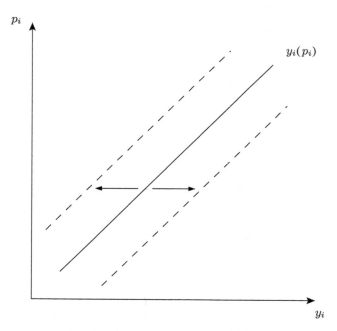

Fig. 4.8 A shift of the supply curve: a change in an explanatory variable that is not displayed in the figure leads to a right- or leftward shift of the curve

venience. There are, of course, other causal mechanisms that are likely to influence individual and, therefore, market supply. Expectations, as already discussed, are one example, but there are other influences as well. For example, a natural disaster may destroy part of the production capacity, which cannot be offset in the short run (supply shifts leftward), good weather conditions may increase the crop (supply shifts rightward), or technological progress may lead to an increase in output per unit of input (supply shifts rightward).

The same arguments can, of course, also be made for the demand side, which implies that a complete analytical description of demand and supply functions should include all those other factors. Call these explanatory variables $\alpha, \beta, \gamma, \ldots$ and suppose that all customers have identical budgets b. Market supply and demand for good i could then be written as $y_i(p_i, r, w, \alpha, \beta, \gamma, \ldots)$ and $x_i(p_1, \ldots, p_n, b, \alpha, \beta, \gamma, \ldots)$. Specifying the most interesting causal mechanisms, therefore, depends on the specific problem.

With these concepts in the back of one's mind, one is now ready to move on and see how demand and supply are coordinated on a competitive market.

4.3 Equilibrium

Have you ever thought about how it is possible that the baker knows that you will buy a bagel when you go to the city? When you enter his store, the bagel is just there, ready for you to buy and eat it. How come? How could the baker have known, even though you never ordered the bagel in advance? If this example seems a little bit underwhelming, to put it mildly, then one had better think twice. The great miracle of the market mechanism is that millions and billions of people are making decisions in an apparently uncoordinated, decentralized way and, despite this fact, there is a great deal of order in market outcomes. How is this possible?

Well, the first hint is that decisions are, of course, not uncoordinated. They are coordinated by market prices that shape individual incentives to buy and sell (and, more generally, to act), so decisions in a market economy are decentralized, but not uncoordinated. The question then becomes: to what extent are prices able to coordinate individual behavior and what does this imply for the functioning of markets?

Economists put a lot of emphasis on the idea of *equilibrium*. To motivate a formal definition, look at the following example: assume that, at a given market price, demand exceeds supply, i.e. customers want to buy more than suppliers are willing to sell. A situation like this has an in-built tension, because some customers have to go home unsatisfied: the decentralized plans of the economic agents are mutually inconsistent. (One reaches the same conclusion in the opposite case of excessive supply.) Therefore, the only situation where all the plans of the economic agents are compatible is at a price where supply equals demand. This situation is called an *equilibrium*. It can be defined for the whole economy with n different goods and associated markets (general equilibrium), or for a single market for good i, leaving the rest of the economy out of the picture (partial equilibrium). For

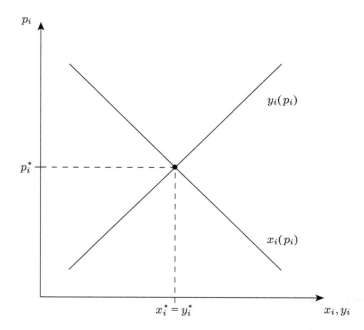

Fig. 4.9 Supply, demand and equilibrium

simplicity, and without significance for the results, suppose that all customers have identical budgets, b.

▶ **Definition 4.7, General equilibrium**   Assume there are n goods with n market prices. A general equilibrium (in goods markets) is a set of prices $p_1^*, p_2^*, \ldots, p_n^*$, such that supply equals demand, i.e. for all $i = 1, \ldots, n$ one has $y_i(p_i^*, r, w) = x_i(p_1^*, \ldots, p_n^*, b)$.

▶ **Definition 4.8, Partial equilibrium**   Assume there are n goods with n market prices. A partial equilibrium on market i is a price, p_i^*, such that supply equals demand on market i: $y_i(p_i^*, r, w) = x_i(p_1, \ldots, p_{i-1}, p_i^*, p_{i+1}, \ldots, p_n, b)$.

Graphically speaking, a partial or general equilibrium is reached at the point where the supply curve intersects the demand curve, as represented in Fig. 4.9.

As this chapter has explained, equilibrium implies mutual consistency of plans and has, therefore, the status of a local property. However, the question is whether it is a good requirement from an empirical point of view. In other words, the question is whether or not "real" markets tend to be in equilibrium. This is a question about the dynamic forces that would act on the variables, if the system were not in equilibrium. Economists usually have two epistemic views on that problem.

Some would argue that equilibrium is nothing more than the requirement of logical consistency, which implies that markets cannot be out of equilibrium, by definition. When one empirically tests these theories, one can only observe equilibrium points. The rest of supply and demand curves are empirically non-accessible. Others have an epistemically less rigid approach and interpret the idea of equilibrium more metaphorically, admitting that markets can be out of equilibrium in the above sense. Equilibrium analysis, in this case, must be complemented by a model that explains the adjustment of prices out of equilibrium. Otherwise, the concept would be analytically arbitrary. This is usually done by the formulation of the following conjecture:

▶ **Assumption 4.1, "Law" of supply and demand** The prices of goods adjust in a way that supply becomes equal to demand.

One may have noticed that there is a tension in the way the "law" of supply and demand is introduced. It has the status of an assumption but, at the same time, it looks like a property of markets that can either be true or false and that could, in principle, be empirically tested. Intuitively, it makes perfect sense that excess demand will have the tendency to drive prices up and excess supply will have the tendency to drive prices down, but mainstream economics is still lacking a full-sized theory that explains out-of-equilibrium behavior in this sense and, at the same time, it unambiguously supports the "law." The conceptual problem is nicely summarized by the following quote from Kirman (1992): "Economists have no adequate model of how individuals and firms adjust prices in a competitive model. If all participants are price-takers by definition, then the actor who adjusts prices to eliminate excess demand is not specified." In other words, the invisible hand of the market has to belong to someone and this someone is absent in standard theory. Therefore, the standard theory is static in nature, and the epistemically most convincing position is to bite the bullet and interpret equilibrium as a logical constraint of the model.

Why are economists interested in a general equilibrium? There are at least five reasons that one should briefly discuss.

- As mentioned before, economists are concerned about the *existence* of an equilibrium, because its existence implies that "the invisible hand of the market" can guide individual decisions in a way that makes them mutually consistent. The modern treatment of the existence problem and general equilibrium theory, in general, goes back to Léon Walras, a French economist from the nineteenth century. His idea, from today's perspective, was of striking simplicity: the equilibrium condition for each market, i, is characterized by the equation $y_i(p_i^*, r, w) = x_i(p_1^*, \ldots, p_n^*, b)$, or $x_i(p_1^*, \ldots, p_n^*, b) - y_i(p_i^*, r, w) = 0$, for every good or service $i = 1, \ldots .n$. Thus, the problem of the existence of a general (partial) equilibrium boils down to characterizing the conditions under which a system of n (1) equation(s) with n unknown variables – the prices – has a solution. Nearly a hundred years later, in the beginning of the 20th century, there were still uncertainties as to whether or not such an equilibrium exists under circumstances that are sufficiently general to be representative of real-world supply and demand decision. One of the major achievements of the so-called *general-*

equilibrium theory is the clarification of these conditions. Digging deeper into the problem would be far beyond the scope of an introductory textbook, but the general consensus is that the conditions that are sufficient to guarantee its existence are relatively mild. Thus, I give a statement of this property without proof:

▶ **Result 4.1, Existence theorem**   A general (partial) equilibrium exists under quite general assumptions.

This statement, of one of the most fundamental results of general-equilibrium theory that has been used, is rather loose, so it makes sense to work on understanding this result by means of an example: assume there is only one market with one price, so that the equilibrium condition boils down to $z(p^*, b, r, w) = x(p^*, b) - y(p^*, r, w) = 0$. This function is called the *excess-demand function* and the equilibrium is the root of this function. Intuition dictates that this function would tend to be positive, if the price approaches zero, because a lot of people would like to buy, but only few are willing to sell. By the same token, this function tends to be negative, if the price approaches infinity, because no one can afford to buy the good, but it is very attractive to sell. (Making this conjecture precise requires a lot of work, but this is the insight of intuition in a nutshell.) If one knows the properties of the excess-demand function for very low and very high prices, the intermediate-value theorem tells one that the function has at least one root, if it is continuous, because continuity makes sure that it cannot "jump" above or below zero. Does it make sense to assume that the excess-demand function is continuous? Continuity implies that demand and supply change only a little bit for small price changes. One's demand for apricots does not "jump," if apricots become a little bit more expensive. This assumption also seems pretty reasonable, but whether it holds true or not is ultimately a question of preferences and production technologies. The art and craft of general-equilibrium theory is to work out conditions based on preferences and technologies that make sure that demand and supply are continuous. If they are, a generalization of the intermediate-value theorem confirms that an equilibrium exists. Interested readers can find more details in graduate-level textbooks on advanced Microeconomics.

- A related question is the *uniqueness* of an equilibrium. It is an important property of any positive theory to make unique and testable predictions. If there are *Multiple equilibria*, then the predictive power of the theory is rather limited. Unfortunately, it turns out that the assumptions that are necessary to guarantee uniqueness are much stronger than the assumptions that are necessary to guarantee existence.

- Coming back to the epistemic question of the "correct" interpretation of the existence property, a lot of economists consider stability a desirable property of an equilibrium. *Stability* refers to the property of a system to return to its initial state after a shock has occurred. Assume, for example, that an economy is not in equilibrium initially. Are there forces at work that will lead the economy towards the equilibrium? The answer is similar to the case of uniqueness: only under strong assumptions.

- The problems of uniqueness and stability are also relevant for the predictive power of the model. Assume, for a moment, that the model of competitive markets is a good description of the "real" economy and that it is used to inform oneself and political decision makers about the likely effects of policy reforms like changes in taxes, integration of markets, and so on. As covered before, comparative-static analysis is the analysis and comparison of equilibria for different states of the explanatory variables like, for example, the equilibrium with high and low income taxes. In order to be able to learn something meaningful about real-world policies, uniqueness and stability are, in fact, crucial. Without uniqueness, it would be very hard to predict the outcome of policy reforms and, without stability, one could not be sure that the "new" equilibrium would ever be reached.
- The basic normative question that results from the theory of comparative advantage is how to design institutions (like competitive markets) to make sure that the gains from trade that are, in principle, possible from specialization and trade can actually be realized. This is the question about economic efficiency: are competitive markets capable of inducing incentives such that the problem of scarcity is alleviated to the largest extent possible, given technological and resource constraints? Chapter 5 will be devoted to an in-depth analysis of this question.
- Last, but not least, one could be interested in questions of economic justice, in the following sense: assume that there are two individuals, Ann and Bill, who can specialize and trade according to the theory of comparative advantage. Let the monetary value of the gains from trade that is made possible by this process be CHF 100 (which will henceforth be called *rents*). The fact that markets are, in principle, able to make sure that these gains from trade can, in fact, materialize does not tell one anything about their distribution. They could, in principle, go exclusively to Ann, exclusively to Bill or could be shared. This raises the question of distributive justice and if society has strong opinions about the distribution of these rents, there might be a tension between economic efficiency and distributive justice.

4.4 Equilibrium Analysis

As one says, the proof of the pudding is in the eating so, the next step is to see how useful the model of perfect competition is for gaining better understanding of the economy. The most important comparative-static exercises for the supply and demand side have already been covered. This subchapter will now put them together to show how they can be used to develop one's intuition for the effects of external shocks or economic policy. The case studies below are intended to give one a basic idea of how to analyze economic problems by means of models. The purpose is not to develop a complete picture, which would be a very demanding task. Even the very simple model of demand and supply gives one a lot of mileage in understanding complex social phenomena.

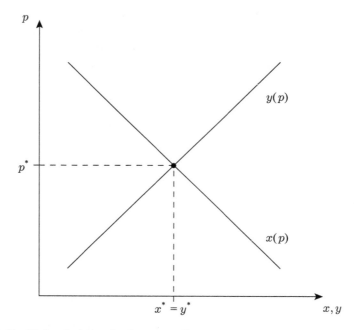

Fig. 4.10 Equilibrium in the market for green coffee

Case study: How bad weather in Brazil affects the Swiss coffee market After oil, coffee is the world's second most important commodity and Brazil is the world's largest coffee producer. In the days before the so-called second and third waves of coffee culture, coffee was essentially run down to a commodity of moderate quality and low prices. Low quality implied that the customers had low willingness to pay, and a low willingness to pay implied that farmers would have a low willingness to invest in quality. Assume that there is a world market for green coffee and that the situation in this market can be summarized by Fig. 4.10.

The supply of green coffee is shown by the upward-sloping supply function $y(p)$, which is determined by the coffee farmers from the different growing regions. On that note, the downward-sloping function, $x(p)$, is the demand for green coffee, which is determined by the coffee roasters, who buy green coffee, roast and package it, and sell it directly to the consumer or to national retailers. Assume that these supply and demand functions reflect the situation in an average year with average harvests. The equilibrium in this market is given by the intersection of the supply and demand curves, which implies a market price of p^* and a trade volume of x^*.

Now, examine the effects of adverse weather conditions in a country, like Brazil, that is responsible for about one third of the world supply of green coffee. These weather conditions reduce the crop by a substantial amount compared to an average year. The effect of this reduction in supply is illustrated in Fig. 4.11: the world

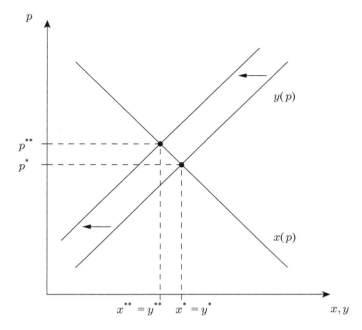

Fig. 4.11 The effect of a reduction in the crop on the equilibrium

supply function for green coffee shifts leftward, because the quantity available at any given price is now smaller than in an average year. The demand function is unaffected by this change, because it is mainly determined by the demand function of the final customers, which leverages onto the demand function of the roasters.

Therefore, the total effect of the shift in the supply function is that the market price for green coffee increases to p^{**}, and the volume of trade drops to x^{**}. What are the implications of this change? At first glance, Brazilian coffee farmers are, of course, negatively effected by the adverse weather conditions, because they suffer a loss in their crop. However, they are at least partly compensated for this loss by the increase in the market price which can even lead to higher revenues of the Brazilian farmers.

The following example illustrates this point: assume the demand for green coffee is given by $x(p) = 1-p$, and supply by $y(p) = a+p$. a is a scaling parameter that can be used to analyze the effect of changes on the crop. Smaller values of a shift the supply function leftwards. These demand and supply functions lead to a market equilibrium of $p^* = (1-a)/2$, $x^* = (1+a)/2$. In addition, the total revenues R of the coffee farmers are given by the total quantity of green coffee times the market price, $R = p^* \cdot x^* = (1-a^2)/4$. Assume that supply in an average year is given by $a = 1/2$, which implies that $p^* = 1/4$, $x^* = 3/4$, and $R = 3/16$. The loss in crop can be analyzed by a change in a from $1/2$ to, say, 0. In this case, $p^* = 1/2$,

$x^* = 1/2$, and $R = 1/4 = 4/16 > 3/16$: total revenues go up, despite the fact that farmers sell less coffee.

It should be noted that the positive effect of the adverse weather conditions on the farmers' revenues depends on the parameters of the model (please check for different values of a). It may also be the case that revenues go down, so it is ultimately an empirical question as to whether a shortage in supply has positive or negative effects on revenues. However, assume for a moment that the effects are positive, because this nicely illustrates the nature of perfect competition. If revenues go up and production costs do not overcompensate this effect, then profits should go up, as well (I will come back to this point in Chap. 8, when I will be in a position to determine profits explicitly). In that case, why do coffee farmers have to rely on bad weather conditions to reduce their crop when they could reduce it voluntarily? Are they irrational, too stupid to understand what is good for them, or is there something more profound going on? The reason why every single farmer has no incentive to reduce his crop is that he is too small to be able to influence the market price by a reduction in crop size. Bad weather conditions, on the other hand, influence approximately one third of the farmers, which has a substantial effect on the market. This explains why prices go up. Therefore, even though each single coffee farmer acts rationally, the total effect is that prices and revenues are low. This is an illustration of the idea of *unintended consequences*, first brought forward by John Locke and Adam Smith.

Other coffee farmers, of course, profit from the increase in price, because their crops are unaffected, so they are able to sell them at higher prices. The effect on roasters is unclear at the moment, because it depends on whether they can pass on the price increase to the retailers or final customers or not. On that note, one can have a look at the downstream market for coffee and, for simplicity's sake, assume that roasters sell directly to customers, bypassing retailers (this simplifying assumption has no qualitative effect on our analysis). The market for roasted coffee looks similar to the market for green coffee, but has a different interpretation. Figure 4.12 shows this market.

The demand function, $X(P)$ (I use capital letters to distinguish this downstream market from the upstream market for green coffee), represents the demand for roasted coffee by customers like students and professors. The supply function, $Y(P)$, for coffee is determined by the coffee roasters. They buy green coffee on the downstream market, which was analyzed before, and use it as an input to produce the different varieties of coffee that customers can find on the shelves. The equilibrium in an average year is given by X^* and P^*. What is the effect of the shortage of green coffee due to the Brazilian crop failure? The increase in the price for green coffee raises the costs of the coffee roasters, which implies that their supply functions also move leftward. Thus, the situation is qualitatively similar to the market for green coffee and is illustrated in Fig. 4.13.

As before, the price for roasted coffee tends to go up to P^{**} and the quantity sold tends to go down to X^{**}. In a situation like this, the roasters are able to pass on part of the increase in input prices to the customers, but only part of it, because customers react to an increase in prices by a reduction in the quantity of coffee

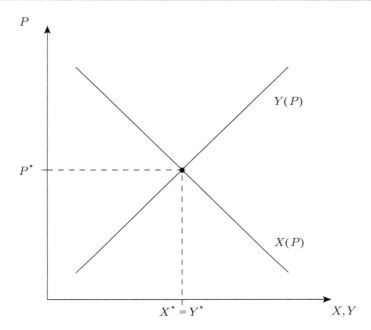

Fig. 4.12 Equilibrium in the market for roasted coffee

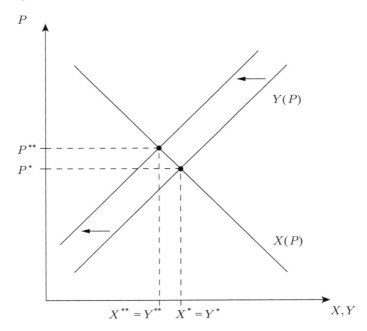

Fig. 4.13 The induced effect of a reduction in the supply of green coffee on the market for roasted coffee

consumed (the demand function is downward sloping). In summary, bad weather in Brazil will ultimately affect Swiss coffee drinkers, because of the tight relationship between the different markets.

Case study: What makes financial markets special? In principle, it is possible to use the same techniques to analyze the functioning of financial markets. There are, however, a few peculiarities of the traded goods that make these markets special and that create an inherent instability. For simplicity's sake, one can restrict one's attention to equity markets, where shares of firms are traded and then start with a representative, potential buyer of stocks. How does she determine her willingness to pay? Contrary to consumer goods, like apples or shoes, shares are not directly useful; people buy them because they want to make money and there are two ways to make money with shares: the first source of income results from the future flow of profits the share will bring to the shareholder, and the second results from a difference between the future selling and the present buying price.

If the flow of future profits were known, the price of a share should be equal to the discounted cash flow (DCF). For example, assume that a share brings a profit of CHF 100 a year from now, CHF 100 two years from now, and nothing from then on. Assume further that she could invest her money in safe government bonds that give her an annual return of 10%. In that case, she should use this interest rate to discount the future profits, which leads to a DCF of $1/(1+0.1)^1 \cdot 100 + 1/(1+0.1)^2 \cdot 100 \approx$ 173.5. With uncertainty about future returns, the price would be reduced by a risk premium. In such a world and with rational expectations, there could only be a difference between discounted selling and buying prices, if unanticipated surprises regarding the future value of the shares take place (for example, because of an unanticipated invention).

The problem with the DCF method is, of course, that no one in the market knows the discounted cash flow for sure, which implies that no one knows if a share is overvalued or undervalued at any given point in time. Take Apple Inc. as an example: its DCF depends on the perceived ability to create "the next big thing." This ability depends on a large number of factors, from the current personnel, to the ability to attract creative employees in the future, and to the general corporate culture and technological constraints and opportunities (the usefulness of devices, like the Apple Watch, depends on the reliability of sensors that allow one to track certain parameters of one's health, and it is unclear *ex ante* if and when the technology will be marketable).

The plethora of different factors that determine future profits make any assessment of DCF risky. If all market participants had equal information, then trade should only occur because of differences in attitudes towards risk. If all market participants also had equal attitudes towards risk, the transactions should only take place between those individuals with more optimistic and those with less optimistic expectations.

The market becomes inherently unstable, because expectations about future profits may become self-fulfilling, if they can spread through the market and influence buyers and sellers. Similarly to viruses that infect a population, rumors can spread and influence the willingness to pay for shares. If market participants, for example, lose their faith in the ability of Apple Inc. to deliver the next big thing, they will reduce their DCF, which can cause a downward price spiral. If share prices go down, however, the ability of a company to finance investments and research and development can be severely impeded, which creates a situation where its ability to create profits in the future is, in fact, reduced: the negative expectations become self fulfilling, even if there was initially no reason for the more pessimistic view. This is the key difference between the share price for Apple Inc. and the price for apples.

Case study: Markets as production technologies The theory of comparative advantage was covered in Chap. 2 and it is now time to come back to those basic ideas to see how they relate to competitive markets. To recapitulate, the sequential integration of ever more trading partners increases the size of the cake, but may, at the same time, produce losers who would be better off with a system of partial integration. The identity of winners and losers is determined by the institutional structure of the economy. I will elaborate on this idea by focussing on policy interventions on competitive markets. There is a striking formulation of the central thought underlying the idea of comparative advantage that goes back to David Friedman. In the words of Steven Landsburg (1995): "There are two technologies for producing automobiles in America. One is to manufacture them in Detroit, and the other is to grow them in Iowa. Everybody knows about the first technology; let me tell you about the second. First, you plant seeds, which are the raw material from which automobiles are constructed. You wait a few months until wheat appears. Then you harvest the wheat, load it onto ships, and send the ships eastward into the Pacific Ocean. After a few months, the ships reappear with Toyotas on them. International trade is nothing but a form of technology. The fact that there is a place called Japan, with people and factories, is quite irrelevant to Americans' well-being. To analyze trade policies, we might as well assume that Japan is a giant machine with mysterious inner workings that convert wheat into cars. Any policy designed to favor the first American technology over the second is a policy designed to favor American auto producers in Detroit over American auto producers in Iowa. A tax or a ban on 'imported' automobiles is a tax or a ban on Iowa-grown automobiles. If you protect Detroit carmakers from competition, then you must damage Iowa farmers, because Iowa farmers are the competition."

This way of thinking allows one to focus on the key aspects of policy interventions in globalized markets. Markets for wheat and markets for cars are connected, because individuals spend the money that they make with wheat for buying cars and *vice versa*. The connection does not have to be direct, but may follow from a complicated maze of interactions that, in the end, make both markets interdependent. Wheat farmers may, for example, sell their harvests to bakeries, which produce and sell bread and pay wages to their workers. The bakeries sell the bread to retailers, who pay their workers, as well. Farmers and bakers use part of their wage income to buy cars and thereby transform, via a complicated chain of events, wheat into Toyotas. The really fascinating insight from this example is that policy interventions in one industry may have – for the layperson – completely unanticipated side effects on other industries. The above example shows that, at the end of the day, the complicated interplay between different markets makes Detroit-built cars and Iowa-grown wheat substitutes. Economists, who are used to think in terms of market interdependencies, can play a very important role in society by carving out these effects.

✍ Here is another example that this chapter does not fully work out, but that can serve as food for thought. Switzerland is currently an immigration country and most people are motivated to move to Switzerland, because of high wages. A couple of years ago, some politicians and economists discussed the idea to skim off part of the 'immigration rent' by the introduction of an immigration tax that has to be paid by the immigrant workers. This may sound like a good idea, however, are we aiming at the right target? In order to understand this question, assume that the labor market for internationally mobile workers is competitive. In this case, immigration will take place up until the point at which workers are indifferent between working in different countries, in the long run. This is called an *arbitrage condition*. The introduction of an immigration tax does not change this logic: arbitrage between labor markets makes sure that the tax-induced reduction of net incomes of mobile workers will be compensated by an increase in gross income, up until the point where the marginal immigrant is again indifferent between working in Switzerland and working abroad. Therefore, the immigration tax cannot skim off rents from mobile workers, because the market reacts in such a way as to compensate them for the tax. Then again, all immigrant workers pay the tax and the state generates revenues.

Hence, there seems to be a puzzle: who pays the immigration tax, in the end? To answer this question one has to analyze the connections between different markets and the most promising candidate is the housing market: immigrants have to live somewhere, so one can expect that immigration would tend to increase rents and the overall price level on the housing market. (A side effect is that immigration triggers a redistribution from Swiss renters to landlords, because rents tend to increase for both Swiss and foreign renters.) What one can conclude is that at least part of the immigration tax is passed on to property owners. Given that an arbitrage condition holds in labor markets for internationally mobile workers, in the long run, it is the group of property owners who profit from immigration, because of the increased values of their property, and it is the same group that, in the end, has to bear the better part of the immigration tax. The phenomenon that the groups that are legally responsible for paying taxes and the groups that, in fact, pay it do not have to coincide is called *tax incidence* in economics and it is, again, one of the most important duties of an economist to create an awareness of the real effects of policy decisions.

Case study: The likely consequences of autonomous vehicles A more speculative example is to tinker with the likely effects of self driving cars, which is no longer exclusively Google's hobby horse. Companies like Mercedes Benz or Volvo have invested a lot of resources into this new technology, and an economic analysis reveals why this is the case. This case study will combine the insights gained from the concepts of both opportunity costs and competitive markets in order to speculate about the economic and social implications of this technological innovation.

It is tempting to think about autonomous vehicles (AVs) in terms of increased convenience, creating the appearance that it is just another innovation. The reality will, however, be quite different, because the new technology will have massive ramifications for cities, insurance markets, labor markets and the way one thinks about mobility.

To begin with, if cars can move around autonomously, then why should they be idle most of the day? It may make much more sense to let a car pick one up and drop one off at work before it continues on to transport someone else. The reason why this option opens up with AVs is a change in opportunity costs. With the current technology, mobility is bound to a human driver, so the opportunity costs of letting one's car transport other people during one's office hours is the wage rate of the car driver (this is the Taxi or Uber model). These opportunity costs drop to essentially zero as soon as the new technology takes over, which makes it more costly (in terms of opportunity costs) to leave a car idle.

These changes will have four likely major consequences:

- They will likely cause a change in ownership structures, because it becomes more and more attractive for an individual user to borrow usership rights than to own a vehicle, because the transaction costs of organizing vehicle sharing are likely smaller with large, specialized sharing companies (that may but do not have to be the car manufacturers themselves; Zipcar, Car2Go and Mobility are good examples). Additionally, one may observe the emergence of new sharing models that do not exist today. People will not stop buying cars altogether, because they might still enjoy the act of driving or the flexibility of owning a car, but the opportunity costs of these aspects of individual ownership will increase, implying that fewer cars will be owned.
  More fundamentally, with AVs forming a large part of transportation, commuting could be thought of as an interlinked system of complementary transport carriers that, together, form a seamless network of mobility opportunities. AVs will be moving through the streets and people will hop on and off as needed. For long-distance rides, AVs will pick one up at home, drop one off at the train station or airport, and another AV will pick one up at the final destination. As a result, the divide between private and public transportation will get blurred.
- If total mobility remains stable, a better utilization of the existing fleet frees up parking space in cities (AVs can park more consistently and closer together humans do now and they can park outside of the city center), which changes the appearance and functioning of cities, leaving more room for pedestrians, but also freeing up valuable properties for better uses than parking space.
- However, if consumers develop from owners of cars to users of mobility services, the functioning of the insurance industry will change, as well. Today, it is mostly a highly regulated business-to-customer market where each individual car driver or car owner is required to insure against accidents. In the future, it will transform into a business-to-business market where insurance companies insure the providers of mobility services. This move is also inevitable, because the main sources for accidents will change from human error to technology failures.

In addition, experts assess that the new technology will be safer than the existing one. McKinsey predicts a 90% drop in accidents, implying annual savings on repair and health-care bills of up to $190 billion in the US alone. This will save lives and drastically reduce insurance premiums.

- The incentives to rethink ownership structures will also likely change the way individuals think about cars and mobility. Cars have been a very important status symbol for the better part of the 20th century. Changes in opportunity costs make this status symbol increasingly costly and maybe even ridiculous in the eyes of a majority of people.

However, AVs will also influence productivity and labor markets. They will, for example, fundamentally reform the logistics industry. The current complementarity between vehicles and human drivers will be replaced by substitution competition between man and machine. The short-run consequence will be downward pressure on wages in these sectors and the long-run consequence will be massive job losses.

Regarding productivity and leisure time, the time spent commuting might not go down but people, freed from the need to drive, can spend their commute working, consuming media or being in contact with friends. These changes will provide opportunities for other industries, for example by creating an in-car media market.

The above examples illustrate the usefulness of the model of perfect competition for an understanding of the functioning and interdependence of markets. They have been exercises in what has been called *positive* economics. A natural question at this point may be if markets are able to coordinate economic activities in a way that is desirable from a *normative* point of view. The next chapter is devoted to answering this question.

References

Arrow, K. J. (1974). General Economic Equilibrium: Purpose, Analytic Techniques, Collective Choice. *American Economic Review, 64*(3), 253–272.

Bourdieu, P. (1983). Ökonomisches Kapital, Kulturelles Kapital, Soziales Kapital. In R. Kreckel (Ed.), *Soziale Ungleichheit* (pp. 183–198). Göttingen: Schwartz.

Kirman, A. P. (1992). Whom or What does the Representative Individual Represent? *Journal of Economic Perspectives, 6*(2), 117–136.

Landsburg, S. E. (1995). *The Armchair Economist*. Free Press.

Further Reading
Bowles, S. (2003). *Microeconomics: Behavior, Institutions and Evolution*. Princeton: Princeton University Press.

Debreu, G. (1972). *Theory of Value: An Axiomatic Analysis of Economic Equilibrium*. Yale University Press.

Frank, R. H. (2008). *Microeconomics and Behavior*. McGraw-Hill.

Kreps, D. M. (1990). *A Course in Microeconomic Theory*. Harvester Wheatsheaf.

Normative Economics 5

Question: How should something be?

e.g. Taxes on raw materials?
reintroduction of study fees?

This chapter covers . . .

- what consequentialist and deontological theories of justice are and how they relate to virtue ethics.
- how mainstream economics is based on a consequentialists' ideas of justice.
- the concept of Pareto efficiency and why competitive markets are efficient.
- how there could be tensions between efficiency and distributional objectives.
- how individuals can fail to do what is good for them.

5.1 Introduction

'It is demonstrable,' said he, 'that things cannot be otherwise than as they are; for as all things have been created for some end, they must necessarily be created for the best end. [. . .] [A]nd they, who assert that everything is right, do not express themselves correctly; they should say that everything is best.'
  'If this is the best of possible worlds, what then are the others?' (Voltaire, Candide)

If one is a utilitarian in philosophy, one has the perfect right to be a utilitarian in one's economics. But if one is not [. . .] one also has the right to an economics free from utilitarian assumptions. (John Hicks)

The analysis of the coffee market in the last chapter showed how the model of perfect competition can be used to better understand economic phenomena. This has been an exercise in what economists call positive economics, which is a very important aspect of economics, as a social science. However, most people are not only interested in the logic of social interaction, but also in normative questions about desirable properties of institutions like markets. Economists, like other social scientists, are not experts in justifying specific, normative criteria, but what they can do is to analyze if or to what extent certain institutions make one's ideas about justice and fairness a reality. There is a division of labor between economists, practical philosophers and the general public in the discourse about the "right" way to

© Springer International Publishing AG 2017
M. Kolmar, *Principles of Microeconomics*, Springer Texts in Business and Economics,
DOI 10.1007/978-3-319-57589-6_5

organize society. The general public has certain (culturally influenced) viewpoints and gut feelings about justice that are scrutinized and systematically analyzed by practical philosophers, and some of these theories are put to the test by economists, who try to figure out how institutions have to be designed to help promote the normative goals of the individual members of society. Under ideal circumstances, this process can lead to a fruitful discourse between philosophers, economists and the general public, because the coherence of one's ethical gut feelings with the implied institutional consequences can thereby become visible and may lead to a process of adjustments in one's ethical views as well as one's ideas of just institutions. John Rawls (1971), a philosopher, called such a state of balance among ethical intuitions and institutions, which is reached through a process of deliberative mutual adjustment among general principles and particular judgments, a *reflective equilibrium*.

   The outlined picture of the division of labor is, maybe, a little bit too optimistic, in the sense that mainstream economics is overwhelmingly concerned with a specific class of normative theories, which are called *welfarism*. Welfaristic theories of just institutions start from the normative premise that individual welfare, and only individual welfare, should matter for an evaluation of institutions. Individual welfare is measured in terms of the (subjective) wellbeing (often called *utility*) the individuals experience (or are supposed to experience) in a specific institutional context. Welfarism is a subclass of a larger class of normative theories that is called *consequentialism*. All consequentialist theories of justice share the view that the consequences of acts are all that matters for normative evaluations. This property has far-reaching implications for the way one perceives the role of institutions: they are basically incentive mechanisms that have to guarantee that individual behavior leads to the socially most desired outcomes. Institutions are like irrigation systems: the flow of water follows the laws of gravity so, to make sure that a garden flourishes, one has to dig the channels in the right way. The same is true for society: individuals follow their interests so, to make sure that individual and social interests are aligned, one has to make sure that individual interests are "channeled" in the right way, by means of adequately designed institutions.

   In looking at the big picture, consequentialism itself is only one of three major classes of normative theories that are debated in practical philosophy, the other two being *deontology* and *virtue ethics*. Deontological theories assert that consequences are irrelevant for the normative evaluation of acts, but rather the focus belongs on certain properties of the procedure, which lead to decisions. A prominent representative of this way of thinking is Immanuel Kant, who famously claimed that good will is the only analysis that counts for the normative evaluation of acts, though there are many more. This view puts much more emphasis on individual moral responsibility and less on institutions. It states that the primary entity that makes sure that individuals behave morally is the law of reason, not the law of the state. The role of formal institutions is, therefore, secondary.

   Another classical proponent for a completely different deontological concept of justice is John Locke, who argued that humans have absolute natural rights. Rights are not assigned because they serve a higher purpose (they are means), but because they are an integral part of what it means to be human (they are ends). According to this view, natural rights are not contingent upon the laws, customs, or beliefs of any particular culture or government, and therefore are universal and inalienable. They are life, liberty, and property. However, if property is a natural right of every human being, then markets get a direct, normative underpinning, because liberty, private property and markets go hand in hand. Disciples of the natural-rights tradition do not support markets because they have desirable consequences, but because they respect property and liberty.

Virtue ethics goes all the way back to at least Aristotle and is a theory that sees the main challenge a human being faces in the quest to perfect his or her virtues. Very similar ideas can be found, in for example, Confucianism and Buddhism. The virtuous moral person, like the virtuosic violin player, acts morally effortlessly, because she trained herself to make it her "second nature." The virtuous person does not act morally in the sense of Kant, because she does not act out of a sense of duty. If a person performs an act, it is because she is inclined to act this way, due to it it "feeling natural" to the virtuous person, Kant calls this act *beautiful*, not moral.

The virtuous person acts in accordance with his or her moral duties, which again changes the view one has on the role of institutions. Contrary to Kant, who puts a lot of trust in the ability of reason to control individuals, institutions play an important role in virtue ethics, because good institutions help individuals to become (morally) virtuous. The good state, according to this view, is the state that helps its citizens become virtuous: "We become just by the practice of just actions, self-controlled by exercising self-control, and courageous by performing acts of courage. [...] Lawgivers make the citizens good by inculcating [good] habits in them, and this is the aim of every lawgiver; if he does not succeed in doing that, his legislation is a failure. It is in this that a good constitution differs from a bad one." (Aristotle, Ethics 1103a30)

There is also a decisive difference between virtue ethics and consequentialism regarding the role of institutions, which can be traced back to Machiavelli. He wrote that "anyone who would order the laws [...] must assume that all men are wicked [...] it is said that hunger and poverty make them industrious, laws make them good." (Machiavelli 1984, 69–70). The task of government for Machiavelli, was not to make citizens moral, but to make them act *as if* they were (Adam Smith's invisible hand that leverages self-interest onto social welfare lurks in the door). Institutions, in this sense, are *incentive mechanisms* and this view made its way via Mandeville and Hobbes into modern consequentialism, with far-reaching consequences for people's ideas about the role of institutions and the balance between individual responsibility, autonomy and the state. A state, whose main purpose is to make selfish people behave as if they were not selfish, is a different state from the one that helps people to develop, for example, the virtue of justice. Both ideas about the role of institutions start from different anthropologies and it is unclear which one describes a human being more adequately.

Mainstream economics has mostly, if not exclusively, focused on welfaristic theories of just institutions and is, in this respect, normative. Insofar as it is not tailored to the specificities of consequentialism, the toolbox could, in principle, be used to analyze the implications of other ethical views, but this is not done in practice.

✍ Economists' self-perception is that they are no experts in normative theories and that they, therefore, focus on what could be seen as a minimum criterion for a just society: the criterion of *Pareto efficiency*. The idea goes back to the Italian economist Vilfredo Pareto. He wanted to understand under which conditions institutions are able to cope with the problem of scarcity in order to avoid waste. Waste, in this sense, is not the peel of a carrot, but a specific property of the allocation of goods, services and resources. An *allocation* is a technical term for the distribution of resources, goods and services among the individuals in a society. The basic idea is that this allocation would be wasteful, if it were possible to redistribute the available goods and resources in a way that makes at least one individual better off without making any other individual worse off. This type of wastefulness will henceforth be called *inefficiency*, and an allocation that avoids waste will be called *efficient*.

The idea of efficiency sounds rather intuitive: an allocation cannot be just in the welfaristic sense, if it is possible to make some people better off without harming others. Therefore, efficiency is, in a sense, a necessary condition for a just allocation of goods and resources. The question as to whether this is sufficient or not will be the topic of later discussion.

In order to make this idea more precise, one can split the production and consumption of goods and services into two classes of activities: production, given resource constraints, and consumption, given constraints on the available goods and services (scarcity).

▶ **Definition 5.1, Efficiency in production** An allocation of given quantities of resources is efficient in production, if it is not possible to reallocate the resources among the producers in such a way as to increase the production of at least one good without reducing the production of some other good.

▶ **Definition 5.2, Efficiency in consumption** An allocation of given quantities of goods and services is efficient in consumption, if it is not possible to reallocate the goods and services among the consumers in such a way as to increase the well-being of at least one consumer without reducing the well-being of another consumer.

▶ **Definition 5.3, Pareto efficiency** An allocation of given quantities of resources, goods and services is Pareto-efficient, if it is efficient in production and consumption.

✍ It is straightforward to extend the above definitions to the concept of a *Pareto improvement*: comparing allocations A and B, if no one is worse off and at least one person is strictly better off in A than in B, then A is said to Pareto-improve B. (Note that two Pareto-efficient allocations can never Pareto-improve each other, but it is not true that a move from an allocation that is not Pareto-efficient to an allocation that is Pareto-efficient is always a Pareto improvement. Assume, for example, that allocation A gives 30 apples to individual i and 30 apples to individual j, allocation B gives 80 apples to individual i and 20 apples to individual j, and allocation C

gives 40 apples to individual i and 40 apples to individual j. The individuals prefer more apples to fewer apples. A is not Pareto efficient, because it is dominated by C, but both, B and C are Pareto efficient. Moving from A to B implies a change from a Pareto inefficient to a Pareto efficient allocation, but it is no Pareto improvement, because j is worse off.)

The concept of Pareto efficiency has some intuitive appeal as a normative principle, but has nevertheless been criticized even by adherents of welfarism. The reason is that Pareto efficiency is "blind" with respect to the distribution of economic rents. Assume that Ann and Bill prefer more money to less money and try to distribute CHF 100 in a Pareto-efficient way. It is straightforward to see that *any* distribution of the money among the two is Pareto-efficient: the only way to make one person better off is by taking money away from the other person, which makes this person worse off. Thus, Pareto-efficient allocations may easily be at odds with one's ethical intuitions about just or fair distributions of goods and services.

On the other hand, it is hard to deny that a plausible normative theory (among the welfaristic ones) would not qualify a Pareto improvement as a general improvement in the well-being of society: if it is possible to improve the lot of at least one person without harming any other, why should one not move in this direction? As long as one is not malevolent, it is hard to justify arguments against Pareto improvements. To summarize, if one considers welfarism to be a convincing class of normative theories, then seeking Pareto improvements is necessary, but may not be sufficient for justice.

5.2 Normative Properties of Competitive Markets

The definition of Pareto efficiency is very general and relies on a concept of individual wellbeing that this textbook has not formally introduced so far. While motivating individual and market demand, Chapter 4 made a vague point that it has something to do with individual preferences that we will formally introduce in Chap. 5. In order to see if one can say anything about the efficiency of equilibria on competitive markets, one has to derive a proximate measure for efficiency. Fortunately, this can be done.

In order to see how to do this, it makes sense to focus on a special example of a market, a market for some good in which the demand of a single customer is typically either zero or one, like refrigerators. The analysis is completely general, though, and extends to all products. Figure 5.1 shows the demand function on the market for fridges.

Each point along the demand function can be associated with a specific individual in society and the individuals are ranked according to their willingness to pay for a fridge. This interpretation allows for a very powerful interpretation of the points along the demand function: they give us the customers' maximum willingness to pay. Look at the individual who is "behind" the first unit of the good. The market-demand function at this point signals a willingness to pay that is equal to CHF 2,000. How does one know? By analyzing the response of this customer to

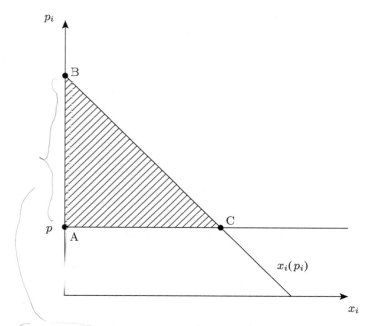

Fig. 5.1 Consumer surplus in the market for refrigerators

different prices. If the market price is below CHF 2,000, the customer is willing to buy, if it is above, she prefers to not buy. Thus, CHF 2,000 is the critical price of the good where the customer is indifferent between buying and not buying, hence it is her willingness to pay.

Assume that the price of the good is equal to CHF 1,200. In that case, the customer will buy one unit of the product. Is it possible to infer anything about the customer's increase in well-being? Under a certain condition that will have to be scrutinized below, yes, because her willingness to pay would have been CHF 2,000 and she pays only CHF 1,200, so a monetary measure for her increase in well-being is CHF 2,000 − CHF 1,200 = CHF 800. The same logic can be applied to all customers, whose willingness to pay exceeds the market price. (All other customers are neither better nor worse off, because they do not buy the good.) Therefore, the aggregate monetary surplus is given by the added differences between one's maximum willingness to pay and one's actual payment. It is equal to the triangular area ABC in Fig. 5.1. This area is called the *consumer surplus.*

In order to define this measure formally, one has to make use of the concept of an *inverse function.* Remember that a function, f, is a mapping from one set A to some other set B that links elements from A with elements from B, so $f : A \rightarrow B$. Assume that the mapping is one-to-one such that, for every element a in A there is exactly one element b in B that is connected with the element in A by f, $b = f(a)$ and *vice versa.* The function, f, answers the question as to which elements in B

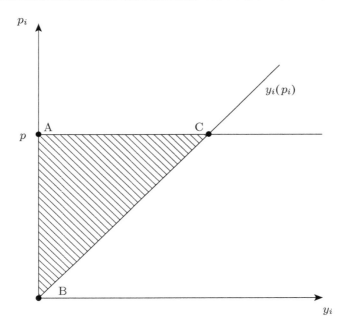

Fig. 5.2 Producer surplus in the market for refrigerators

are associated with the elements in A. One can also ask the opposite question: take an arbitrary element of B; which element of A is associated with it? Given that the mapping is on-to-one, the answer is given by the inverse function that is usually denoted by f^{-1} and which is a mapping from B to A.

▶ **Definition 5.4, Consumer surplus** Given a market demand function for some good i, $x_i(p_i)$, and a market price p_i, let $P_i(x)$ be the inverse demand function and define as $x(p_i)$ the demand where the price equals the willingness to pay. The consumer surplus is the aggregate difference between the customers' willingness to pay and their actual payment,

$$CS(x(p_i)) = \int_{x=0}^{x(p_i)} (P_i(x) - p_i)dx.$$

One can develop a similar argument for the supply side. Figure 5.2 gives one the supply function for refrigerators.

Assume, for simplicity, that each seller sells either one or no fridge. Then, each point along the supply function can be associated with a specific seller in society and the sellers are ordered according to the minimum price they want to receive in order to be willing to sell the fridge. In order to understand why, look again at the firm that is "behind" the first unit of the good. The market supply function at this point signals a minimum price that is equal to CHF 100. How does one know? Again, by analyzing the response of this firm to different prices. If the market price is below CHF 100, the firm prefers not to sell the good; if the price exceeds CHF 100, it is willing to sell. CHF 100 is the critical price where the firm is indifferent between selling and keeping the good, hence it is its willingness to sell (which is also sometimes called the reservation price). Formally, this price is equal to a point on the inverse of the supply function. Assume that the price of the good is equal to CHF 1,000. In this case, the firm will sell one unit of the product. This increases its (monetary measure of) well-being by CHF 1,000 − CHF 100 = CHF 900.

Again, the aggregate monetary surplus of all firms that sell at a given market price is given by the added differences between market price and willingness to sell. It is equal to the triangular area ABC in Fig. 5.2. This area is called *producer surplus*.

▶ **Definition 5.5, Producer surplus** Given a market supply function for some good i, $y_i(p_i)$, and a market price p_i, let $Q_i(y)$ be the inverse supply function and define as $y(p_i)$ the supply where the price equals the willingness to sell. The producer surplus is the aggregate difference between the market price and the firms' willingness to sell,

$$PS(y(p_i)) = \int_{y=0}^{y(p_i)} (p_i - Q_i(y))dy.$$

Combining supply and demand in the same figure, one can now calculate a measure for the aggregate rent on this market, see Fig. 5.3.

What one can see in this figure is the sum of consumer and producer surpluses as the total area between the supply and demand function up to the equilibrium quantity x^*. This sum of consumer and producer surpluses is a measure for the gains from trade that are made possible by this market.

How do the concepts of consumer and producer surplus relate to the concept of Pareto efficiency? If one identifies the willingness to pay and the willingness to sell as expressed on the market with the individual's "true" willingness to pay and sell, then one can identify the allocation that maximizes the sum of consumer and producer surplus with a Pareto-efficient allocation: the only way to make sellers better off is by increasing prices, which makes customers worse off, and *vice versa*. By the same token, selling more than the equilibrium quantity requires both a price below the market price, to induce a buyer to buy, *and* a price above the market price, to induce a seller to sell, which boils down to saying that one would destroy rents. This observation is one of the most profound findings of the theory of competitive markets and, therefore, has a very prominent name.

▶ **Result 5.1, First Theorem of Welfare Economics** Every equilibrium on competitive markets is Pareto-efficient.

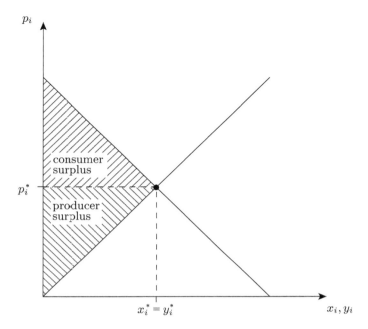

Fig. 5.3   Consumer and producer surplus in the market for refrigerators

The First Theorem of Welfare Economics is a strong result in support of competitive markets, because it implies that markets have a tendency to avoid socially wasteful activities. Under conditions of scarcity, when people would always prefer a larger slice of the cake, competitive markets make sure that the cake is as large as it can be, given the available resources. This is why many economists have a lot of confidence in market economies and competition.

It gets even better. The previous subchapter explained that Pareto efficiency is only a necessary but, for many people, not a sufficient criterion for distributive justice, because the resulting allocation may be highly unequal. Can one say anything about the distribution of welfare? The so-called Second Theorem of Welfare Economics gives a clue.

▶ **Result 5.2, Second Theorem of Welfare Economics**   Assume there are endowments of goods and resources, and that demand and supply fulfill certain conditions of regularity. Then, every Pareto-efficient allocation can be reached as a competitive equilibrium by means of reallocating the endowments.

Again, the statement of this result is not very precise, but it is sufficient for working on the economic reasoning underlying the theorem. Building this reasoning is important, because the theorem became very influential for the way economists think about redistribution. For simplicity, assume that one looks at an economy without production, where individuals are endowed with certain goods. They can decide to consume their endowments (autarky), or they may enter the market and trade their endowment for some other goods. For example, Ann and Bill are endowed with apples and pears and can try to do better than what they can expect from their endowments, by trading apples for pears with each other. The total endowment of apples and pears is 10 and 10, and both want to consume as many apples as pears. Assume that Ann has all the apples and pears in her endowment and Bill has nothing, so the endowments are $e^A = (10, 10)$ and $e^B = (0, 0)$. In this case, there is nothing to trade and the allocation is Pareto-efficient, but highly unequal. Next, assume that the endowments are $e^A = (2, 8)$ and $e^B = (8, 2)$. In this case, it makes sense to trade and a plausible candidate would be to trade three apples for three pears, allowing Ann and Bill a consumption of five apples and five pears each. This trade would lead to a market price of apples in terms of pears that is equal to 1 (one gets one apple for one pear), and the resulting allocation is the egalitarian one.

Now, assume that one is a social planner or politician, who is leaning towards egalitarian outcomes, and one is confronted with initial endowments $e^A = (10, 10)$ and $e^B = (0, 0)$. The Second Theorem of Welfare Economics tells one what to do: in order to reach a more egalitarian outcome, one should redistribute the endowments of the individuals in roughly the desired direction and let the market do the rest. Therefore, if a social planner, "the state", or politicians have sufficient coercive power to administer this type of redistribution, then there is no tension between efficiency and equity.

One should devote some more effort to deeply understanding the meaning of the welfare theorems. A modern economy is an unbelievably complex social arrangement, where millions and billions of decisions are made every day. Each decision has a tiny influence on the way goods and resources are distributed among individuals. If I decide to spend CHF 150 for a new pair of sneakers, I am revealing that the pair of sneakers are worth more to me than their price and, at the same time, they must be worth less to the producer, because the purchase is voluntary. Thus, trading the sneakers is efficiency-enhancing. On that note, if there are people who are willing to sell sneakers at the given market price, they will enter the market. Given that this process only stops when the willingness to pay of the "last" buyer equals the willingness to sell of the "last" seller, markets are Pareto-efficient and goods and resources are directed towards their most efficient uses. There is no centralized planner with information about the willingness to buy and sell of billions of individuals to get to this point: the only thing that is needed is that individuals have information about the prices that are relevant for them.

At the same time, I am revealing that CHF 150 for a pair of sneakers is worth more to me than any other alternative use of the money, including saving the money for my future (concept of opportunity costs). This creates a link between the market for sneakers and *all other* markets. This complicated network of markets makes sure that signals about relative scarcity are transmitted in a way that guides resources towards their most efficient uses. If, for example, a technological innovation in the IT sector (for example a new accounting software) creates a substitute for a traditional job, like an accountant, which has capital costs lower than the wage rate, then firms will start replacing accountants with software. If software is cheaper to use, it reduces the costs of production, which reduces the firm's willingness to sell. For given market prices, profits increase, but the firm will ultimately be pressured on market prices, because high profits will encourage market entry. Hence, the technological innovation influences the price of the goods that are produced with this technology and makes them relatively cheaper compared to other goods. This effect, again, redirects consumer behavior: if the good is ordinary, consumers will buy more of the cheaper goods, increase consumption of their complements, and reduce consumption of their substitutes, which has effects on these markets, as well. Therefore, the effect of a relatively local technological change will ultimately be spread over the whole economy, leading to adjustments in all kinds of markets.

How about the accountant? The technological innovation created a substitute for his job, making him compete with a new technology. The only way for the accountant to keep his job is to be willing to reduce his wage to the point where the employer is indifferent between using the new computer software and human labor. In this sense, wage rates also signal relative scarcity: the emergence of new technologies makes this specific type of labor less scarce, leading to lower prices (wages). In the long run, this reduction in wages is an important signal, because it discourages people from becoming accountants, making labor available for more valuable uses. Thus, wage rates are also an important signal of scarcity that support individuals in their decisions to qualify for certain jobs. However, this knowledge may be of little help for a fifty-year old accountant with two young kids and a mortgage to pay, who becomes unemployed.

Should one trust the theorems of welfare economics? There are three points that should be mentioned before one can reach a conclusion:

- The reason why there is no tension between efficiency and equality in the example is that redistributing exogenous endowments has no adverse incentives for the individuals. The amount of ingredients that are available for baking the cake do not depend on the initial property rights of the ingredients. If this were the case, redistribution might have adverse incentive effects. For example, if the state levies an income tax, people are likely to be discouraged from working. In this case, there is a tension between efficiency and equity, because moving into the direction of more egalitarian outcomes shrinks the pie. Therefore, the policy advice that follows from the Second Theorem of Welfare Economics is to look for "tax bases" that do not react to redistributive policies. However, such tax bases are rather limited. The only ones that come to mind are land plus the natural resources in the ground (but, even in this case, the willingness to extract them may depend on the tax system), potential ability of the people, like IQ (but there is a lot of evidence that IQ is, to a certain extent, a function of effort), or the individual himself (which is called a *head tax*). All other tax bases may react to changes in redistributive policies. Hence, the range of applicability of the theorem, in its pure form, is rather narrow, but the general insight is very important: if one wants to minimize the efficiency costs of egalitarian policies, one should try to identify tax bases that are as independent as possible from the redistributive policies.

- In order to be able to impose and enforce redistributive policies that are in line with the Second Theorem of Welfare Economics, the agency that is in charge needs sufficient independence and sufficient coercive power to be able to enforce the policies. *Independence:* coming back to the apple-pear example, it is likely that the endowment-rich Ann will oppose redistributive policies and she has at least two channels to be effective in this respect. First, she can try to influence the agency's decisions, for example by lobbying. Putting politicians on the payroll of the rich is a very effective way to prevent even worse redistributive policies (from the point of view of the rich). Therefore, the quality of political institutions becomes important in determining whether redistributive policies can be implemented or not, if one cannot rely on the intrinsic motivation of the politicians and bureaucrats to execute them. *Coercive power:* a second problem, which has to do with the quality of political institutions, is the ability of the agency that is responsible for redistributive policies to actually enforce them. Ann, for example, could try to shield her fortune by complicated tax-avoidance strategies, trusts, etc. If the agency has only limited means to enforce its policies, then it has to rely on the voluntary cooperation of the "rich."
- The third point worth mentioning is more methodological. In the apple-pear example, the "state" would like to enforce the egalitarian solution $(5, 5), (5, 5)$. However, if this is the case, why do they choose the detour $(2, 8), (8, 2)$ and rely on markets, instead of choosing the desired allocation directly? Looking at the problem from this angle shows that the second theorem is, of course, correct, but it does not provide us with a strong argument in favor of competitive markets, because it is unclear why markets are needed in the first place.

5.3   Is One's Willingness to Pay One's Willingness to Pay?

The argument about the efficiency of market equilibria relies heavily on a rather innocuous-looking, implicit assumption about the relationship between the willingness to pay and the "true" willingness to pay of individuals. Research, which has been primarily conducted by so-called "behavioral" economists, neuroscientists and psychologists has increasingly scrutinized whether one can always identify the expressed willingness to pay or sell with the "true" willingness to pay or sell.

The identification of both is an example of what economists call the *theory of revealed preference*, which makes the point that the true, normatively relevant preferences of a person can be elicited from his or her (market) behavior. This conjecture has strong implications for the normative evaluation of individual choices, because it implies that individuals make no mistakes when they choose among different alternatives. This does not mean that they never regret their choices, but that any regret is a necessary consequence of resolved uncertainty: I caught a virus during my trip to a foreign country so, *ex-post*, I would have preferred to have stayed at home. However, *ex-ante*, before the trip, and given my subjective assessment of the risks, it was still the right decision.

Whether or not the observed willingness to pay is a reliable measure for the actual preferences of the individuals is a highly controversial and disputed question, because much is potentially at stake. If one assumes that people sometimes do not know what is best for them, then the door is wide open for paternalistic interventions that undermine individual freedoms. However, at the same time, not interfering with individual freedoms implies that those who understand those weaknesses and design products and pricing strategies to their advantage can exploit systematic weaknesses in the ability to make correct decisions. I will come back to this point in Chap. 10 when discussing pricing strategies.

A comprehensive overview of so-called behavioral biases, which point towards a gap between the actual and the revealed interests of people, would be far beyond the scope of this book, but this subchapter will use two examples to illustrate the point:

- There is a lot of experimental evidence that decisions can depend on apparently arbitrary "anchors." *Anchoring* describes a process from behavioural economics, based on which one can influence people's estimates with arbitrarily suggested associations – even if the association, the so-called anchor, is completely un-related. For example, if one asks people whether Gandhi was older than 114 when he died and then ask them for his age of dying, one will get higher esti-mates then if one asks whether he was older than 35. In a famous experiment, researchers demonstrated how arbitrary and irrelevant information can influence the willingness to pay. MBA students could buy a bottle of wine. In a first step, they were asked if they would be willing to pay an amount that equals the last two digits of their social-security number. In a second step, they were asked how much they would actually be willing to pay for the bottle of wine. According to standard theory, the social security number should have no influence on their willingnesses to pay for the wine. In practice, however, it turned out that stu-dents with a social-security number that ended with a number below 50 were willing to pay significantly less than those whose social-security number ended with a number above 50. The average willingness to pay in the former group was €11.62, whereas the other group was willing to pay €19.95, on average. Bringing the social-security number to mind makes it an anchor from which the subjects develop their estimates. It implies that completely irrelevant informa-tion can influence one's willingness to pay, even for relatively ordinary products like wine, which challenges the idea of revealed preference, because decisions to buy or sell are likely highly context dependent and the specificities of the con-text that will determine decisions are hard to anticipate. People are especially prone to anchoring effects when they make financial decisions and it can explain a number of marketing strategies like arbitrary rationing: customers will, on av-erage, buy more items in sales promotions if one sets a (high) limit than if one sets no limit at all.

- Another effect is called *ego depletion*. A number of studies have shown that people who are confronted with a challenging cognitive task and a temptation (like eating chocolate) are more likely to give in to the temptation than people are who do not have to solve the cognitive task. The term *ego depletion* reflects the fact that the cognitive task exhausts important aspects of the personality: motivation and self-control. Ego depletion has a lot of behavioral consequences, from aggressive responses to status-oriented behavior. However, from the point of view of willingness to pay, the most interesting consequences are the following: people with depleted egos are more prone to overspending and impulsive purchases (people, for example, are more prone to impulsive purchases after a long day of work, which partly explains why some companies concentrate internet adds during this part of the day), and they have a harder time abiding their diet. Therefore, economic decisions, which are made with a depleted ego, will likely be regretted and one cannot infer the "true" preferences from the observed behavior.

What are the areas where it is very likely that individuals do not consistently act according to their true interests? Loewenstein, Haisley and Mostafa (2008) give an overview: "There are areas of life [...] in which people seem to display less than perfect rationality. For example, although the United States is one of the most prosperous nations in the world, with a large fraction of its population closing in on retirement, the net savings rate is close to zero and the average household has $8,400 worth of credit card debt. Fifty percent of U.S. households do not own any equities, but the average man, woman and child in the U.S. lost $284 gambling in 2004, close to $85 billion in total. Many workers don't max out' on 401k plans despite company matches (effectively leaving free money 'on the table') and what they do invest often goes undiversified into their own company's stocks or into fixed income investments with low long-term yields. At lower levels of income, many individuals and families sacrifice 10–15 percent of their paycheck each month to payday loans, acquire goods through rent-to-own establishments that charge effective interests rates in the hundreds of percent, or spend large sums on lottery tickets that return less than fifty cents on the dollar. Worldwide, obesity rates are high and rising rapidly, and along with them levels of diabetes and other diseases, and people with, or at risk for, life-threatening health conditions often fail to take the most rudimentary steps to protect themselves."

If one takes this list at face value, a pattern becomes visible: the decisions that require a minimum degree of financial literacy, far-sightedness and commitment seem to be the ones where people struggle the most. Maybe our evolutionary past did not shape our brains in a way that makes it easy for us to handle these problems, because they have not been very relevant for the better part of the history of our species.

If one agrees that there are economic decisions where it is uncertain whether an individual is acting according to his or her well-understood interests, then the revealed-preference paradigm is hard to defend and, if one cannot defend it, then one can no longer be certain that consumer and producer surplus are an adequate measure of welfare, which – finally – leaves the relevance of the welfare theorems up in the air. This assessment does not imply that competitive markets are not efficient, if the revealed-preference paradigm cannot be defended in a substantial number of market contexts. What it implies, however, is that one cannot build one's understanding of Pareto efficiency on the welfare theorems.

References

Haisley, R., Mostafa, R., Loewenstein, G. (2008). Subjective Relative Income and Lottery Ticket Purchases. *Journal of Behavioral Decision Making, 21*, 283–295.
Machiavelli, N. (1984). *Discorsi sopra la prema deca di Tito Livio.* Milano: Rizzoli (first published in 1513–1517, translation by Samuel Bowles).
Rawls, J. (1971). *A theory of Justice.* Cambridge (Ma.): Harvard University Press.

Further Reading
Caplin, A., & Schotte, A. (Eds.) (2008). *The Foundations of Positive and Normative Economics: A Handbook.* Oxford University Press.
Fleurbaey, M. (2008). Ethics and Economics. *The New Palgrave: Dictionary of Economics.*
Hausman, D. M., & McPherson, M. S. (1996). *Economic Analysis and Moral Philosophy.* Cambridge University Press.
Sen, A. (1970). *Collective Choice and Social Welfare.* North-Holland.

Externalities and the Limits of Markets

6

This chapter covers ...

- the implicit assumptions underlying the assertion that competitive markets are efficient.
- the concepts of interdependency and externality and how they contribute towards understanding the problem of how to organize economic activities and the role of< markets.
- the concept of transaction cost and why it is important to not only understand limitations of markets, but also the firms and the state as alternative means to organizing economic activities.
- how to apply the concept of transaction costs to understanding how specific markets have to be regulated.
- the relationship between externalities, common goods and public goods, and why these types of goods may justify state interventions beyond property rights enforcement, contract law and market regulation.
- a lot about climate change, why status concerns make one unhappy, and the social responsibilities of firms.

6.1 Introduction

It is not possible to add pesticides to water anywhere without threatening the purity of water everywhere. Seldom if ever does Nature operate in closed and separate compartments, and she has not done so in distributing the earth's water supply. (Rachel Carson (1962))

The release of atom power has changed everything except our way of thinking... If only I had known, I should have become a watchmaker. (Attributed to Albert Einstein)

The last chapter showed that competitive markets are a very effective way to organize economic activity, because they are able to coordinate the behavior of economic actors in a Pareto-efficient way, as long as one sticks to the assumption implicit in mainstream economics that true and revealed preferences coincide (First

© Springer International Publishing AG 2017
M. Kolmar, *Principles of Microeconomics*, Springer Texts in Business and Economics,
DOI 10.1007/978-3-319-57589-6_6

Theorem of Welfare Economics). This finding has potentially far-reaching consequences for one's perception of the economic role of institutions and, especially, of the state: if competitive markets can alleviate scarcity efficiently, and if efficiency is a convincing normative ideal, then the role of the state is restricted to that of a night watchman. The *night-watchman state* is a metaphor from libertarian political philosophy that refers to a state whose only legitimate function is the enforcement of property rights and contracts and whose only legitimate institutions are, therefore, the military, police, and courts.

This concept of a state, whose monopoly on violence is restricted to the enforcement of property rights and contracts, is sometimes also called a *minimum state*. According to this view, a state that extends its functions beyond this role needs a different normative legitimization by, for example, including distributive objectives. However, even in this case, the Second Theorem of Welfare Economics guides directions of government intervention: distributive objects can best be achieved by redistributing exogenous endowments and, if they are not subject to redistribution, one should look for the closest substitutes.

The purpose of this chapter is to scrutinize this narrative by extracting the implicit assumptions underlying the claims about competitive markets. The idea is to put the conclusions into perspective in order to allow one to better understand the reasons for the efficiency of competitive markets, as well as their potential and their limitations in the organization of economic activities. In summary, there are three qualitatively different lines of reasoning suggesting that the First Theorem of Welfare Economics cannot be the final say in the debate about the best way to organize economic activities:

- The first line of reasoning has already been mentioned in the last chapter: irrespective of the functioning of the price mechanism, it is unclear whether the revealed-preference paradigm is adequate for all types of goods and services.
- The second point that one has been able to tackle, is the relationship between the mode of production in a given industry, summarized by the production technology, and the viable market structures. Not all market structures are compatible with all technological modes of production, but there is a close link between the two. For example, perfect competition requires a specific production technology to be sustainable. I only briefly mention this point in this chapter, for completeness, but will dive into the details in Chap. 8.
- Last, but not least, there can be contractual limits to the establishment of markets, and these limits will be the focus for now. As I argued in the last chapter, any exchange of goods and services has two dimensions: a real one that focuses on the physical aspects and a legal one that focuses on the transfer of rights. This is why the majority of arguments, which will be developed below, are also relevant in legal contexts and, in fact, *law and economics* as an interdisciplinary field of research evolved along some of the lines demonstrated here.

I will start with some observations that should puzzle one if one looks at them from the perspective of the First Theorem of Welfare Economics.

First, the welfare theorems provide only necessary, but not sufficient, arguments for a night-watchman state, because one has not looked for the efficiency-properties of alternate solutions yet. It may be that, under the conditions of the welfare theorems, other organizational structures would also turn out to be efficient. As I will show in later chapters, both monopolistic and oligopolistic markets can turn out to be Pareto-efficient as well, and one does not have any *a priori* reason to assume that centralized planning is not efficient, even though the big historical experiment in centralized planning called socialism can, in all fairness, be called a failure. However, maybe a comparison between "capitalism" and "socialism" is too bold and ideologically charged to allow for a constructive view on institutions. A strong and complete argument in favor of markets has to close this gap.

Second, there is a big theory-reality gap that one has to approach. Assume that $\mathcal{L}, \mathcal{B}$ the First Theorem of Welfare Economics is a correct characterization of competitive markets, for all types of goods and services. What one should expect, in this case, is a strong tendency of real economies to evolve into the direction of perfect competition, because such an organization would outcompete others. What would such an economy look like? Every transaction would take place in markets and, for example, corporations and other institutional entities would not exist. The assembly of, for example, cars would be organized by a complex chain of bilateral contracts between all the persons who contribute to the manufacturing of the car and the customers. Everyone would act as "You Inc.'s" on atomistic markets without any firms as hierarchical organizations, which replace the market place by a system of hierarchical command and control mechanisms. However, this is not what we observe.

A lot of economic activities are revoked from markets and are, instead, organized according to the different logic of corporations. Basically, what happens if a firm hires a worker is that the worker accepts, within a certain scope, to comply with the instructions of one's principal, which is a hierarchical and not a market interaction. The first step into the corporation is of course a market transaction (signing the job contract), but it is exactly with this contract that one agrees to simply follow the orders of one's boss without further negotiations about prices and so on. A firm can be interpreted as an institution that replaces markets with hierarchies. Yet how could this ever be beneficial, if a market is a reliable instrument for achieving Pareto-efficiency? Taking the First Theorem of Welfare Economics at face value, firms should not exist. But they do. Here are two potential reasons why: first, because people are not sufficiently smart to figure out how efficient markets are, so they make mistakes by withdrawing so many transactions from the market place; second, there is something missing in the theory.

One can also turn the question on its head: if one infers, from the existence of firms, that there must be good reasons (in efficiency terms) for their existence, why does one not organize all economic activities within a firm? Why does one organize some transactions with the use of markets? This question has been baptized the "Williamson puzzle" after one of the founding fathers of contract theory and the theory of the firm, Oliver Williamson. Here is the idea: if one has a set of transactions that are organized on markets, one could just as well organize them under the roof of one big firm. If markets are efficient, the manager leaves everything as it is, so the performance of the firm must be equal to the performance of the market. However, if the market is, for some reason, inefficient, then the manager can correct this inefficiency by a centralized, selective intervention. Hence, the firm should be able to outperform the market. However, if one thinks about it, this one big firm, which is organizing all economic activities under its roof, comes close to a system of centralized planning. Thus, again, the puzzle shows that there must be something missing in the theory thus far.

Additionally, to further increase the confusion, why do some firms replace the market mechanism for a set of transactions and then hurry to imitate its functioning in mimicking its mechanisms internally by, for example, the introduction of cost and profit centers, where the inter-center transfer of goods and services is organized by centrally administered transfer prices?

6.2   Transaction Costs

There is a plethora of different institutions in modern economies: markets, profit-oriented firms, non-profit organizations and government agencies. These are all responsible for the mediation of the production and distribution of goods and services, all with their own distinctive logic for providing and distributing goods and services. Any economic theory that aims to understand the reasons for the existence and boundaries of these different ways, in order to organize economic activity, must go beyond the First and Second Theorem of Welfare Economics.

Therefore, the challenge is to identify the missing idea that explains institutional diversity. In order to do so, it makes sense to look at the logic of the First Theorem of Welfare Economics from a different perspective. This allows one to reach a deeper understanding of the reasons why markets can be efficient, but also points towards possible explanations for the limitations of markets.

On a very basic level, scarcity implies that individual acts and consequences are interdependent. My decision to drink this glass of wine implies that no one else can drink it. My decision to wear a blue sweater implies (a) that no one else can wear this sweater at the same time and (b) that everyone passing along my way has to see me wearing it. In a world without scarcity, acts would be independent from each other and, therefore, individual goals would not compete with each other. Therefore, what scarcity does is to make individual acts interdependent. As a result, my decisions have repercussions on some other peoples' well-being and the question is whether I take these consequences into consideration when I make a decision. Efficiency, from this perspective, requires exactly this: that each and every person takes the effects of his or her decisions on others into consideration and behaves accordingly. The technical term is that the person *internalizes* the effects of his or her behavior on others.

However, if I am selfish or ignorant, or both, then I do not care about the effects of my behavior on others. This is the point where markets step in: if I own a car and I consider driving it myself, I am also aware of the fact that I could alternatively sell it on the market. What I am doing in this situation is comparing the monetary value of using the car myself with the market price. If the market price is higher, I want to sell my car; otherwise, I prefer to use it myself.

What does this almost trivial observation have to do with other people? Remember what one has found out about equilibrium prices thus far. The market price in a competitive market is equal to the willingness to pay of the consumer, who is just indifferent between buying and not buying. Thus, prices reflect the willingness to pay of other market participants: my ability to sell the good makes me implicitly internalize the effects that my choices have on others, with the consequence that I only use the good if my willingness to pay exceeds the willingness to pay of other potential users. This is the deeper meaning behind Adam Smith's famous remark on self-interest mediated by the market: "It is not from the benevolence of the butcher, the brewer, or the baker that we expect our dinner, but from their regard to their own interest. We address ourselves, not to their humanity but to their self-love, and never talk to them of our own necessities but of their advantages." The self-love of the baker leverages onto one's well-being, because one pays him to do so. Prices, in this respect, have two very powerful functions in an economy: they motivate the selfish to care about the effects of their actions upon others, and they also help the benevolent, because prices considerably reduce complexity.

6.2.1 An Example

Assume a firm produces some good (bread) by means of capital and labor. The capital (oven) is debt-financed and labor (the baker's time) is employed. This economic activity has three effects. First, the bread makes those people eating it better off (it is crispy, tasty bread). Second, it ties capital to the specific use, which has opportunity costs in the sense that it cannot be used elsewhere. Third, the baker spends some time baking bread, which also has opportunity costs either in the form of forgone alternative earnings or in the form of forgone leisure time. With competitive markets for capital, labor, and goods, there will be market prices for both inputs and the output. The owner of the bakery has to decide how much bread to bake, how much to invest in capital, and how much labor to hire. The price for bread signals the social value of an additional loaf of bread, which implies that he correctly internalizes the additional welfare that he creates with his bread. The price for capital (the interest rate) signals the opportunity costs of the next-best use of capital, which implies that the owner correctly internalizes the "damage" that he creates by detracting capital from alternative uses. Additionally, the price for labor (wage) signals the opportunity costs of labor, i.e. the loss in welfare that results because the baker cannot do anything else during the time he is baking bread. This example illustrates not only that decisions are interdependent, but also that markets make sure that they are made in an efficiency-enhancing way.

So far, so good, but one still is not at the point where it becomes apparent how markets are *not* efficient. In order to reach this point, I will modify the above example. In the first modification, the production of the product now has sewage as a necessary byproduct, which is dumped into a nearby lake. This reduces the profit of a fisherman. Can one still count on markets doing their magic and leading the economy towards efficiency? The answer is that it depends, and this is where the legal side of the problem enters the picture. There are three possible scenarios:

1. The firm has the legal right to dump sewage.
2. The fisherman has the legal right to prohibit the dumping of sewage.
3. The existence and allocation of rights is unclear.

The first and second cases are qualitatively identical to the first example: property rights are completely assigned, which is a prerequisite for bilateral negotiations between the firm and the fisherman. Assume that the reduction of sewage by 10% reduces the profits of the firm by CHF 1000 and increases the profits of the fisherman by CHF 1500. In this case, there are gains from trade between the fisherman and the firm, and the fisherman can buy "sewage-abatement rights" from the firm, in case the firm owns the rights. Any price for a 10% reduction between CHF 1000 and CHF 1500 increases the profits of both, the firm and the fisherman, and it is *a priori* not clear why negotiations should not be successful. However, the same holds true if the fisherman is the initial owner of the rights. In this case, the firm can buy "sewage rights" from the fisherman. There is no reason to assume that one assignment of rights is better than the other, from the point of view of efficiency but, of course, both scenarios lead to different distributions of economic rents, because the owner of the right gets paid. This is no different from the case of, for example, apples: ownership rights, of course, have a value, but they are irrelevant with respect to the efficiency of the resulting allocation.

It is only case three where markets cannot do their magic. If there is no "owner" of the lake, the fisherman and the firm can haggle until eternity without ever reaching a legally binding agreement. Therefore, what one can take away from the example is that markets can only be established if property rights are well defined. These findings motivate the following definition:

> **Definition 6.1, externality** An institution is inefficient if not all interdependencies caused by the individuals are internalized. These non-internalized interdependencies are called *externalities* or *external effects*.

Definition 6.1 is sufficiently general to include non-market as well as market institutions. In a market context, the institution is, for example, a system of competitive markets and the internalization takes place by means of market prices. A situation where externalities exist in a system of markets is sometimes also called a *market failure*. If the institution is a firm, the internalization could take place by means of internal transfer prices between divisions or by means of wage contracts for employees. It is important to stress that the concept of external effects refers to the institutional framework in which transactions take place; externalities are not properties of goods and services *per se*.

The above example has shown that incompletely specified property rights can lead to externalities, because markets cannot emerge. This is an example of what is called *incomplete markets*, and the key question is whether markets are necessarily incomplete, because it is impossible to assign property rights, or if the problem can be fixed by "closing the gaps" and assigning previously unassigned property rights.

The narrative of the example has purposefully been developed around an environmental problem, because many people think that there is something inherent in environmental goods that prevents markets from being efficient. This is a profound misunderstanding, as the example shows. The fact that the interdependency between the fisherman and the baker is caused by sewage is inconsequential for the ability of markets to steer incentives efficiently; what is relevant is the existence and enforcement of property rights and contracts. The same type of problem, as in case 3, would occur if the property rights for bread were not assigned or unclear. If everyone were to enter the bakery and take as much bread as he or she could, the allocation of bread would likely be inefficient and the owner of the shop would lose any incentive to continue production. Thus, why is it that, especially environmental goods are prone to market inefficiencies? There are several reasons, but none of them is causally linked to the "environmental quality" of a good or service. One reason is that, for a long time in human history, a lot of environmental goods have not been scarce. Fresh air and water became only scarce in a lot of areas over the last century or so. However, without scarcity, it is not necessary to think about efficient uses and there is, therefore, no need to assign property rights to these goods. Thus, part of the problem is a lagging behind of the assignment of property rights when scarcity finally kicks in. For the better part of human history, humans simply did not have the technology to completely deplete fishing grounds, so there was no need to regulate access, and the same goes for other natural resources. However, these problems are relatively easy to solve because, in principle, one can assign rights.

In addition to incompletely assigned property rights, there is another reason why markets may fail. Assume, in the above example, that property rights are completely assigned, so that either the firm or the fisherman have the user rights for the lake. Therefore, in principle, it should be possible to set up a contract that specifies the quantity of sewage the firm is allowed to emit into the lake. The problem may then be that the contracting parties are not able to verify if the other party sticks to the terms of the contract. There may be emissions by the firm that cause a reduction in the population of fish that is not easily detectable or even impossible to detect. In a situation like this, setting up a contract that specifies emissions may be insufficient to reaching efficiency because the contract cannot be enforced if neither party can verify a potential breach of it (in front of a court, for example).

However, there are other reasons why markets may fail. In order to get to this point, consider a further variation of the example. In this case, the firm no longer produces sewage that impedes with a single fisherman, but pollutes the air with negative consequences for all the residents in the nearby city. Now, one can look into what property rights and markets can do in this example.

1. The firm has the right to emit.
2. The residents have the right to prohibit emissions.

In the first case all the residents have to find an agreement with the firm. However, given that there are a lot of them, reaching such an agreement is likely to be very costly (think of the opportunity costs of time the residents and firm representatives have in reaching an agreement). Thus, it is very likely that negotiations will break down. The same is also true, if the residents hold the rights. Here is a numerical example: assume, as before, that the reduction of pollution by 10% reduces the profits of the firm by CHF 1000 and has a monetary value for each of the 10,000 residents of CHF 2. So there are huge gains from trade (CHF 20,000 − CHF 1000 = CHF 19,000), but each resident is only willing to negotiate up to the point where his or her opportunity costs of time are smaller than CHF 2, which is if they take no longer than, say, five minutes.

These opportunity costs are an example for a type of costs that turned out to be the key for understanding the economic role of institutions:

✍ ▶ **Definition 6.2. transaction costs** Transaction costs are the costs of economic activity that are caused by the institutional framework.

Transaction costs are, therefore, the costs of organizing economic activities, of measuring and policing property rights, of lobbying and rent seeking, or of monitoring performance, to mention a few potential sources.

One can check if the above-mentioned type of opportunity costs prevents successful negotiations. Assume that mutual negotiations take longer than five minutes for each resident. In this case, the potential gains from trade are more than consumed by the transaction costs of negotiations (the transaction costs of five minutes of negotiations are CHF 2 times 10,000 residents = CHF 20,000), so it is very unlikely that negotiations will be successful and, even if they are, they create a negative net value.

The fact that markets will likely lead to inefficient outcomes is not an argument against markets *per se*. The question is if alternative institutions exist that economize on transaction costs. In the above case, the residents could, for example, delegate the authority to negotiate with the firm to a single representative. Even if some residents have to accept a compromise in the negotiations because of their very specific preferences, this compromise may be better than the externalities that would result from decentralized negotiations. To be more specific, assume that the opportunity costs of reaching an agreement to delegate authority to a representative are CHF 1 per resident, and that the subsequent negotiations between the representative incurs additional opportunity costs of CHF 1000 but reach an efficient agreement. In order to calculate the "net" gains from trade one has to subtract the sum of transaction costs from the gains from trade, i.e. CHF 20,000 − CHF 1000 − CHF 10,000 = CHF 9000. In this case, delegating authority consumes part of the gains from trade, but dominates the decentralized market outcome, because it reduces the transaction costs. Note that the resulting arrangement can no longer be described as a decentralized market mechanism, but more closely resembles what might be called "representative democracy."

Digression 11. Class Action $\mathcal{L}$

A class action is an element of the U.S. legal system that allows a group to sue another party. It is a way to overcome the collective-action problem that exists, if many people are harmed by the actions of one party. The problem, in cases like these, is often that the small recoveries that can be expected by any individual do not provide an incentive to sue individually, despite the fact that the aggregate recoveries may be very high. Such a situation creates an incentive for parties to take disproportionately high risks, because the likelihood that they will be brought to court in case of harm is inefficiently small without class action. This problem leads to externalities.

Class action is a means to internalize these externalities. This argument has been explicitly used by the United States Court of Appeals. In Mace v. Van Ru Credit Corporation (1997), the court argued that "[t]he policy at the very core of the class action mechanism is to overcome the problem that small recoveries do not provide the incentive for any individual to bring a solo action prosecuting his or her rights. A class action solves this problem by aggregating the relatively paltry potential recoveries into something worth someone's (usually an attorney's) labor."

This point is also stated in the preamble to the Class Action Fairness Act of 2005: "Class-action lawsuits are an important and valuable part of the legal system when they permit the fair and efficient resolution of legitimate claims of numerous parties by allowing the claims to be aggregated into a single action against a defendant that has allegedly caused harm."

Swiss law, on the contrary, does not allow for class action. When the government proposed a new Federal Code of Civil Procedure in 2006, replacing the cantonal codes of civil procedure, it rejected the introduction of class actions. In the message to Parliament on the Swiss Code of Civil Procedure (Federal Journal 2006, p. 7221) it has been argued that "[it] is alien to European legal thought to allow somebody to exercise rights on the behalf of a large number of people if these do not participate as parties in the action."

6.2.2   Analysis of Externalities on Markets

Externalities in markets can easily be analyzed using the supply and demand diagram that was introduced in Chap. 4. One describes the demand function as a function that measures the customers' willingness to pay, and the supply function as a function that measures the producers' willingness to sell. If interdependencies between individuals remain uninternalized, there is a gap between the individual and the social valuations of economic transactions, implying that individual demand and supply do not adequately reflect the social value of the transaction. Take the emissions problem from above as an example and assume that the residents do

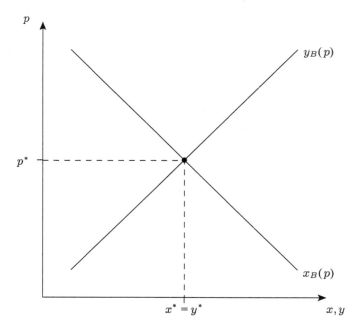

Fig. 6.1 Supply and demand in the bread market

not figure out ways to organize collective action. In this case, markets are incomplete, because a market for emissions does not come into existence. If one wants to analyze this problem using standard supply and demand diagrams one, therefore, has to focus on the market for bread, which is given in Fig. 6.1.

The figure shows the demand and supply of bread in this economy, with an equilibrium quantity x^* and equilibrium price p^*. Focus on the case where the baker pays for pollution first. One knows, from the above reasoning, that there must be a difference between the baker's willingness to sell, if he does not have to pay for pollution, and the willingness to pay, in case he has to pay. If pollution is proportional to the quantity of bread, the supply curve with internalized interdependencies must be *above* the supply curve with uninternalized interdependencies. Making the baker pay for pollution increases his opportunity costs of production, which should influence his willingness to sell any given quantity of bread. Production becomes more expensive, so his willingness to sell should be higher than with uninternalized interdependencies. This situation is given in Fig. 6.2.

Given that the upward-shifted supply curve has been derived for the hypothetical case of complete markets, where the baker has to pay for pollution, it reflects the true social opportunity costs of bread production. Hence, the intersection between the demand curve and the "truncated" supply curve (point O in Fig. 6.2) represents the Pareto-efficient solution, and one sees that uninternalized interdependencies lead to inefficiently high levels of production at disproportionately cheap prices:

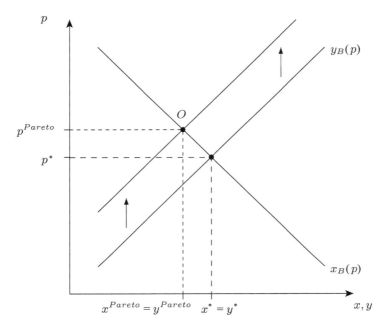

Fig. 6.2 Bread market, if the baker has to pay for pollution or if the fisher has to pay for the omission of pollution

too much for too little. A situation like this is also called a *negative externality in production*.

Digression 12. Externalities, "Polluter-Pays Principle," and the "Principle of Minimum Harm" $\mathcal{L}$, Φ

In environmental law, the "polluter-pays principle" makes the party responsible for producing pollution responsible for paying for the damage. It has support from the Organization for Economic Co-operation and Development (OECD) and European Community countries and it seems to make a lot of sense intuitively: in the above example, the baker is responsible for the pollution of the lake, so why not making him pay for cleaning up his mess?

Before one rushes to this conclusion, however, it makes sense to hold on for a second. It is correct to say that the baker causes the pollution, but this does not mean that he also causes the externality. This claim seems odd at first and it is one of the many counterintuitive insights from Ronald Coase to stress that externalities, necessarily, involve more than a single party. The externality exists only because both, the baker and the fisherman, are located on the same lake. If one of them would move away, the externality would cease to exist. In other words, externalities must be treated as a *reciprocal* problem. The polluter-pays principle ignores the fact that externalities are jointly caused by all involved parties: to avoid harm to a pollutee necessarily inflicts harm on the polluter.

If one is still not convinced, because it *is* the baker who *pollutes* the lake, think about a situation where a dynasty of bakers has been living at the lake for generations. Then, from one day to the other, a fisher decides to settle and set up his business. A few days later, he starts complaining about the pollution. Is it still so obvious that the baker causes the *externality*?

The polluter-pays principle is one way to assign rights, because it implies that one party, and not the other has to pay and, with adequately set payments, the externality gets internalized. One has, however, also seen that the same type of solution can be reached if the baker has the right to pollute and the fisherman pays for reductions in pollution. Such a "pollutee-pays principle" may be at odds with one's intuitions of fairness but, from an efficiency point of view, one has no reason to assume that it is better or worse than the polluter-pays principle. If one sticks with efficiency as a normative principle, it makes sense to replace the principle with a "cheapest cost avoider principle." The idea behind this principle is that it cannot be assumed, in general, that both assignments of rights are equally efficient. With differences in transaction costs, however, it makes sense to assign the rights in a transaction-costs minimizing way.

The above discussion was exclusively concerned with the normative criterion of efficiency, which is an example of an anthropocentric ethic. The reason why the normative problem of externalities vanishes, if the fisherman moves away, is because there is no human being left to be harmed. Environmental ethics like "deep ecology" make the point that such an ethic is too narrow, because the lake, as an ecosystem, still gets harmed and the only way to solve this problem is to reduce pollution. If one includes considerations like this, the polluter-pays principle requires a different interpretation, because it is the only one that respects the integrity of nature. From this perspective, it can be seen as a special case of the more general *principle of minimal harm* or *ahimsa* that is a fundamental moral position of Jainism, Hinduism, and Buddhism. A very popular proponent of the principle of ahimsa was Mahatma Gandhi, and it also shaped Albert Schweitzer's principle of "reverence for life."

$\mathcal{L}$ As previously stated, there are always two ways of internalizing interdependencies (that may differ in transaction costs), depending on which side of the market holds the property rights. Therefore, the alternative in the example above would be to analyze the effects on the bread market, if the residents pay the baker. Is the effect on the bread market identical to the example above or can one expect something different? If the previous analysis is correct, then the assignment of property rights should not influence the efficiency of the solution (without transaction costs), so both scenarios had better yield the same effects on the bread market. In order to check this, assume that the baker gets paid for the reduction in emissions and emissions are again proportional to bread production. For simplicity, assume that one loaf of bread produces one unit of emissions. Let the price of bread be p^b and the price for each unit of omitted emissions be p^e. In this case, an additional loaf of bread has two effects on the baker: it increases his revenues by the market price for bread, p^b, and he reduces his revenues because of the additional emissions by p^e. The total effect on the baker's revenues is therefore $p^b - p^e$, whereas it had been p^b with uninternalized interdependencies. The effect of this change is that the supply curve moves *upwards* as in Fig. 6.2: the only way to convince the baker to sell as much bread as with uninternalized interdependencies is to pay him *more*. This finding verifies the conjecture that, in the absence of transaction costs, it is irrelevant which side of the market pays for the interdependency: it is only relevant that one side does. This insight plays an important role in the literature on law and economics that tries to understand the behavioral consequences of different legal rules.

If there are negative externalities in production, it should not be too surprising that there can also be positive externalities in production, negative externalities in consumption, and positive externalities in consumption:

- **Negative externality in production:** The behavior of an individual causes non-internalized interdependencies, the internalization of which would increase the opportunity costs of production. An example is the above-mentioned problem of uninternalized environmental interdependencies.
- **Positive externality in production:** The behavior of a firm causes uninternalized interdependencies, the internalization of which would reduce the opportunity costs of production. An example is the pollination of fruit trees by bees. The presence of a beekeeper in proximity of a fruit farmer increases the crop of the farmer, because more blossoms get pollinated. If there is no market for "pollination services," the resulting equilibrium is inefficient with too few bees, honey and fruits. Such a situation can, for example, be analyzed in the market for honey, where the individual's willingness to pay falls short of the social value of honey, because the quantity of honey is positively correlated with the pollination services provided, for which the beekeeper is not paid. (This is an assumption to illustrate how this type of interdependency can be analyzed using supply and demand diagrams if, in fact, the interdependency causes an externality. In practice, farmers and beekeepers are likely to figure out ways to pay for the services). Alternatively, one can focus on the fruit market, in which the individual's willingness to sell is higher than the efficient one, because the same quantity of fruit is more costly to produce, if there are not enough bees around.

Digression 13. Pollination Services
The first reaction of a lot of people when they first hear about pollination services is to discard them as a slightly idiosyncratic curiosity, without much

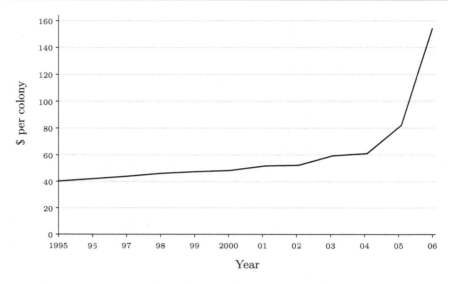

Fig. 6.3 Price level for pollination services. (Source: Sumner & Boriss, 2006, p. 9)

economic relevance. The truth is that pollination services are the backbone of agriculture and are also a very important economic factor.

Pollination makes a very significant contribution to the agricultural production of fruits, vegetables, fiber crops and nuts. Estimates show that pollination services contribute between US $6 and US $14 billion to the US economy per year (Southwick and Southwick, 1992; Morse and Calderone, 2000). The United Nations Environment Programme (UNEP) estimated that pollination services are worth more than US $215 billion globally.

Given the economic importance of pollination services, it should not come as a surprise that commercial pollination services have emerged, mostly provided by honeybees through a long-standing and well-organized market. Californian Almonds are a good example to study the functioning of this market. Almonds are one of the most profitable agricultural products. Recently, honeybee pests and other problems have reduced available bee supplies. At the same time, the high profit margins led to an expansion of almond acreage. Standard supply-and-demand analysis predicts that this trend – shortage in supply and increase in demand for pollination services – leads to an increase in the price. Figure 6.3 shows that this has, in fact, been the case: the average price per colony almost tripled between 1995 and 2006.

Pollination services are an example of what is called an *ecosystem function*, which is defined as "the capacity of the ecosystem to provide goods and services that satisfy human needs, directly or indirectly" (De Groot 1992).

These services are not only provided by bees, but by a wide variety of insects, birds and mammals (like bats). A study for the UK found that insect-pollinated crops have become increasingly important in UK crop agriculture and, as of 2007, accounted for 20% of UK cropland value. Bees account for only about 34% of pollination services, down from 70% in 1984 (Breeze, Bailey, Balcombe, and Potts 2011). Unlike with bees, it is very difficult to create markets for pollination services provided by other species, which leads to externalities. One of the consequences is that the conservation status of pollinating bird and mammal species is deteriorating. Eugenie Regan of UNEP's World Conservation Monitoring Centre makes the point that this trend may be negatively impacting the global pollination services.

- **Positive externality in consumption:** The behavior of an individual causes non-internalized interdependencies with other individuals, which increases the value of their consumption. An example is the decision to buy a product that is interconnected in a network, like a specific type of software or cellphone. The more users coordinate on a given standard, the more valuable the standard becomes for others. For example, the more people use the same text editor, the easier it becomes to exchange text documents. This means that the social value of consumption exceeds the individual. In other words, the Pareto-efficient demand function lies above the market demand with uninternalized interdependencies. Other examples are individual education decisions that raise individuals' qualifications, but also have an effect on the average literacy of a community, or maintenance of housing that not only increases the value of the individual property, but also the attractiveness of the neighborhood.
- **Negative externality in consumption:** The behavior of an individual causes un-internalized interdependencies with other individuals, which reduces the value of their consumption. An example is noise from gardening that annoys the neighbors. In this case, the Pareto-efficient demand function for gardening activities lies below the market demand with uninternalized interdependencies. Another example is vaccination. The World Health Organization (WHO) estimates that vaccination averts 2–3 million deaths per year (in all age groups) and that up to 1.5 million children die each year due to diseases that could have been prevented by vaccination. The individual decision to vaccinate against a pathogen creates a positive interdependency, because it makes the spreading of pathogens more difficult, reducing the risk of other people getting infected. By the same token, the individual decision not to vaccinate creates a negative interdependency. The transaction costs of internalizing these interdependencies on markets are prohibitive, leading to negative externalities in consumption.

The literature on externalities is very inconsistent in its terminology, mixing the physical properties of activities, which one calls interdependencies, together with

the institutional properties, which either lead to an internalization of these interdependencies or do not. It should therefore be stressed again that, in a market context,
the term externality relates to missing or imperfect markets. An analysis that makes
the *assumption* that an externality exists does not ask for the deeper reasons for the
externality and, therefore, risks that one will draw the wrong policy conclusions,
which could have been derived from a more thorough analysis. The baker-fisherman
problem is a good example. If one starts the analysis with the premise that there is
a negative externality between the two, one *assumes* that the two gentlemen cannot
figure out ways to fix the problem. However, it would be in their best interest to
find a solution, because there are gains from trade. Therefore, one must explore
the deeper reasons for this failure and the institutional alternatives. This points one
towards a detailed analysis of transaction costs.

Economists can sometimes be blinded by their own theories. It was, for example,
a staple in the profession that lighthouse services are a good example for positive
externalities in production, because ships cannot be excluded from the insurance
provided by the lighthouses. The obvious policy implication from this analysis
would be that the state has to step in to provide these services, because markets must
fail. However, a more detailed empirical analysis revealed that there are numerous
examples for the provision of lighthouse services without government interventions
and the key to understanding this 'curiosity' was the realization that port owners
have an incentive to provide these services to make their ports more attractive. The
situation is similar to today's free TV or free services on the Internet. Content
providers give away content for free because users allow those firms to make money
on other markets, like advertising or data collection.

6.2.3   The Bigger Picture

It is now time to use these examples to develop a more comprehensive view on
institutions and transaction costs. The idea that something may be missing in standard theory, which helps explain institutions, goes back to a paper by Ronald Coase
that he wrote as early as 1937. Standard theory models firms simply as technological phenomena transforming inputs into outputs, and make a behavioral assumption
that they seek to maximize profits. This "black-box approach" to the firm had the
advantage of simplicity and it allowed for generating a lot of deep insights into the
functioning of markets, some of which the last chapter covered. However, the standard approach turned out to be ill-suited to answering the question of why firms
exist in the first place, given the apparent efficiency of markets. Ronald Coase's
major insight was that transaction costs are at the heart of the problem of optimal
institutional design. Unfortunately, transaction costs are a vexed concept, because
they turned out to be very difficult to define in a precise and useful way.

Much effort has been devoted to better understanding the exact conditions under which the invisible hand can leverage self-interest onto social welfare and the
most useful insight, for this purpose, goes back to another paper by Ronald Coase
(1960). If society is interested in promoting efficiency, then every institution that

is compatible with this goal must share the same structure: it has to make sure that individuals fully internalize the effects of their behavior on others.

As suggested above, internalization of externalities can be achieved by a complete set of competitive markets. The completeness of the markets implies the absence of an important category of market-related transaction costs. The term "transaction costs" is closely related to institutions, since transaction costs can be used to assess the relative "imperfectness" of different institutions. This understanding allows it to put the First Theorem of Welfare Economics into perspective. It was clear from the work of theorists of socialist planning like Oskar Lange (1936, 1937) that, under the conditions of the First Theorem of Welfare Economics, a central planning mechanism is efficient, as well. In order to find the market clearing prices, "the market" needs information that, in the hands of a central planner, would be sufficient to implement the efficient allocation directly without the detour of market transactions. This implies that, under ideal circumstances, the institutional structure does not matter for the efficiency of the resulting allocation.

Coase (1960) generalized this idea by creating the awareness that it is neither the complete set of markets nor the idealized planner mechanism that is responsible for the result, but two other, implicit assumptions, namely the rationality of economic actors and the absence of transaction costs. The insight is of striking simplicity: if individuals are rational and no transaction costs exist, they should always end up in a situation where gains from trade are completely exhausted; it would simply not be rational to leave them unexploited. This result is called the "Coase Irrelevance Theorem." In an ideal world, without transaction costs, potential externalities would be fully internalized, whether through market prices, centralized planning in either the form of centrally determined transfer prices or direct quantity control, or other institutional arrangements.

The implication of this finding is, of course, not that institutions do not matter in reality, but that one has to identify the institution-specific transaction costs, if one wants to understand the relative efficiency of, for example, markets, firms and government agencies. The transaction-costs-free economy plays the role of the frictionless pendulum in physics: it is not a good description of reality, but a benchmark that allows one to understand the role of frictions (or, for that matter, transaction costs) better.

A number of important research areas emerged from this benchmark over the last decades, all of them unified by their attempt to understand transaction costs and their implications for efficiency and the organization of economic activity. The following are some of the most important types of transaction costs:

1. Transaction costs due to the formation of contracts: as shown in the above example, contracts are not just "there" but have to be negotiated, which requires investment of scarce time and effort. Thus, contracts will only be written (and market transactions will only be performed), if the gains from trade exceed the (opportunity) costs of negotiations. Even buying a smoothie requires that one enter the shop, check the price and pay.

A very dramatic example of market failure, due to the impossibility to cope with interdependencies by means of contracts, is the interdependencies between generations. Most of one's present decisions are likely to have long-term consequences far beyond one's own planning horizon. However, they will likely affect the well-being of future generations. The most prominent examples are anthropogenic climate change and nuclear power. In both cases, there surely are intergenerational interdependencies and they cannot be internalized by the use of markets, because one side of the contracting table has not yet been born when the relevant decisions have to be made. Markets must create externalities almost by definition. On the other hand, if markets must fail, what other means does one have to include the interests of future generations? Given that unborn people cannot be part of *any* decision procedure, be it market-based or political, there is only one alternative left: the literal internalization of interdependencies by means of moral concerns of contemporaries. If the present generations are willing to think and act according to the legitimate claims of future generations, then and only then is it possible to internalize the otherwise existing externalities. Even if political decisions to, for example, raise the price of fossil fuels constrain individual behavior, the decision to implement these regulations is not a result of some kind of bargaining between all the affected parties. It is a commitment mechanism by contemporaries that makes it easier for them to follow their moral standards.

Φ   **Digression 14. Is There Someone to be Harmed? The Non-Identity Problem of Intergenerational Justice**
There is an aspect of the problem of intergenerational justice that makes it different from standard allocation problems between contemporaries. There is a debate in practical philosophy about the normative status of unborn people that focuses around the question, of whether unborn people have the same rights as contemporaries and whether and in what sense contemporaries can harm unborn human beings (Parfit 1984). One of the key obstacles is the so-called *non-identity problem*, which argues that apparently trivial changes in one's plans are likely to change the identity of the future people (for example, because the egg is fertilized by a different sperm).

Thus, changes in the political environment are likely to have some influence on the identity of future generations but, if this is the case, it cannot be argued that anybody is worse off in the future because one is comparing different people. A pragmatic view would accept this problem as it is and declare the specific identity of a future human being to be morally irrelevant. The only fact that counts, one could argue, is that future generations will come into existence and that they can profit or can be harmed by present generations' choices. Plausible as this approach may sound, it implies a major deviation from standard welfarism, which builds on the idea that the welfare of actual people is normatively relevant.

2. Transaction costs due to the enforcement of contracts: even in a night-watchman state, property rights and contractual arrangements have to be backed by the police and courts. The capital and labor costs of maintaining these agencies must be considered part of the transaction costs of markets. From an efficiency point of view, the police is only indirectly productive, because its presence creates the necessary environment in which people feel save to invest and trade but, if police were obsolete (for example because individuals behave cooperatively out of an intrinsic motivation), capital and labor would be freed for other directly productive purposes.

3. Transaction costs due to the incompleteness of contracts: An extensively studied problem is the role that information plays in contract design and in the performance of institutions. There are several strands of literature that I will briefly discuss in turn.

   (a) Asymmetric information: Asymmetric information refers to a situation where one of the contracting parties is aware of information that is relevant for the contract and that of which the other contracting party is not aware. This situation is, of course, the rule rather than the exception, because the parties, in almost any buyer-seller relationship, are unaware of the willingness to pay or sell of the other party. Here is an example that highlights the specific problems that may be caused by asymmetric information. Assume a market for used cars, where the sellers are better informed about the quality of the cars than the buyers are. The representative buyer's willingness to pay depends on her assessment of the *average* quality of the car, which implies that the price is not attractive for high-quality sellers. These sellers will withdraw from the market. If the buyers anticipate this incentive, they will further reduce their expectations about average quality and, therefore, their willingness to pay. In the end, the market can completely unravel, leaving only cars of poor quality for sale. George Akerlof (1970), one of the pioneers of information economics, called this type of market a *market for lemons* (a lemon is an American slang term for a car that is found to be defective only after its purchase).

   It turns out that this informational incompleteness is especially relevant on insurance markets and explains why unregulated insurance markets are likely to be inefficient. Specific forms of regulation, like mandatory insurance and obligation to contract (plus some form of price regulation that is necessary to prevent insurance companies from levering out the obligation to contract by charging high prices), reduce these inefficiencies. This kind of regulation works on insurance markets, but not generally on other markets because the standard for efficiency is easy to set. If individuals want to avoid risk (they are *risk averse*), an efficient solution is one where everybody gets full insurance. Such a standard is relatively easy for a government to regulate.

   (b) Non-verifiable contracts: Some contractual arrangements may refer to properties of the good or service that are observable for both contracting parties, but are not verifiable, for example, in front of a court. An example would be

a labor contract, where both parties know that the employee is cheating, but
the employer is unable to prove it.

(c) Imperfect foresight: Many contracts expand into the future, which makes
the anticipation of future consequences of contractual arrangement crucial.
However, in a number of cases, the future cannot be foreseen with sufficient
precision to allow for efficient contracts. An example is a different labor
contract where a person is hired to conduct research for a company. By
definition, the terms of the contract cannot be specified contingent on the
outcome of the research project, because it would contradict the nature of research and development. Something completely and qualitatively new may
come out of a research project, which makes contracts necessarily incomplete.

Φ From the perspective of transaction-cost economics, climate change is maybe the
worst problem someone could have invented to challenge humanity, because it com-
bines a lot of elements that human beings are ill-prepared to solve. First of all, the
very nature of intergenerational interdependencies makes it impossible for everyone
who is influenced by a decision to participate in a market or any other form of ne-
gotiation. Therefore, the only way to incorporate the interests of future generations
into today's decision-making is by means of the morality of the present generations.
Second, even if one is sufficiently morally motivated to care for future generations,
one has imperfect foresight about the future consequences of one's behavior. Third,
humanity evolved as a species that had to solve small-group problems for the better
part of its history. One's "hard wired" moral instincts are restricted to one's kin
and tribe. Problems on a global scale require going beyond one's moral intuitions
and caring for the lot of all human beings, not only one's relatives and fellow tribe
members. However, reason is a lazy and easily exhausted companion. The executive summary of the so-called Stern Review (2007) makes this point in all clarity:
"The scientific evidence is now overwhelming: climate change presents very serious global risks, and it demands an urgent global response. [...] Climate change
presents a unique challenge for economics: it is the greatest and widest-ranging
market failure ever seen. The economic analysis must therefore be global, deal with
long time horizons, have the economics of risk and uncertainty at center stage, and
examine the possibility of major, non-marginal change. [...] The effects of our
actions now on future changes in the climate have long lead times. What we do
now can have only a limited effect on the climate over the next 40 or 50 years. On
the other hand what we do in the next 10 or 20 years can have a profound effect on
the climate in the second half of this century and in the next. No one can predict the
consequences of climate change with complete certainty; but we now know enough
to understand the risks. [...] For this to work well, policy must promote sound
market signals, overcome market failures and have equity and risk mitigation at its
core."

The following subchapter will cover the examples of traffic congestion and environmental problems in order to illustrate how the concept of transaction costs can
be used to understand the organization of economic activity better and to design
solutions for externality problems.

6.2.3.1 Externalities in Traffic

> A society sufficiently sophisticated to produce the internal combustion engine has not had
> the sophistication to develop cheap and efficient public transport?
>   Yes, boss ... it's true. There's hardly any buses, the trains are hopelessly underfunded,
> and hence the entire population is stuck in traffic. (Ben Elton (1991), Gridlock)

The most common feeling of car drivers who are locked in a traffic jam is anger, but these psychological costs are only the tip of the iceberg of the economic costs caused by crowded streets and overburdened infrastructure. The main causes of traffic jams are accidents, poor infrastructure, peak-hour traffic and variable traffic speeds on congested roads. The Centre for Economics and Business Research and INRIX (a company providing Internet services pertaining to road traffic) has estimated the impact of such delays on the British, French, German, and American economies. Here are some of the main findings (US data):

- The costs of congestion summed up to $124 billion in 2013. This cost is (ceteris paribus) expected to increase 50 percent to $186 billion by 2030. The cumulative cost over the 17-year period is projected to be $2.8 trillion.
- The annual cost of traffic for each American household is $1700 today. This cost is expected to rise to $2300 in 2030, with huge regional variations (the cost is $6000 in the Los Angeles area). To put these numbers into perspective, the median household income was $51,939 in 2013.
- The monetary value of carbon emissions caused by congestion was $300 million in 2013. By 2030, this is expected to rise to $538 million, totaling $7.6 billion over the 17-year period.

Congestion costs of traffic can legitimately count as an externality, because the main causes of these costs are (a) opportunity costs of time, (b) costs of carbon and other emissions and (c) price effects of higher transportation costs. In order to understand this conjecture, it makes sense to look at a car driver's decision problem. When deciding if, when or where to use streets, she takes individual costs and benefits into consideration. However, the lion's share of costs and benefits spills over onto other traffic participants and the general public. Emissions cause either regional or global effects, which are not included in the individual's decision problem, and other drivers' wasted time is also neglected. The reason is that decentralized negotiations about when and where to use the streets would lead to prohibitive transaction costs.

What else can one do to make traffic more efficient? What are the institutional alternatives? Solving congestion is not easy. Building more roads, or widening existing ones, can encourage people to drive even more. Charging road users for travelling at busy periods can help to solve the efficiency problem, but it may cause other problems. To highlight them, one can focus on the *London Congestion Charge*. The standard charge in 2016 was £11.50 on most motor vehicles operating within the Congestion Charge Zone (Central London) between 07:00 a.m. and 08:00 p.m., Monday through Friday. In theory, the charge should be set such that the individual driver pays a price that is equal to the costs caused by his decision to use a specific network of streets during a given time period. Hence, if the charge is calculated correctly, one can infer that the externality caused by a single driver is approximately £11.50. If the price of going to central London goes up, demand should go down and one gets the desired increase in efficiency, because congestion is reduced. What makes this instrument problematic is that is has distributional consequences, because the fee is especially burdensome for the relatively poor, who are disproportionately deterred from coming to the city center by car.

Φ **Digression 15. The Role of Public Space in Democracy**
Congestion charges or road prices not only have distributive consequences, which one might find objectionable, but also have more profound effects on how one thinks about the societal role of public space. In a democracy, public spaces have an important role in the expression of political opinions, as locations for spontaneous gatherings and, more generally, places where a representative profile of people comes together and has the right to do so. A public space is a site where democracy becomes possible. Henri Lefebvre (1974) made this point quite poignantly: "(Social) space is a (social) product [...] the space thus produced also serves as a tool of thought and of action [...] in addition to being a means of production it is also a means of control, and hence of domination, of power." Charging high prices for the access to public space, which makes it more difficult for specific groups to access them is, therefore, politically questionable. A narrow economic view, which focuses on efficiency gains, easily loses sight of the bigger context in which the instruments are embedded.

A good example for the relationship between democracy and public space is the *Landsgemeinde* (cantonal assembly). This is a Swiss institution where eligible citizens of the canton meet on a certain day in a public space and debate and decide on laws and public expenditures. Another example is the *Speakers' Corner*, an area for unrestricted public speaking, debate and discussion, which became a symbol for the importance of unrestricted access to public space in a democracy. An interesting, yet unresolved, question is whether virtual public space on the internet can take over the role of physical public space, thereby overcoming physical and legal boundaries.

6.2.3.2   Environmental Externalities

Climate change is a result of the greatest market failure the world has seen. (Nicholas Stern)

The metaphor is so obvious. Easter Island isolated in the Pacific Ocean – once the island got into trouble, there was no way they could get free. There was no other people from whom they could get help. In the same way that we on Planet Earth, if we ruin our own [world], we won't be able to get help. (Jared Diamond (2005), Collapse: How Societies Choose to Fail or Succeed)

Oil spills that waste beautiful beaches and wilderness areas are only the tip of the iceberg of environmental externalities. The following are some examples of environmental externalities in production that lead to social costs that are not internalized by market prices. Unregulated air pollution from burning fossil fuels becomes a problem, if no market for pollutants exists. Anthropogenic climate change, as a consequence of greenhouse gas emissions, involves future generations. Negative effects of industrial animal farming include, for example, the overuse of antibiotics that results in bacterial resistance and the contamination of the environment with animal waste. Another problem is the cost of storing nuclear waste from nuclear plants for very long periods of time.

There is a broad consensus among scientists that the rate of species loss is greater now than at any time in human history. In 2007, the German Federal Environment Minister acknowledged that up to 30% of all species would be extinct by 2010. The Living Planet Report 2014 comes to the conclusion that "the number of mammals, birds, reptiles, amphibians and fish across the globe is, on average, about half the size it was 40 years ago." If one follows the scientific consensus and assumes that part of the loss in biodiversity is a consequence of the economic system, the question is whether this loss is a result of externalities. Is it possible that mass extinction of species can be Pareto-efficient? This is a tough question, because it requires information about the role of biodiversity in supporting human life on this planet and it relies on assumption about the way humans value biodiversity *per se*. If one starts with the conservative assumption that biodiversity has only instrumental value in supporting human life and if one admits that intergenerational externalities exist, because current generations do not adequately take the interests of future ones into consideration, then one can make a case for the existence of an externality. This is if one assumes that a more diverse biosphere is more likely to support human life than an impoverished one. This latter conjecture, however, is built on deep uncertainty of the complex role of the biosphere in supporting human life. The deeper problem is that the concept of Pareto-efficiency, as seen before, is blind with respect to the distribution of gains from trade, and, more generally, economic welfare. A policy where the present generation has a big 'party' and uses up most of the natural resources, leaving a devastated planet where future generations scrap along at the subsistence level, is Pareto-efficient as long as there is no alternative policy to make future generations better off without harming the present ones.

The concept of Pareto-efficiency has a lot of shortcomings when it comes to long-term problems, which is why is has been supplemented, and even replaced, by the concept of *sustainability* in the normative social and natural sciences and in politics. The most popular definition of the concept of sustainable development goes back to the so-called Brundtland Commission of the United Nations (1987): "sustainable development is development that meets the needs of the present without compromising the ability of future generations to meet their own needs." This concept implicitly acknowledges the right of future generations to live a decent life and is, therefore, stronger than the Pareto criterion. However, it still suffers from the need to understand the complex role of ecosystems and it is anthropocentric in nature. I will come back to this latter point at the end of this subchapter.

Returning to a less complex externality, the example of an oil spill illustrates the basic problems and solutions. Assume that a company operates a fleet of oil tankers, which move large quantities of crude oil from its point of extraction to the refineries. The environmental risk of this business model is that oil spills, due to accidents, affect the (marine) environment and may also affect the fishing industry. One can divide the discussion into two parts. Part one assumes that it is possible to attach a meaningful monetary value to the damage caused by oil spills and to ask for institutional arrangements that lead to efficient outcomes. Part two scrutinizes this assumption and takes a closer look at the normative issues that are involved when attaching price tags to oil spills.

The risk of an accident can be influenced by the shipping companies' investment in safety technology. A profit-oriented company faces a tradeoff between the costs and benefits of such investments and the question is whether it adequately reflects the social costs and benefits when it makes its decisions. In an unregulated market, with only property rights and contract law, this is very unlikely, because many people are potentially influenced by an oil spill, so decentralized negotiations cannot solve the problem efficiently. Therefore, safety standards are presumably inefficiently low in an unregulated market. How can one internalize these externalities? I will discuss three different instruments:

- A very direct and crude way of enforcing safety standards is by setting and enforcing mandatory standards. This instrument is effective, if enforcement is guaranteed, but not necessarily efficient. It becomes the more efficient, the more homogeneous the global fleet is because, in this case, the costs and benefits of a reduction in the risk of accidents are the same for all tankers. Unfortunately this is not the case and, the more heterogeneous the ships are, the less efficient a homogenous regulation will be. One could argue that this is not a problem, as long as regulation can be fine-tuned to the specific characteristics of the tanker, but regulations that are more complicated are more difficult to enact and enforce. Therefore, it is very likely that, in practice, standards would lead to some efficiency losses.
- It is also possible to tax activities that are positively correlated with risks and to offer state subsidies for activities that are negatively correlated with risks. Taxes and subsidies change the perceived prices, either making risky activities more costly or making risk-avoiding activities cheaper. The effect is that one creates incentives to influence investments into safety in a socially desirable way. The major advantage of this solution is that, unlike with standardization, this instrument works selectively for different types of tankers and it is, in principle, able to avoid inefficiencies that result from the one-size-fits-all approach of standards. However, a tax-subsidy system has to be administered, which causes transaction costs of its own.
- Last, but not least, one can react with the introduction of liability law. Liability law makes shipping companies pay in case of damage. Liability law increases the costs of the firms in case of an accident and is, therefore, a theoretically promising instrument for internalizing externalities. When it becomes

more costly to have an accident, the company will be more prudent and invest in higher safety standards. However, this legal instrument can conflict with other legal instruments, which have legitimizations of their own. For example, most countries have an insolvency law that restricts the risks of firms and individuals. If such a law is in place, the worst that can happen to a firm is for it to become insolvent, which effectively restricts its monetary risks. Since oil spills are usually big events, liability law can, therefore, be a toothless tiger, if the owners of the company are protected by insolvency regulations.

The above discussion has shown that there are several tools for coping with environmental externalities in the economist's toolbox and it depends on the case at hand which tool (or combination of tools) will work best.

The second aspect of the problem, which one should at least briefly consider, is the question of whether it is possible to attach a price tag to environmental damages. It is relatively uncontroversial that it is possible to get reasonable estimates of damages to the local fishing or tourism industries, because the goods and services they provide have market prices and past experience gives a good proxy for the loss in revenues and profits that result from environmental damages. The question becomes more involved, if one tries to estimate the non-economic costs to human beings that result from the depletion of resources such as air, water, and soil, the destruction of ecosystems and the extinction of wildlife. What is the value of a species of beetle to humankind, which is threatened to become extinct?

However, a radical position would even go beyond the evaluation of non-economic (in a narrow sense) damages and scrutinize the implied anthropocentrism implicit in the normative values underlying Pareto efficiency (or more generally welfarism) and also in the idea of sustainability in the sense of the Brundtland report. According to, for example, the *deep ecology movement*, heavily influenced by the Norwegian philosopher Arne Næs, animals, wildlife and biosystems have intrinsic value, whereas the mainstream approach is to see them exclusively as means for human ends. The latter approach would deny wildlife a right to existence, if it does not serve any needs of human beings. The deep ecology movement would reject the characterization of non-human life as a means to an end. The core principle is the belief that the living environment, as a whole, should be respected and regarded as having certain inalienable legal rights to live and flourish, independent of its utilitarian instrumental benefits for human use. This has far-reaching consequences for normative economics, which are based on welfaristic ideas about good and bad, right and wrong. From the perspective of deep ecology, classifying a meat market as being efficient is completely off the mark, because animals are ends and not means to human needs. A comparison to slave marketsx is illuminating: trading slaves on markets can be classified as Pareto-efficient, as long as one denies slaves human and civil rights and does not see them as ends, but rather as means for the needs of the class of "non-slaves." Hence, it is a meaningful problem to discuss the efficiency properties of slave markets in such a society. As soon as one extends basic human and civil rights to all human beings and declares them unalienable, however, there is no meaningful way to discuss the efficiency of such a market, because the traded "resources" are no longer means, but rather ends in themselves. One gets the same fundamental transformation if one grants rights to non-human species.

It would be far beyond the scope of an introductory textbook to dig deeper into the thorny issues of environmental ethics and the consequences for one's perception of economic systems. What the above discussion should have made clear, however, is that our perception of markets relies on normative principles that are – despite their widespread acceptance – far from obvious and innocuous.

There are other, less obvious, ways to cope with externalities and also other, less obvious, sources of externalities in markets. The next two examples focus on business ethics and, especially, the concept of *corporate social responsibility* (CSR), status concerns and relative-performance measurement as illustrations.

6.2.3.3   Morality and Corporate Social Responsibility

> Globalization makes it clear that social responsibility is required not only of governments, but of companies and individuals. (Attributed to Anna Lindh (2002))

In the realistic case that the institutional structure of a state is imperfect, in the sense that it does not always provide incentives for (Pareto-)efficient behavior, the question is how the people within society do or should deal with these inefficiencies. An example of this is when property rights are imperfectly enforced because of high transaction costs. The better part of everyday transactions is, for example, formally but not materially protected by property rights, because it would be too costly to enforce them. If a customer buys a bottle of orange juice at a kiosk and the retail clerk refuses to give back the change, the opportunity costs of calling the police, verifying the tort (which is difficult, if the retail clerk refuses to confess), etc. are likely prohibitive. Alternatively, on that subject, it is equally unlikely that the retail clerk can do much to prevent the customer from saying thank you and walking away with the bottle of juice without paying for it. Property rights cannot explain the fact that the overwhelming number of these transactions take place smoothly and efficiently.

There must be other mechanisms at work, and I will briefly discuss two of them. First, the interaction may not be singular but rather repeated and, if there is always a probability that the customer and the retail clerk will meet again in the future, it would be rather shortsighted to sacrifice future trades for the (relatively small) present gain. Repeated interactions can, therefore, be used to build up a reputation as a reliable trading partner, which can stabilize transactions, even in situations where formal property rights cannot be protected by the state. Second, the trading partner may have an intrinsic motivation to play fair. There is broad, scientific consensus by now that individuals are, for good evolutionary reasons, not always selfish, but have the ability and also (sometimes) the desire to act morally. The willingness to keep one's promises, to pay one's bills, etc., however, depends very much on the perception of the situational context. If people have the feeling that

– by and large – society gives everyone his or her fair share, their willingness to cooperate, to act fairly and to voluntarily follow certain moral standards of behavior is much larger than in a situation that is considered unfair from the beginning. Social norms and the intrinsic desire to act morally are then substitutes for formal property-rights enforcement. The more porous the system of property-rights enforcement is, the more important moral behavior becomes.

How relevant is the above observation? Is moral behavior, as the example suggests, only necessary for small-scale transactions, like buying soft drinks at kiosks, or is there more to the story? Here is an example. As one has seen, imperfect and asymmetric information is potentially a major cause of transaction costs. Therefore, in all cases where the better-informed party can exploit the other party, moral behavior can reduce transaction costs and facilitate trade. This view has been nicely expressed by Kenneth Arrow (1971): "In the absence of trust [. . .] opportunities for mutually beneficial cooperation would have to be forgone [. . .] norms of social behavior, including ethical and moral codes (may be) [. . .] reactions of society to compensate for market failures." What could be scrutinized in this quote is the implied supremacy of markets. It is too narrow of a view to see morality only as a repair shop for market failures. However, the general point is irrespectively valid: if specialization, exchange and trust go hand in hand, it is much easier for a society to flourish.

As one will see below, the existence of public goods, like infrastructure, basic research or defense, is a reason why the state can improve efficiency by playing a role beyond the enforcement of property rights. In order to be able to do so, the state needs to have access to finances, which are primarily collected as taxes. The process of globalization has, however, created opportunities for (multinational) firms and (mainly) wealthy individuals to minimize their tax burden by ever more complicated financial constructions. It may be a good deal for small countries to attract big companies by low tax rates, but the result is a global tax structure and provision of public goods that is inefficient. The point is that the international system of sovereign national states and international tax treaties creates loopholes and leads to discretionary power for firms and wealthy individuals and, despite the OECD initiatives, it is unrealistic to close those loopholes by means of enforceable treaties. As a consequence, one can either accept the resulting inefficiencies or appeal to the moral responsibilities of these firms or people. This is what the former Swedish politician Anna Lindh had in mind in the quote from the beginning of this section and international tax evasion strategies, of course, do not exhaust the number of challenges imposed by globalization.

In the field of Business Ethics, corporate social responsibility (CSR) emerged as $\mathcal{B}$ a separate field of research, exactly because current trends in international markets led to a redistribution of power from the institutions of the traditional state into the hands of corporations. One of the key questions, in this literature, is whether this increase in power goes hand in hand with the moral responsibilities of the managers and the corporation as an institutional actor.

6.2.3.4   Status

> From whence, then, arises that emulation which runs through all the different ranks of men, and what are the advantages which we propose by that great purpose of human life which we call bettering our condition? To be observed, to be attended to, to be taken notice of with sympathy, complacency, and approbation, are all the advantages which we can propose to derive from it. It is the vanity, not the ease, or the pleasure, which interests us. (Adam Smith, The Theory of Moral Sentiment, Chapter II.)

> Comparison brings about frustration and merely encourages envy, which is called competition. (Jiddu Krishnamurti (2005))

Social comparisons and the urge to outperform others seem to be deep motivational factors for human beings. Humans are, sometimes, a cooperative but, even more regularly, they are a competitive species and there are good evolutionary reasons for why relative performance is an important factor in mate selection. The human metabolism also requires that some absolute standards are met to stay healthy (for example daily caloric intake) but, apart from that, relative (status) concerns are an important factor for the explanation of human behavior across cultures and times.

However, this powerful drive has a dark side to it and almost all spiritual traditions from Christianity to Buddhism warn people that the concern for relative status is the road to unhappiness and suffering and that the way towards a fulfilling live is to free oneself from social comparisons. Mark Twain has the, perhaps, shortest account of this fact: "Comparison is the death of joy."

May this be as it is; is there anything that one can say, as an economist, about the functioning of markets, if demand is driven by status concerns? The first observation that one can make is that scarcity works differently for status than for other goods. Say one is eating apples for nourishment. An additional apple makes one better off, irrespective of what the other people are doing. Thus, if everybody in society eats twice as many apples, everybody is better off. This is not true for status goods. If cars are acting as a status symbol (or for that matter, smartphones or Prada shoes) and one buys a bigger car, while no one else in one's neighborhood does the same, then one gains in status and prestige. However, if everyone buys a bigger car, the effects neutralize and one ends up in the same status position as before, when everyone had the smaller car. It is like running a race: if everyone trains harder and runs faster, the odds of winning remain the same but, if one is the only one who "goes the extra mile," then one can tip the balance in one's direction.

What this example shows is that technological progress or an increase in material well-being can alleviate scarcity for ordinary goods (people live healthier, longer lives, are better nourished, etc.), but not for social status. This is why status is called a *position good*. It is the relative ranking in the pecking order that determines one's position and, if everyone works twice as hard to improve, then no one will be better off in the end. It might even be the case that a point is reached when everybody is worse off, when people start paying tribute for working longer hours. However, do status concerns create externalities in market economies? To understand this, one can use the extended supply and demand analysis introduced in Sect. 6.2.2. Assume

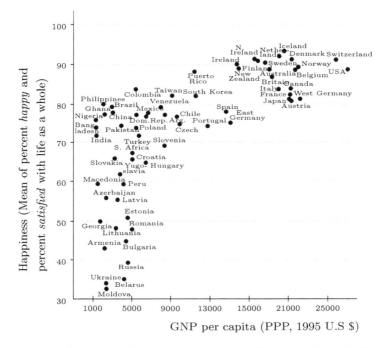

Fig. 6.4 Average subjective happiness and average income. (Source: Inglehart & Klingemann, 2000, p. 168)

that x measures supply and demand for a status good (mechanical watches), which means that part of the reason for buying this good is to impress the neighbors. Assume further that this interdependency between one and one's neighbor cannot be internalized directly (think about it: "how much would you pay me to not buy that Rolex?"). In that case, the individual value of the status good is, in general, higher than the social value.

One may wonder if *positional externalities* are a mere theoretical curiosity, or if anything more significant is going on. One way to approach this problem is to look for empirical evidence that is anomalous, given the predictions of standard theory (without status concerns), but that can consistently be explained, if one accounts for status. In fact, such evidence exists and it became famous as the "happiness paradox." However, it is still highly contested whether the empirical findings are valid.

The happiness paradox refers to patterns in empirical research on happiness. Two findings are key for it. The first relates to the relationship between average subjective happiness and average income. The findings are summarized in Fig. 6.4.

The figure depicts *average* happiness levels and *average* income levels in different countries. It shows that there is a positive association between average happiness and average income up to an annual income of about $12,000. However, there is

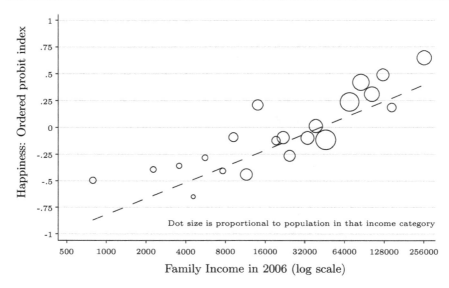

Fig. 6.5 Individual happiness as a function of individual income. (Source: Stevenson & Wolfers, 2008; General Social Survey, 2006)

no positive association between average happiness and average income for higher income levels. This "flatness" is sometimes also referred to as the "hedonic treadmill" where one runs faster and faster without moving forward. This finding is difficult to square with the idea that individuals are mutually unconcerned or selfish, because this assumption would imply that increases in material well-being (and average income should be a proxy for this) increase subjective well-being (i.e. happiness). This is apparently not the case. Here is a nice summary of this aspect of the paradox: "People in the West have got no happier in the last 50 years. They have become much richer, they work much less, they have longer holidays, they travel more, they live longer, and they are healthier. But they are no happier." (Layard, 2005)

The second finding refers to *individual* happiness levels as a function of *individual* income within countries, see Fig. 6.5.

The figure reveals that richer individuals are happier than poorer individuals, in a given society, and this is the puzzle: how is it possible that individual income is a good proxy for individual happiness while, at the same time, these effects net out for the whole society as soon as the average income exceeds a certain minimum?

Status preferences provide a missing link for resolving this puzzle: in poor countries, where individuals have to fight for subsistence, the relative importance of non-status-related compared to status-related consumption and production is high. However, the richer a society gets, the more important status concerns become. Thus, in these societies, there is no longer a positive association between *average* income and *average* happiness, because status effects "net out." If one climbs up on the status ladder, someone else necessarily has to climb it down. However, increases in *individual* income make a difference, because one climbs higher on the status ladder, and the negative happiness effects on others that result from this improvement are irrelevant for one's individual happiness.

As I have mentioned before, the findings and the interpretation of the happiness paradox are contested, but they are in line with a lot of evidence from other fields, like evolutionary biology, where *relative* fitness is key for mating and therefore evolutionary success of ones' genes. Additionally, it comes as no surprise that all the major spiritual traditions humans have created attach large warning signs to individual comparisons. However, even if one takes the interpretation at face value, the policy implications are complicated. Should one infer from the hedonic treadmill that the state has an active role in the internalization of status externalities that is similar to its role in the internalization of, for example, environmental externalities (taxation of status goods, etc.), or should one leave it to the individual to overcome the attachment to status? These are deep questions and they are even more pressing because, as long as social norms have declare that social status is a function of material well-being, one straps oneself to the wheel of consumerism and materialism, which is, at least partly, responsible for the environmental externalities mentioned above.

6.3 Four Boundary Cases

[T]hey devote a very small fraction of time to the consideration of any public object, most of it to the prosecution of their own objects. Meanwhile each fancies that no harm will come to his neglect, that it is the business of somebody else to look after this or that for him; and so, by the same notion being entertained by all separately, the common cause imperceptibly decays. (Thucydides (2013), The Peleponnesian War, Book 1, Section 141)

Coming back to the variations of the bakery example from the last subchapter, the distinctive difference between the two types of environmental interdependencies (sewage and air pollution) was the physical "reach" of the interdependency-causing activity. In the sewage-case, there was only one person, the fisherman, who caused the interdependency with the bakery whereas, in the air-pollution case, the bakery influenced all the residents. These differences in the number of people, who are influenced by economic activities, are an important element in the classification of goods and services and in developing an understanding of the functioning of markets.

The implicit assumption behind the model of competitive markets discussed in Chap. 4 was that the interdependency is *bilateral*. A typical example for a bilateral interdependency is an apple. Either one or the other person can eat an apple (one cannot eat the same apple twice), so Ann's decision to sell an apple to Bill has no direct physical consequences for third parties. The same was true in the sewage example. However, the bilateralism of the interdependency was a result of the fact that only one fisherman made his living from the lake. If two fishermen had have cast their nets into the lake, the interdependency would have been trilateral, because the emissions by the factory would have reduced the catches of both fishermen. In the air-pollution example, the reach of the interdependency was even larger, covering all residents of the area. This observation motivates the following definition.

✍ ▶ **Definition 6.3, reach** The reach of an economic activity is the set of people directly influenced by the activity.

The two meaningful boundary cases are the minimum and the maximum reach. The use of a good with minimum reach has an effect on only one person, and the use of a good with maximum reach has an effect on all individuals in an economy. A good with minimum reach is called *rival in consumption*, and a good with maximum reach is called *non-rival in consumption*. Combined with scarcity, goods with minimum reach create a bilateral interdependency. One has already seen that an apple is an example for a rival good and it is either person A or B who gets nourished by the apple. An example for a global non-rival good is a fossil-fuel combustion increasing CO_2 levels which, in turn, contributes to anthropogenic climate change, which has an impact on all individuals on the planet. Finally, an example for a national, non-rival good is the protection against foreign aggressors due to national defense.

A good part of the goods and services fall in-between these extremes. The reach of national defense, for example, is the boundaries of the nation-state. A live sports event or a music concert has a reach that is limited to the visitors of the stadium or concert hall. Additionally, a piece of music uploaded on YouTube has everyone with internet access within its potential reach. Even though reach can vary widely in range, it is customary to start with a discussion of the two extreme forms of rival and non-rival goods and this book will stick to this custom here, keeping in mind that the understanding that one can develop from these cases must be somewhat modified, when applied to intermediate cases.

What one has also seen from the air-pollution example is that different types of transaction costs exist that have an impact on the functioning of markets, as well as on other institutions. In order to be able to use markets to allocate goods and services, one relies on the ability of the owner of the good to exclude others from its use. Without excludability, people would freely use the goods and services provided by others with the consequence that market transactions would never take place. Excludability of goods and services is, therefore, a necessary condition for the establishment of markets and the (opportunity) costs of exclusion are a major source of transaction costs in the market mechanism. This motivates the following definition:

✍ ▶ **Definition 6.4, exclusion costs** The transaction costs that are necessary to exclude third parties from the appropriation and use of goods and services owned by a person are called exclusion costs.

The reach of an economic activity and the exclusion costs span a two-dimensional map where goods can be pinned down according to their specific characteristics. Figure 6.6 illustrates this point.

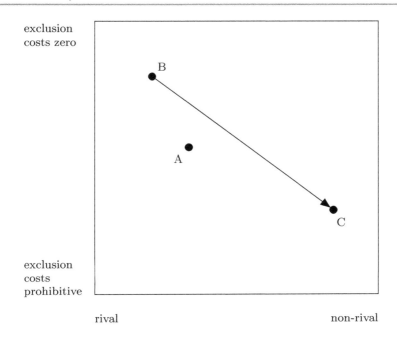

Fig. 6.6   Types of goods according to reach and exclusion costs

The four corners of this "map" define the boundary cases of minimum and maximum reach and zero and prohibitive exclusion costs. In reality, all goods are located somewhere in the middle. A point like *A* could, for example, be a car. Its reach is to carry up to five persons and exclusion costs are given by the price for locks and the alarm system.

Exclusion costs can vary over time. Take music as an example. In the good old days of the phonograph record, excluding third parties from the illegal consumption of music was relatively easy: in the absence of technologies for copying, exclusion required investments to prevent the theft of the physical record. The piece of music, as such, was non-rival in consumption, but the specific physical "carrier", the record, made it *de facto* rival (a point like *B* in the figure). With the invention of music cassettes, copying music became easier, which had an impact on the way property rights had to be protected. However, the big change came, of course, with the digitalization and distribution of music via the internet. This technological change essentially transformed music from a rival to a non-rival good and had a huge impact on the ability of the owner to exclude people from the illegal use of music (a shift from point *B* to point *C* in the figure). It took the music industry years to cope with this problem and to develop new business models. Technological inventions like copy and data-storage devices can, therefore, cause externalities for other products, like music or software.

Table 6.1 A taxonomy of goods

	Rivalrous	Non-rivalrous
Excludable	Private goods	Club goods
Non-excludable	Common goods	Public goods

Again, custom has it that one focuses on the two most extreme manifestations of exclusion costs. If exclusion causes zero transaction costs, then the good or service is called *(perfectly) excludable*. If exclusion causes prohibitive transaction costs, then the good or service is called *(perfectly) non-excludable*. Perfect excludability is, obviously, a simplifying assumption. To quote Madison in the Federalist Papers No. 51, "If men were angels, no government would be necessary," because mankind would never steal, which is the only way perfect exclusion is possible without any costs. Otherwise, shop owners protect their shops by locks, security systems and guards, all of which contribute to transaction costs. The same is true for the general public that protects its flats, houses and cars against theft. However, some goods come relatively close to the ideal of perfect excludability, for example the above-mentioned apple. An essential good that is non-excludable is oxygen in the air. Just try to enforce any property rights on a specific molecule.

The extreme cases of rivalry and excludability give rise to a two-by-two matrix of goods that is useful for a first discussion of the different types of challenges that have to be overcome, if one wants to organize economic activities. Table 6.1 gives an overview. The four boundary cases are called private goods, common goods, club goods, and public goods, and I will discuss them in turn.

Private goods One does not have to devote much time and attention to private goods, because they are the ones whose efficient production and distribution can be organized relatively easily, at least in principle. They are also the type of good that is implicitly assumed in the theory of competitive markets, which is analyzed in Chap. 4. Their minimum reach makes the interdependence bilateral under conditions of scarcity and market prices induce efficient incentives to produce and exchange these goods. If it is, furthermore, costless to exclude others from the use of these goods, without consent from the owners, there is nothing standing in the way of establishing markets.

Common goods Things are getting much more involved when it comes to common goods. These goods share the minimum-reach property, but it is not possible to allocate them using market mechanisms, because the owner of these goods cannot prevent others from their use, which is a prerequisite for the functioning of market transactions. The ability to exclude others from the use of resources, goods, and services depends very much on the state's ability to function properly. Even the night-watchman state needs laws and law enforcement to support the development of markets and, with weak state institutions (insufficient funding of the police, corrupt officials, etc.), excludability is far from guaranteed, which prevents markets from functioning efficiently. Irrespective of the quality of institutions, though, there

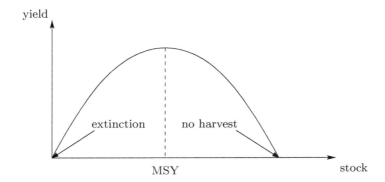

Fig. 6.7   Maximum sustainable yield

are some goods and resources, whose inherent qualities make exclusion very costly. Examples are migratory species, like fish and birds, or oxygen (see above). In comparison to cattle, where the assignment of property rights to specific animals is possible and effectively enforceable in principle, it is very hard to assign and enforce property rights to individual fish. This need not be an impediment to effective exclusion, as long as close substitutes to property rights for fish exist, and a close substitute could be property rights over the part of the sea where a shoal of fish lives. For example, the United Nations Convention on the Law of the Sea assigns exclusive economic zones (EEZ) to states. These zones grant special rights regarding the exploration and use of marine resources to nation states. They stretch from the baseline out to 200 nautical miles from its coast (as long as there are no overlaps between different countries). Territories beyond this stretch are international waters without exclusively assigned user rights.

Exclusion is, therefore, possible for all fish that migrate only within the boundaries of a given EEZ. However, for fish that migrate beyond or across the EEZs or in international waters, property rights over the sea are no effective substitutes for property rights over the fish themselves. The result is often overexploitation, because sustainable shoal management is not in the interest of the states or the fishery fleets: they have to bear the (opportunity) costs of sustainable management, but part of the revenues spill over to other nations or fleets. In order to understand this problem better, it makes sense to dig a little deeper into the economics of renewable resource management. For all renewable resources, there is a causal relationship between the size of the stock and the yield. If the stock size is zero, obviously, the yield is zero, as well. Increasing stock size makes positive yields possible and the yield increases with stock size up to a certain point, where larger stocks require smaller yields again, up to the point of maximum stock size, which can only be sustained if the yield is zero. Figure 6.7 shows this relationship.

The *maximum sustainable yield* (MSY) is the largest yield (or catch for the fishery example) that can be taken from a species' stock over an indefinite period. It is given by point MSY in Fig. 6.7. Given this biological law, it is in the interest of a long-term business to adjust the yield around MSY. Underexploitation would leave money on the table and overexploitation would trade long-term for short-term profits. However, if the stock is not excludable, the incentives to act according to the long-run interests are diminished, because no user of the stock can be sure that the stock will still be there tomorrow. There is a tension between the logic of individual and the logic of collective action. I will come back to this point in Chap. 11.

Φ

Digression 16. Cod

One of the most 'famous' examples for the overexploitation of marine resources is *gadus morhua*, or cod. Cod has been a very important commodity for about 600 years and dried cod (also called stockfish or clipfish) was an essential food for mariners. During the Middle Ages and the Age of Discovery, it was one of the most important commodities that made seafaring possible, because dried cod was one of the world's first non-perishable foods. It also became a popular food in Europe and, for about 250 years, 60 percent of all the fish eaten in Europe was cod. As early as 1620, cod fishing was at the center of international conflict, because various nations attempted to monopolize rich fishing grounds. Even the King of Spain married off his son to the royal house of Portugal, because of fishing rights. By the late 1700's, codfish made New England an international commercial power.

For a very long time, it was beyond imagination that human activity could negatively impact the species, because it was famous for its reproduction rates. In the words of Alexandre Dumas (1873), "It has been calculated that if no accident prevented the hatching of the eggs and each egg reached maturity, it would take only three years to fill the sea so that you could walk across the Atlantic dryshod on the backs of cod." Human imagination proved to be too limited. Since the late 1950ies, technological advances, which have made fishing more effective, have heralded the start of a period of overfishing, which led to a first partial collapse of Atlantic northwest cod fishery in the 1970ies and a complete collapse in the 1990ies. In the summer of 1992, the Northern Cod biomass fell to 1% of its earlier level, see Fig. 6.8.

Cod is only a very prominent example of the problem of overfishing: the Peruvian coastal anchovy fisheries crashed in the 1970s after overfishing, the sole fisheries in the Irish Sea and the west English Channel have become hopelessly overfished and many deep-sea fish are at risk, as well as a number of species of tuna. A 2008 UN report asserts that the world's fishing fleet could be halved with no change in catch. Even more fundamental is the impact on the whole marine biosystem. Scientific evidence regarding the impact of humans on marine life is nicely summarized in a recent paper by McCauley et al. (2015): "Three lessons emerge when comparing the marine and terrestrial defaunation experiences: (i) today's low rates of marine extinction may be the prelude to a major extinction pulse, similar to that observed on land during the industrial revolution, as the footprint of human ocean use widens;

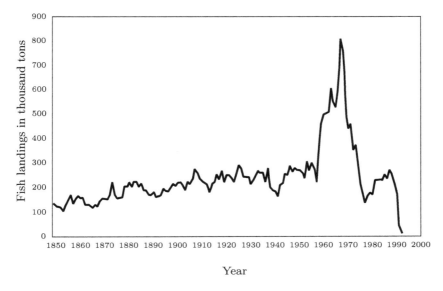

Fig. 6.8 Collapse of the North Atlantic cod fishery. (Source: Millennium Ecosystem Assessment, 2005, p. 12)

(ii) effectively slowing ocean defaunation requires both protected areas and careful management of the intervening ocean matrix; and (iii) the terrestrial experience and current trends in ocean use suggest that habitat destruction is likely to become an increasingly dominant threat to ocean wildlife over the next 150 years. [...] Human dependency on marine wildlife and the linked fate of marine and terrestrial fauna necessitate that we act quickly to slow the advance of marine defaunation."

Unresolved commons problems even led to the collapse of whole societies throughout human history. As far as we know today, examples are the Greenland Norse, Easter Island, the Polynesians of Pitcairn Island, the Anasazi of southwestern North America and the Maya of Central America. There are, of course, always several factors that contribute to the collapse of a society, but overexploitation of natural resources plays a very prominent role.

Markets are only sufficient, not necessary, to reach efficiency. Humans have developed other effective means to cope with commons problems and property rights plus trade, interestingly, is not one of the most common forms of resource management, as is stressed by Elinor Ostrom who systematically analyzed solutions to the commons problem in different societies. She came to the conclusion that the absence of private property and markets need not be an impediment to the efficient and sustainable use of common goods. On the contrary, evidence suggests that well-maintained systems of resource and ecosystem management can, in fact, yield better results than markets can. These findings are of considerable importance for the way one should think about institutions, because they point towards the blind spots of the standard model in economics, which puts (too?) much emphasis on markets. What makes Ostrom's findings difficult to integrate into this discussion about common goods is, however, a tension between this definition of common goods and the way she uses the term. This book defines them by the "technological" property of non-excludability, which excludes certain institutions *by definition*. Ostrom (2005) starts from a different perspective, focusing on goods and resources for which common property exists *de facto*. The set of admissible institutions, therefore, remains unclear and some of her criteria for successful institutional principles show that they rely on excludability. With this caveat, one can briefly discuss the basic principles of successful management of common goods that have been identified.

- Pretty much in line with the standard model, precise delineations of the resources and effective exclusion of externals is important. Hence, even if exclusion is not practiced within the group, it is important to exclude outsiders.
- One needs rules regarding the appropriation and provision of the common goods and these rules have to be adapted to local conditions. This property shows that institutional diversity is key, because there are close ties between the norms and cultures of groups and the environment.
- Rules for collective decision-making play an important role by giving voice to as many users of the common resource as possible and allowing the management system to adapt to changing environmental and social conditions.
- Monitors maintain compliance effective and are part of or are accountable to the users.
- One needs a scale of graduated sanctions for resource appropriators, who violate community rules.
- One needs mechanisms of conflict resolution that are cheap and are easily accessible for the conflicting parties.
- The self-determination of the community is recognized by higher-level authorities.

This list shows that institutions, which can effectively manage common goods, are diverse, but share common patterns. One-size-fits-all solutions that rely on property rights and markets should, therefore, be considered with caution, because they are only one means to cope with commons problems and may even be maladapted to local norms and traditions. However, given that the above principles have been identified in mostly stable and small communities, it remains an unresolved question whether they can be "scaled up" to cope with large or even global commons problems. Trust and sanctions are relatively easy to establish in small and stable communities and small-scale communities are also the environments in which human beings evolved and developed their intuitions about fairness and justice. A suggestion about how to deal with larger common-goods problems, which comes out of this line of research, is to organize them in the form of multiple layers of nested organizational units.

Club goods If exclusion is possible and the good is non-rival, then it is a club good. The name sounds strange at first, but it will become clear as I discuss the implications of this combination of factors. Think of a live music concert or sports event. In order to be able to enjoy it, one has to enter a stadium or concert hall and this physical barrier can be used to exercise exclusion and to force one to buy an entrance ticket. Further examples for club goods are Pay TV, lectures, music and software or – to a certain extent – roads. I will briefly discuss them, to see if there are interesting patterns to be found.

Lectures at universities, for example, are pretty much like live music and sports events club goods, because one can, in principle, exclude people from attending and thereby enforce the payment of prices. These prices are sometimes also called entrance or user fees. Moreover, if the primary motive for attending lectures is a grade certificate, one can enforce the payment of user fees by withholding the certificate. Given that exclusion is possible, in principle, it is mainly a political decision of whether access should be regulated by the price mechanism and whether it shall be complemented by other mechanisms (like making a high-school degree mandatory). A lot of public universities in Europe charge only moderate or no tuition fees, whereas private universities and also public universities in countries like the USA charge substantial amounts. MIT, for example, charged its undergraduate students \$45,016 annually (2014/15), which is pretty much in line with other top US universities. The University of Cambridge (for its undergraduate computer science program) discriminates tuition fees between UK and EU and other international students and the fees are £9,000 and £22,923 for both groups, respectively.

Another aspect of lectures, music and sports is that the "live event" has a limited reach defined by the capacity of the lecture room, concert hall or stadium. Therefore, the maximum supply is defined by this capacity. To make sure that supply meets demand, one can rely on the price mechanism, adjusting user fees accordingly or one can use alternative rationing schemes. Universities, for example, screen students by means of entrance tests, and so on.

Why is there a difference in the way demand is rationed between, for example, music events and university programs? Profit-oriented universities face a tradeoff between short- and long-term profits. Assume that, at given tuition fees, demand exceeds supply, such that entrance tests must be used to ration. In the short run, the university could increase its revenues by increasing tuition fees, but this may have a negative impact on the selection of the student body, which may have a negative effect on the future reputation of the university, which – as the last step in a complex causal chain – has a negative effect on future entrance fees. This is not the case with other commercial events, like concerts and sports, because the talents and motivation of the audience has only a very limited effect on the quality of the event.

One has already seen that live events face certain capacity constraints, which limits the reach of a club good. These limitations can, in principle, be overcome by "going digital." Broadcasting sports events or live music and selling studio music via, for example, iTunes markedly extended the reach of these goods such that, at the maximum, everyone with access to the internet can get access to the product, which creates huge profit potentials for firms. However, every distribution channel has its own enforcement costs and the music industry had to learn this the hard way during the early days of the internet, when it was almost impossible to prevent illegal downloads. Thus, digital products somehow oscillate between the characteristics of a club good and a public good. "Going digital" is also a recent trend in education, where so-called MOOC's (massive open online courses) are mushrooming.

Last, but not least, roads are an example of a good for which regulating access via price mechanisms is becoming increasingly popular, partly because of changes in the available technologies of exclusion, and partly because of other trends. Access to most roads in the majority of countries is free and traffic infrastructure is financed by taxes. One of the reasons is that road pricing and the investments in the setup and maintenance of the necessary exclusion technology is very costly in general. Furthermore, there is a lot of evidence that, as long as congestion is not an issue, a region's traffic infrastructure creates huge positive externalities, because it facilitates trade. For example, Paris experienced a boost in its economic development after the abolishment of bridge tolls by Baron Haussmann in the 19th century. However, positive externalities caused by traffic infrastructure can easily be compensated by negative externalities, if traffic gets congested. The current trend to (re-)introduce tolls on highways, bridges and other major roads is, to a certain extent, a reaction to the increasing economic costs of congested traffic infrastructure, combined with more efficient technologies for the enforcement and collection of tolls that bring down the transaction costs of enforcing fees.

If one looks at club goods from a slightly different angle, one observes an interesting property because, as long as no capacity constraints are binding (there are still empty seats in the lecture room), an additional user of the good causes no additional costs. This property has two interesting implications.

First, from an efficiency perspective, it makes sense to increase the number of users to the largest extent possible, because each additional user increases the gains from trade (no additional costs, but additional consumer surplus). It follows that actually excluding people from using the club good can never be efficient. Exclusion is a mechanism that can be used to establish a market and, therefore, has to be distinguished from the act of actually excluding people. The threat of exclusion makes the enforcement of prices (like tolls or tuition fees) possible, but it depends on the actual prices whether potential users will be excluded or not.

Second, the fact that firms can serve additional customers at approximately zero additional costs creates a tendency towards the monopolization of markets for club goods. Take software as an example. From the point of view of a software developer, the lion's share of the costs she has to invest is caused by the development of the product. As soon as the product is on the market, each additional user causes approximately zero additional costs. Hence, the more users there are the better, for the software developer. The fact that the minimum price that is necessary to break even falls as the number of users increases creates an inherent tendency for market concentration: firms with larger market shares can outcompete their smaller competitors, because they can charge lower prices without running a deficit. This is the reason why club goods are sometimes also called *natural monopolies*. I will come back to this point when I discuss production costs in Chap. 8 and monopoly pricing in Chap. 10.

Public goods   The last type of good is non-rival in consumption and exclusion is impossible. If exclusion is impossible, markets cannot be used to incentivize the production and allocation of these goods, so one has to look for alternative ways to organize economic activity.

Examples for public goods are fireworks, basic research, national defense, avoidance of climate change and legal systems, and the following paragraphs will discuss all five examples in turn. Fireworks are an example for a (local or regional) public good, because no one in a city can be effectively excluded from the spectacle and it is also non-rival. Arguably, the other examples are more important than fireworks.

Basic research is non-rival, because the fact that I understand a mathematical theorem does not make it impossible for another person to understand it, as well. All knowledge, in this sense, is non-rival. The difference between basic and applied research is, therefore, not the degree to which goods are rival, but the ability to exclude. Applied research usually builds on basic research and "brings it to the market." A good example is quantum physics. Without quantum mechanics, there would be no transistor and hence no personal computer and no laser. Therefore, the development of quantum mechanics made the development of a large number of products possible, without which today's world would be impossible. Products or components thereof, like transistors or computers, can be effectively protected by patent law. However, the protection of property rights for the mathematical formulation of quantum physics, like the Schrödinger equation or the uncertainty principle, is not as easy to do. Even if a formal property right exists, one cannot sell the Schrödinger equation directly and it is, in general, very hard to establish a causal link between abstract physical principles and marketable goods, such that potential property-rights infringements would be hard to detect.

The public-goods character of basic research requires alternatives to the market mechanism and one can find essentially two different ways to organize the production process. One is public financing. Major resources for basic research at universities and research institutions are provided by the state and financed by taxes, and career incentives for scientists have the form of a contest, where the relatively most successful qualify for professorships and research money. The alternative is to interpret education and research as complementary bundles where basic research is, at least partly, financed by tuition fees and students profit indirectly from the direct access to a research-intensive environment, because new ideas disseminate earlier, which gives them a competitive edge in developing new, marketable products. A good example is the synergistic relationship between Stanford University and Silicon Valley startups and companies.

A staple example for public goods is national defense. It is relatively obvious that, within certain geographic limits, a military of a given size provides a non-rival service to its citizens. By its very nature, the reach of national defense is the people living within the territory of a nation state (protecting people living abroad is much more difficult). The non-excludability of national defense becomes apparent, if one distinguishes between an actual military conflict and the insurance against attacks provided by the military. In case of an act of aggression, exclusion of specific citizens is, in principle, possible. One could escort them to the border and hand them over to the enemy. However, it is virtually impossible to exclude people within the territory from the insurance provided by the existence of the military, which results from the fact that one is not attacked at first place.

Last, but not least, the avoidance of climate change has important properties of a truly global public good: CO_2 emissions have global effects on the climate, so measures to slow down climate change cause non-rival effects. Similarly, no one can be excluded from the effects of climate change (or the effects from slowing it down). The global nature of climate change is what makes the problem so difficult to solve. The expected costs and benefits of climate change are unevenly distributed between countries and regions and international negotiations take place within the holey network of international law. International agreements are difficult to reach and they are even more difficult to enforce. If one would ask a group of social scientists and psychologists to design a problem that is hard to solve for human beings, I am pretty sure that it would look very much like climate change.

The last example for a public good that I will discuss is the legal system of a country, because it allows one to focus one's attention on the fact that excludability need not be a physical characteristic of a good. The legal system of a country is clearly nonrival. If A uses contract law to set up contracts, it does not impede B from using the contract law himself. Things get more complex when it comes to excludability. Technically, it is no problem to exclude people from contract law because the courts could decide not to apply it to contracts signed by specific people. However, contract law is embedded within the rest of the legal system, which makes such restrictions illegal. It can (and, in practice, usually does) specify that all laws apply equally to all citizens of a country. Such a norm creates a legal non-excludability and the system depends on levels of analysis to determine whether such constraints are taken as a given or if they are subject to scrutiny.

References

Akerlof, G. (1970). The Market for Lemons: Quality Uncertainty and the Market Mechanism. *The Quarterly Journal of Economics, 84*(3), 488–500.

Arrow, K. (1971). *Essays in the Theory of Risk-Bearing*. Amsterdam: North-Holland.

Breeze, T. D., Bailey, A. P., Balcombe K. G., & Potts S. G. (2011). Pollination services in the UK: how important are honeybees? *Agriculture, Ecosystems and Environment, 142*(3–4), 137–143.

Carson, R. (1962). *The Silent Spring*. Mariner Books.

Coase, R. (1937). The Nature of the Firm. *Economica, 4*(16), 386–405.

Coase, R. (1960). The Problem of Social Cost. *Journal of Law and Economics, 3*, 1–44.

De Groot, R. S. (1992). *Functions of Nature: Evaluation of Nature in Environmental Planning, Management and Decision Making*. Groningen: Wolters-Noordhoff.

Diamond, J. (2005). *Collapse: How Societies Choose to Fail or Succeed*. Penguin Books.

Dumas, A. (1873). *Grand Dictionnaire de Cuisine*. Hachette.

Elton, B. (1991). *Gridlock*. Sphere Books.

Inglehart, R., & Klingemann, H. D. (2000). Genes, Culture, Democracy, and Happiness. In E. Diener, & M. Eunkook (Eds.), *Culture and Subjective Well-Being*. MIT-Press.

Layard, R. (2005). *Happiness: Lessons from a New Science*. Penguin Books.

Lefebvre, H. (1974). *The Production of Space*. John Wiley.

Lindh, A. (2002). Speech by Anna Lindh in the Helsinki Conference.

Krishnamurti, J. (2005). *Life Ahead: On Learning and the Search for Meaning*. New World Library.

Lange, O. (1936). On the Economic Theory of Socialism: Part One. *The Review of Economic Studies, 4*(1), 53–71.

Lange, O. (1937). On the Theory of Economic Socialism: Part Two. *The Review of Economic Studies, 4*(2), 123–142.

McCauley, D. J., Pinsky, M. L., Palumbi, S. R., Estes, J. A., Joyce, F. H., & Warner R. R. (2015). Marine Defaunation: Animal Loss in the Global Ocean. *Science, 347*(6219, 1255641).

Millennium Ecosystem Assessment (2005). *Ecosystems and Human Well-Being: Synthesis*. Island.

Morse, R. A., & Calderone, N. W. (2000). The value of honey bees as pollinators of U.S. crops in 2000. *Bee Culture, 128*, 1–15.

Ostrom, E. (2005). *Understanding Institutional Diversity* Princeton University Press.

Parfit, D. (1984). *Reasons and Persons* Oxford University Press.

Southwick, E. E., & Southwick, L. (1992). Estimating the economic value of honey-bees (Hymenoptera, Apidae) as agricultural pollinators in the United States. *Journal of Economic Entomology, 85*, 621–633.

Stern, N. (2007). *Stern Review on the Economics of Climate Change*. Cambridge University Press.

Stevenson, B., & Wolfers, J. (2008). Economic Growth and Subjective Well-Being: Reassessing the Easterlin Paradox. Discussion paper, National Bureau of Economic Research.

Sumner, D. A., & Boriss, H. (2006). Bee-Conomics and the Leap in Pollination Fees. *Agricultural and Resource Economics Update, 9*(3), 9–11.

Thucydides (2013). *The History of the Peleponneasian War*. CreateSpace Independent.

World Wildlife Fund (2014). Living Planet Report 2014. http://www.footprintnetwork.org/images/article_uploads/LPR2014summary_low_res.pdf.

Further Reading

Arrow, K. J. (1969). The Organization of Economic Activity: Issues Pertinent to the Choice of Market versus Non-market Allocations. In *Analysis and Evaluation of Public Expenditures: The PPP System*. Washington, D.C: Joint Economic Committee of Congress.

Coase, R. (1998). The New Institutional Economics. *American Economic Review, 88*, 72–74.

Furuborn, E. G., & Richter, R. (2005). *Institutions and Economic Theory: The Contribution of the New Institutional Economics*, 2nd ed. University of Michigan Press.

Laffont, J. J. (2008). Externalities. *The New Palgrave: Dictionary of Economics*.

Marney, G. A. (1971). The 'Coase Theorem:' A Reexamination. *Quarterly Journal of Economics, 85*(4), 718–723.

Ostrom, E. (1990). *Governing the Commons: The Evolution of Institutions for Collective Action*. Cambridge University Press.

Swiss Code of Civil Procedure. (2006). https://www.admin.ch/opc/en/classified-compilation/20061121/201601010000/272.pdf. Accessed 08/04/2017

Voltaire (1984). *Candide*. Bantam Books.

Williamson, O. E. (1985). *The Economic Institutions of Capitalism: Firms, Markets, Relational Contracting*. Free Press.

Part III
Foundations of Demand and Supply

So far, I have motivated the structure and interpretation of the individual supply functions heuristically. I have argued that it is *plausible* that the supply of a good increases with its price and decreases with the prices of the inputs that are used for its production. I have argued that it is *plausible* that a point along the supply function can be interpreted as the willingness to sell this particular unit and the area between the market price and the supply curve can be interpreted as a measure for the gains from trade the firm is able to appropriate. By the same token, I have argued that it is *plausible* that the demand of a good decreases with its price.

This approach was a good first approximation, because it allows one to discuss the functioning of markets without large detours into the bits and pieces of firm decision making. However, it has (opportunity) costs (as one should know by now) as well, because it leads one no further than the current point. One's understanding of the demand and supply curve, the interpretation of consumer and producer surplus and the functioning of markets remain necessarily limited as long as one does not move on and look for a deeper understanding of how decisions are made by consumers and within firms.

This is the purpose of the following chapters. Remember that microeconomics is the study of decision making in households and firms and their interactions. Thus, according to this paradigm, the understanding of individual decision making must be at the core of one's discipline and the field that specializes on decisions is called *decision theory*. Models of decision making have two building blocks:

- They have to specify a *set of alternatives*, from which a decision maker can choose.
- They have to specify an *objective function* of the decision maker that specifies a ranking of the alternatives and a motivational assumption, which allows one to make predictions about the alternative she or he will choose, if confronted with a set of alternatives.

This specification is very generic and can be applied to all kinds of decision problems, ranging from demand decisions of consumers, to supply decisions of firms to voting behavior in elections.

The above approach allows one to develop a road map for the endeavor to more deeply understand demand and supply decisions: one first has to specify the set of alternatives from which a firm can choose. Afterwards, one has to specify the objectives of consumers and firms and bring those two building blocks together in order to derive hypotheses about consumer and firm behavior (i.e. demand and supply decisions).

From the point of view of a consumers, admissible alternatives are those consumption bundles that are affordable given prices and income. Economists assume that consumers can rank the affordable consumption bundles according to their tastes or preferences, and it is assumed that they try to select the best consumption bundle available. The formal model that allows to cope with decisions like these is introduced and analyzed in Chap. 7 where we will in a first step introduce the general preference model that will in a second step applied to the problem of consumer choice in competitive markets.

The set of alternatives from which a firm can choose is, as one will see, influenced by the costs it faces. I will, therefore, continue with an analysis of the different types of costs (Chap. 8). With a thorough understanding of the determinants and structures of costs, one can then go back to goods markets and see how they can be linked with objectives of the firm to better understand supply decisions. In order to do so, I assume that firms want to maximize profits (there will be more about this assumption in Chap. 8). Then I will return to the competitive environment to have a second look at firm behavior under perfect competition (Chap. 9). Costs will also be a crucial building block for firm behavior in monopolistic and oligopolistic markets that are analyzed in Chaps. 10 and 12.

The decision-theoretic perspective makes it possible to turn to the managerial perspective in greater detail. Chapters 9, 10 and 12 will move back and forth between a more "economic" perspective, which focuses on the functioning of markets, and a "management" perspective, wich focuses on the decisions firms have to make in order to be successful in certain market environments, the information that is needed to make these decisions and the implications for the internal organization of the firm. Economics and business administration cannot be dealt with separately. Economists can only understand the functioning of markets, if they understand how managers make decisions, and managers can only make sound decisions if they understand the market context in which they act. Economics and business administration are, therefore, complementary and an in-depth exchange between both disciplines can only be fruitful for both of them.

Decisions and Consumer Behavior 7

This chapter covers ...

- the concepts of preferences and utility functions and how these are related.
- how the assumption that individuals maximize preferences can be used to determine the individual demand functions on a competitive market.
- the strengths and weaknesses of this approach as a foundation of choice and decision-making in general and the structure of demand functions specifically.

7.1 Basic Concepts

> The theory of Economics must begin with a correct theory of consumption. (William Stanley Jevons)

Individual and market demand are the consequences of decisions made by individuals. Until now I have taken a shortcut and skipped a more detailed analysis of the way individuals make decisions, because I wanted to keep the focus on the functioning of markets. For that purpose, it was sufficient to heuristically explain how prices, income and other factors influence demand. However, this shortcut's cost is preventing one from developing a deeper understanding of the structure of individual and market demand. Additionally, the way I related the idea of Pareto efficiency and the demand function was also pretty clumsy.

Reduced to its essential core, economic decision theory is very simple: one assumes that individuals choose the best alternative from a set of admissible alternatives. In a market context, the admissible alternatives are the goods and services a consumer can afford, given prices and income. It is more difficult to model the meaning of individuals choosing the best alternative, though. This chapter is devoted to making these ideas precise and to seeing how they can help us to gain a better understanding of market behavior and of behavior in general.

© Springer International Publishing AG 2017
M. Kolmar, *Principles of Microeconomics*, Springer Texts in Business and Economics,
DOI 10.1007/978-3-319-57589-6_7

7.1.1 Choice Sets and Preferences

In order to develop a decision theory, one needs two conceptual ingredients. First, a set of alternatives from which an individual can choose. Call it a *choice set* and denote it by $X = \{x^1, x^2, \ldots, x^n\}$, in which $x^i, i = 1, \ldots, n$ is one of the possible alternatives and assume, for simplicity, that the total number of alternatives n is finite. The idea of a choice set is very general. If one goes to a café, one's choice set is a subset of all of the items on the menu. This implies that an alternative can be a list of individual items, like "one cup of tea, two scones and one portion of orange jam." Mathematically speaking, this type of list is called a tuple. If x^i is the above-mentioned alternative, it could be denoted as $x^i = \{$quantity of tea, number of scones, quantity of orange jam$\} = \{1, 2, 1\}$. If one goes to vote, one's choice set is the set of all admissible parties or candidates and if one is deciding what to do after high school, one's choice set is the set of all potential professions.

Second, the individual may prefer some alternatives to others, which is an expression of her taste or preferences. Assume that she is able to make pairwise comparisons of all the alternatives in X to make statements like, "I prefer alternative x^i to alternative x^j," or "I am indifferent between alternative x^i and alternative x^j." In order to have a lean notation, economists use the following symbols for these statements: "I prefer alternative x^i to alternative x^j" is denoted by "$x^i \succ x^j$" and "I am indifferent between alternative x^i and alternative x^j" by "$x^i \sim x^j$."

It is important to understand the exact meaning of the terminology. Mathematically speaking, I am taking two arbitrary elements of X, x^i and x^j, and comparing them to each other. This comparison is called a *binary relation* on X. The *strict preference relation*, "$\succ$," and the *indifference relation*, "$\sim$," can therefore be denoted as a subset of the Cartesian product of X, $X \times X$. (I am slightly abusing the notation by using the symbols as names for both the relation and for indicating the binary comparison of alternatives.)

Here is an example: assume that Ann can choose between an apple, x^1, an orange, x^2, and a cherry, x^3. In this case, the choice set is equal to $X = \{x^1, x^2, x^3\}$ and the Cartesian product is the set of all ordered pairs $X \times X = \{(x^1, x^1), (x^1, x^2), (x^1, x^3), (x^2, x^1), (x^2, x^2), (x^2, x^3), (x^3, x^1), (x^3, x^2), (x^3, x^3)\}$. Assume that Ann prefers apples to oranges and is indifferent between oranges and cherries, $x^1 \succ x^2$, $x^2 \sim x^3$. If one reads a pair (x^i, x^j) as "x^i stands in relation R to x^j," one can represent her preferences, "$\succ$," by the subset of pairs $\{(x^1, x^2)\}$ and her preferences, "$\sim$," by the subset of pairs $\{(x^1, x^1), (x^2, x^2), (x^2, x^3), (x^3, x^3)\}$. Note that the pairs (x^i, x^i) are elements of the subset, because Ann is indifferent between an alternative and itself. This property is not self-evident from a purely mathematical point of view and, therefore, sometimes is stated as an assumption of the preference relation that is known as *reflexivity*.

As stated, relation "$\succ$" is called the *strict preference relation* and relation $\sim$ the *indifference relation*. It turns out that it is easier to work with a third type of relation that is called the *weak preference relation*, which is denoted by "$\succsim$." It contains all of the pairs from $X \times X$ that either belong to the strict preference or the indifference relation. In this example, it is the set $\{(x^1, x^1), (x^1, x^2), (x^2, x^2), (x^2, x^3), (x^3, x^3)\}$. The strict preference and indifference relations can easily be reconstructed from the weak preference relation by the following operations:

- $(x^i \succ x^j) \Leftrightarrow (x^i \succsim x^j) \wedge \neg (x^j \succsim x^i)$,
- $(x^i \sim x^j) \Leftrightarrow (x^i \succsim x^j) \wedge (x^j \succsim x^i)$,

in which $\wedge$ and $\neg$ stand for the logical operations "and" and "not."

In order for the concepts to have predictive power, one has to make additional assumptions on the structure of the weak preference relation.

▶ **Assumption 7.1 (Completeness)**   For every $x^i, x^j \in X$, either $x^i \succsim x^j$ or $x^j \succsim x^i$ or both are true.

Assumption 1 implies that the individual can compare any two pairs of alternatives. This assumption may sound innocuous, because it seems obvious that one should either be better off with one alternative or the other. However, critiques point out that, depending on the context, alternatives can exist that cannot be compared in a meaningful way. Think, for example, of the alternative "destruction of human life by means of nuclear weapons" and "destruction of human life by means of a lethal virus." It is argued that there is a meaningful difference between being indifferent between two alternatives and not being able to compare them. If one has to choose between alternatives whose consequences are beyond our imagination, it is not clear that an inability to compare and an indifference are the same.

▶ **Assumption 7.2 (Transitivity)**   For every $x^i, x^j, x^k \in X$, if $x^i \succsim x^j$ and $x^j \succsim x^k$, then $x^i \succsim x^k$.

Transitivity implies that there are no "cycles" in the relation. The main justification for this assumption stems from the so-called "money-pump" argument, which rests on the idea that a person with intransitive preferences can be exploited by some other person. In order to understand this, assume that there is a "cycle" $x^i \succ x^j \succ x^k \succ x^i$ and that the individual is willing to pay at least one cent for the next best alternative. In that case, she would be willing to give up x_i plus a small amount of money in exchange for x_k, x_k plus a small amount of money in exchange for x_j and – attention money pump – x_j plus a small amount of money in exchange for x_i. Now she is back where she started, with the exception that the individual has lost three cents. Continuing this process would, in the end, separate the individual from all her money.

However, a lot of empirical experiments have shown that transitivity cannot be taken for granted. Here is an example. Procrastination describes the tendency to delay uncomfortable duties until later. A tendency to procrastinate may have very adverse consequences and the intransitivity of inter-temporal preferences seems to be playing an important role. This is why: assume that it is Monday and you have a report due on Thursday. Overall, you would like to hand in a high-quality report.

However, starting to work on Monday is less preferable to starting to work on Tuesday ("you know, I had a stressful day anyway"). However, when Tuesday comes, it is preferable to delay and start working on Wednesday ("I need the pressure to get things done"). However, from Wednesday's perspective it seems better to delay another day ("well, I simply cannot do it"). However, on Thursday it is too late to prepare and hand in a report of decent quality.

A weak preference relation that fulfills Assumptions 1 and 2 is also called a *preference ordering*. What do they imply in this little example? One already knows that Ann's preferences are $x^1 \succsim x^2$ (because $x^1 \succ x^2$ and $\succsim$ is a weaker condition than $\succ$) and $x^2 \succsim x^3$ (because $x^2 \sim x^3$ and $\succsim$ is weaker than $\sim$). Completeness is guaranteed by assumption (there are only three alternatives in the example) and transitivity implies that one can infer $x^1 \succsim x^3$ from $x^1 \succsim x^2 \succsim x^3$. Hence, the completed preference ordering is given by $\{(x^1, x^1), (x^1, x^2), (x^1, x^3), (x^2, x^2), (x^2, x^3), (x^3, x^3)\}$.

Completeness and transitivity are usually taken for granted in almost all economic applications. However, depending on the specific context that is analyzed additional assumptions have to be imposed. I list three of them in the following paragraphs.

▶ **Assumption 7.3 (Continuity)** For any $x^i \in X$ the set of all $x^j \in X$ is such that $x^i \succsim x^j$ and the set of all $x^k \in X$ is such that $x^k \succsim x^i$ are closed sets in X.

Continuity is less obvious from an economic point of view, but it still has some intuitive plausibility. It implies that the preference relation does not "jump" in the following sense. Assume that an individual is comparing two alternatives, x^1 and x^2, and she weakly prefers x^1 to x^2, $x^1 \succsim x^2$. For example, if one modifies x^1 a tiny bit to $x^1 + \epsilon$, in which ϵ is a very small quantity, then the preference ordering does not suddenly reverse, $x^1 \succsim x^2 \Rightarrow x^1 + \epsilon \succsim x^2$.

▶ **Assumption 7.4 (Monotonicity)** For any $x^i, x^j \in X$, $x^i \geq x^j$ and $x^i \neq x^j$ implies that $x^i \succ x^j$.

Assumption 4 needs a few words of clarification. The specification of X is completely general: elements can be arbitrarily complex or very simple alternatives. However, in some cases the alternatives can be quantitatively measured and compared, for example the quantity of a good like milk. In that case, x^i could be two liters of milk and x^j one liter. In all of these cases, an expression like "$x^i \geq x^j \wedge x^i \neq x^j$" makes sense. It makes no sense, however, to compare smartphones with ice cream. Assumption 4 is, therefore, only applicable for those alternatives that can be measured and quantified on an absolute scale. It then implies that the individual prefers larger quantities to smaller quantities.

▶ **Assumption 7.5a (Convexity)** For any $x^i, x^j \in X$, such that $x^i \succsim x^j$ and for all $t : 0 \leq t \leq 1$, it follows that $t \cdot x^i + (1 - t) \cdot x^j \succsim x^j$.

▶ **Assumption 7.5b (Strict Convexity)**   For any $x^i, x^j \in X$, such that $x^i \sim x^j$ and for all $t : 0 < t < 1$, it follows that $t \cdot x^i + (1 - t) \cdot x^j \succ x^i$.

Assumptions 7.5a and 7.5b are similar in spirit. What they imply is that individuals prefer balanced over extreme alternatives. However, in order to illustrate this idea, one has to restrict one's attention to the alternatives that are quantifiable and measurable in the same way as one has assumed in Assumption 7.4.

Assumptions 7.5a and 7.5b will play an important role in the theory of consumer choice on competitive markets, which is why it makes sense to discuss them in greater detail. Assume that the alternatives from which the individual can choose are quantities of two different goods, like bread and water. Denote two alternatives by $x^1 = \{10, 0\}$ and $x^2 = \{0, 10\}$. In alternative 1 the individual gets 10 units of water and no bread, while in alternative 2 she gets no water and 10 units of bread. In this example, convexity as well as strict convexity imply that an individual would, for example, prefer the more balanced alternative $x^3 = 0.5 \cdot x^1 + 0.5 \cdot x^2 = (5, 5)$ to the extreme ones.

In the example, Assumptions 7.5a and 7.5b seem to make perfect sense, but looking at the alternative example, in which the first good is Miso soup and the second is vanilla ice cream, few people would like to eat them together.

The above assumptions are usually not all imposed simultaneously. As the theory of consumer choice on competitive markets will show, economists try to establish properties of choice behavior with minimal assumptions about preferences, because every additional assumption constrains the admissible behavior of individuals, thus making the theory less general.

One can now define the concept of rationality as used in economics. It has two different aspects. First, if individuals have a preference ordering, a well-defined subset of alternatives $X^o \subset X$ exists that defines the best or optimal alternatives given the preferences. This would not necessarily be the case, if preferences were not complete and transitive. Hence, a preference relation is called *rational*, if it is complete and transitive. Second, it is not sufficient that individuals are able to consistently order the alternatives according to their preferences; they must also *act* according to them. Hence, individual *behavior* is rational, if the individual *chooses* a best alternative given the choice set and the preference ordering. This idea of rationality is at the heart of the concept of *homo oeconomicus*.

▶ **Definition 7.1 (Homo Oeconomicus)**   An individual behaves as homo oeconomicus, if (i) she perceives a choice situation as a choice set X, (ii) has a preference ordering over this choice set and (iii) chooses one of the best alternatives from this choice set, given her preferences.

Two statements are helpful to understand this. First, the concept of rationality is purely instrumental. It only requires that the preferences are structured in a manner that makes it possible to talk about better and worse alternatives in a meaningful way and that individuals act according to their preferences. It does not scrutinize the individual's taste or value judgments that caused her preferences. A debate that allows one to distinguish between better and worse preference orderings would build on a different concept of rationality, which is called value-based rationality. Mainstream economists accept a philosophical position called *subjectivism*, a value judgment that leads to an acceptance of all types of preference orderings. Second, note that no such thing such as selfishness enters this definition of homo oeconomicus. Selfishness is not an integral part of what economists consider rational behavior, even though selfish behavior is added as an *additional* assumption in a lot of analyses. The reason is that concepts such as selfish, altruistic, sadistic etc. preferences refer to *motives of action* and, as I have just said, mainstream economists do not scrutinize such motives but take them as given. It would, therefore, be alien to the idea of instrumental rationality, if it required any specific motive to act. (See Digression 3, Chap. 1, for a more detailed analysis of this topic.)

In order to get started with an analysis of decision making, one needs a few more concepts.

▶ **Definition 7.2 (Not-Worse-Than-x Set)** The Not-Worse-Than-x Set, for an alternative $x \in X$, $NW(x)$, is given by the set of all $x^i \in X$, such that $x^i \succsim x$.

▶ **Definition 7.3 (Not-Better-Than-x Set)** The Not-Better-Than-x Set, for an alternative $x \in X$, $NB(x)$, is given by the set of all $x^i \in X$, such that $x \succsim x^i$.

▶ **Definition 7.4 (Indifferent-To-x Set)** The Indifferent-To-x Set, for an alternative $x \in X$, $I(x)$, is given by the intersection $NW(x) \cap NB(x)$.

7.1.2 Indifference Curves

Thus far, I have introduced the concept of a preference ordering in the simplistic case of a finite set of alternatives X. However, the concepts can be readily generalized to allow for infinitely many different alternatives, which is usually done if the theory is applied to market contexts. In this case, if there are n different goods, then the choice set is a subset of the n-dimensional set of positive real numbers, $X \subset R^n_+$. In this case, one can illustrate the indifferent-to-x set by a graph. Assume that there are two goods whose quantities are represented by the two axes of Fig. 7.1.

The downward-sloping graph represents the indifferent-to-x set for an alternative that one calls a consumption bundle $\bar{x} = (\bar{x}_1, \bar{x}_2)$. It is called *indifference curve*. Hence, Ann is indifferent between this consumption bundle and any other consumption bundle on the indifference curve ($\tilde{x} = (\tilde{x}_1, \tilde{x}_2)$ and $\hat{x} = (\hat{x}_1, \hat{x}_2)$ are two examples for such bundles in the figure), $\bar{x} \sim \tilde{x} \sim \hat{x}$. Please note that if the curvature of this curve is representative for the whole preference ordering, "$\succsim$," then the ordering is both convex and strictly convex. The continuity of the curve reveals that preferences are continuous.

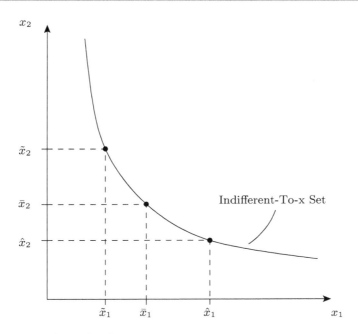

Fig. 7.1 Ann's Indifferent-To-x-Set

The indifference curve in Fig. 7.1, of course, only partially represents the individual's preference ordering. There exists an indifferent-to-x set for every consumption bundle x that can in principle be represented by an indifference curve.

The slope of an indifference curve has an important economic interpretation. Suppose that one not only wants to reallocate the consumption goods but also wants to ensure that the individual is neither better nor worse off. This is only possible if one chooses consumption bundles that lie on the same indifference curve. Now, suppose that at some point x one takes $dx^2 < 0$ away from the individual. Given that the indifference curve is downward sloping, one has to compensate the individual by some extra quantity, $dx^1 > 0$, to ensure that one stays on the indifference curve. See Fig. 7.2 for an illustration of this.

If one looks at infinitesimal changes, $dx^2 \to 0$, then the exchange rate between the two goods is given by the slope of the tangent to the indifference curve at the point $\bar{x}$. The absolute value of this exchange rate, dx^2/dx^1, is called the *marginal rate of substitution* (*MRS*) between good 2 and good 1. It is an expression of the idea of opportunity costs in the context of the individual's decision problem: if one takes a little bit of one good away, how much of the other good does one have to give the individual to make her indifferent?

Figure 7.3a–d illustrate the shape of indifference curves for different types of preference orderings.

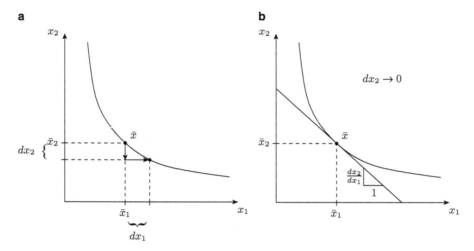

Fig. 7.2 Marginal rate of substitution

Figure 7.3a illustrates so-called *perfect substitutes*. Indifference curves are straight, parallel lines. The outward-pointing arrow indicates that the individual prefers larger quantities to smaller ones (monotonicity). If indifference curves are straight lines, then the MRS is independent of the consumption bundle. This means that the individual is always willing to substitute one good for the same quantity of the other good, hence the name perfect substitutes. Whether two goods are perfect substitutes to each other or not ultimately depends on the perception of the individual, but plausible examples are different brands of toothpaste, yoghurt, shoes, etc. Perfect substitutes are preference orderings that fulfill continuity, monotony and convexity, but not strict convexity.

Figure 7.3b illustrates so-called *perfect complements*. Indifference curves are L-shaped with a kink. L-shaped indifference curves imply that the individual wants to consume the two goods in a fixed ratio. This fixed ratio is given by the slope of the straight line through the origin that connects the kinks. Examples could be left and right shoes (which is why they are sold as pairs), computer hard- and software, coffee and cream, etc. Perfect complements are preference orderings that fulfill continuity, monotony and convexity, but not strict convexity.

Figure 7.3c illustrates strictly convex preferences. Indifference curves bend inwards, but not as extremely as it does for perfect complements. Perfectly convex preferences are somewhere in between perfect substitutes and perfect complements. An individual with such preferences is willing to substitute one good for the other, but has a *ceteris paribus* preference for more balanced bundles.

Finally, Fig. 7.3d illustrates another type of strictly convex preferences, however, with a point of saturation. As the arrows indicate, such preferences are not monotonic, because a globally optimal consumption bundle exists. If consumption falls short of this point, then increasing it makes the individual better off. If consumption

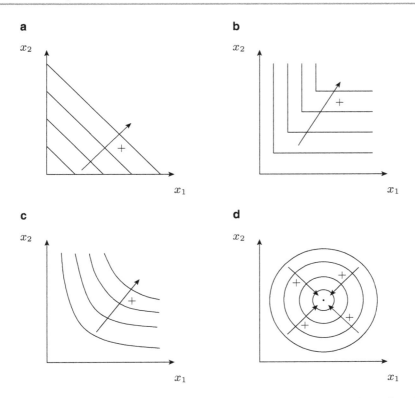

Fig. 7.3 Indifference curves for different preference orderings. **a** Perfect substitutes, **b** perfect complements, **c** strictly convex and **d** strictly convex preferences

exceeds this point, then the individual is better off if she can reduce consumption. Preferences like these are plausible in situations in which goods are not storable and there are physical limits to consumption. Think of ice cream as an example: the first scoop is very good, the second still good, the third is ok, but a fourth, fifth, or sixth scoop makes you sick. It is important to note, however, that if it were possible to produce goods in quantities such that individuals are on or beyond their points of satiation (and the excess can be freely disposed of), society would have overcome scarcity. Hence, the assumption that economics is the science that studies the allocation of scare goods and services implies that one implicitly assumes that one is *not* beyond these points of satiation, either because no such points exist (monotonicity), or because our technological means to production are insufficient to reach these points for all goods in X. In this latter case, however, the indifference curves in Fig. 7.3d look qualitatively similar to the indifference curves in Fig. 7.3c.

7.1.3 Utility Functions

The decision problem of an individual can be completely analyzed by the use of the concept of preference orderings. However, it has turned out that it is sometimes more convenient to represent an ordering by a function, because it allows one to use different and more standard tools from mathematics. This kind of a functional representation of a preference ordering is called a *utility function*. This subchapter will first introduce the concept and then describe some of the potential pitfalls and misunderstandings that come with it.

Economists use the following convention when they represent preference orderings by a function $u(x)$, in which x refers to an arbitrary alternative that can itself be a tuple. They assume that the function assigns a larger number to strictly preferred alternatives, $x^i \succ x^j \Leftrightarrow u(x^i) > u(x^j)$, and the same number to indifferent alternatives, $x^i \sim x^j \Leftrightarrow u(x^i) = u(x^j)$. Any function that meets these requirements qualifies as a utility representation, $u(x)$, of a preference ordering "$\succsim$." More formally, this means:

▶ **Definition 7.5 (Utility Function)** A function $u : X \to \mathbb{R}$ is called a utility function for a preference ordering "$\succsim$" if and only if $x^i \succ x^j \Leftrightarrow u(x^i) > u(x^j)$ and $x^i \sim x^j \Leftrightarrow u(x^i) = u(x^j)$ for all $x^i, x^j \in X$.

This definition of a utility representation or function leaves a lot of freedom when assigning numbers to alternatives or, to put it differently, a given preference ordering has not only one utility representation but many. Here is an example: assume that an individual must choose from a choice set $X = \{x^1, x^2, x^3\}$ and has preferences $x^1 \succ x^2 \succ x^3$. In this case, the following three assignments of numbers to alternatives u_A, u_B, u_C are all utility representations of this preference ordering: $u_A : u_A(x^1) = 3, u_A(x^2) = 2, u_A(x^3) = 1$, $u_B : u_B(x^1) = 354, u_B(x^2) = 7.65, u_B(x^3) = 0$, $u_C : u_C(x^1) = -1, u_C(x^2) = -2, u_C(x^3) = -3$. However, Function D does not represent the preference ordering: $u_D : u_D(x^1) = 3, u_D(x^2) = 1, u_D(x^3) = 2$, because it assigns a larger number to the worst alternative x^3 rather than to the second-best alternative x^2 (2 compared to 1).

An implication of this definition of a utility function is that the absolute values that it assigns to alternatives are meaningless. By the same token, the differences in utility levels for different alternatives are meaningless, as well. The only thing that counts is that preferred alternatives are assigned larger numbers. This is why it is called an *ordinal* concept (absolute values and cardinal differences have no economic meaning).

An immediate implication of this concept is summarized with the following result: assume that $u : X \to \mathbb{R}$ is a utility representation of preference ordering "$\succsim$" and assume that $f : \mathbb{R} \to \mathbb{R}$ is a monotonic and increasing function. In that case, the composite function $v = f \circ u$ is also a utility representation of "$\succsim$." In order to show this, I assume that $u : X \to \mathbb{R}$ is a utility representation of "$\succsim$," which implies, by the definition of a utility function, that

$$u(x^i) > u(x^j) \Leftrightarrow x^i \succ x^j \quad \wedge \quad u(x^i) = u(x^j) \Leftrightarrow x^i \sim x^j.$$

If $f(x)$ is a monotonic increasing function, then one knows that

$$f(u(x^i)) > f(u(x^j)) \Leftrightarrow u(x^i) > u(x^j) \quad \wedge$$
$$f(u(x^i)) = f(u(x^j)) \Leftrightarrow u(x^i) = u(x^j).$$

However, this implies that

$$f(u(x^i)) > f(u(x^j)) \Leftrightarrow x^i \succ x^j \quad \wedge \quad f(u(x^i)) = f(u(x^j)) \Leftrightarrow x^i \sim x^j,$$

And, thereby, that

$$v(x^i) > v(x^j) \Leftrightarrow x^i \succ x^j \quad \wedge \quad v(x^i) = v(x^j) \Leftrightarrow x^i \sim x^j.$$

The transfer from preference orderings to utility functions bears some risk of misinterpretation. Because utility functions assign numbers to alternatives, it is tempting to use these numbers and perform all types of operations with them, like calculating differences ($u(x^i) = 10, u(x^j) = 7$, hence $u(x^i) - u(x^j) = 10 - 7 = 3$ and thus the individual must be three units better off) and comparing them between different individuals (individual A has 8 units of utility, whereas individual B only has 3 units, which makes individual A 5 units better off than individual B). These calculations are mathematically well defined, but economically meaningless, because absolute values of utility or differences in utilities have no meaning if the underlying, primary concept is a preference ordering. What remains, however meaningful, is the marginal rate of substitution MRS, because it is independent of the exact utility representation used. To see this, return to the two representations used above, $u(x)$ and $v(x) = f(u(x))$, and use the following notation: alternative x^i consists of the quantities x_1^i and x_2^i of the two goods 1 and 2. One can express the marginal rate of substitution dx_2^i/dx_1^i by the total differential of the utility function. One can start with the representation $u(x)$ to get the total differential

$$du = \frac{\partial u}{\partial x_1^i} dx_1^i + \frac{\partial u}{\partial x_2^i} dx_2^i.$$

If one wants to stay on the same indifference curve, one has to set $du = 0$, which implies that

$$\frac{dx_2^i}{dx_1^i} = -\frac{\partial u/\partial x_1^i}{\partial u/\partial x_2^i}.$$

For infinitesimal changes in the quantities of the goods, the marginal rate of substitution is equal to the inverse ratio of marginal utilities $\partial u/\partial x_i^k, k = 1, 2$. If one does the same exercise with the representation $v(.)$ instead of $u(.)$, one gets

$$\frac{dx_2^i}{dx_1^i} = -\frac{\partial v/\partial x_1^i}{\partial v/\partial x_2^i} = -\frac{(\partial f/\partial u)(\partial u/\partial x_1^i)}{(\partial f/\partial u)(\partial u/\partial x_2^i)} = -\frac{\partial u/\partial x_1^i}{\partial u/\partial x_2^i}.$$

The MRS is independent of the utility representation that is used. It is the same, irrespective of the exact utility function used, as long as it represents the underlying preference ordering. Hence, the MRS is an economically meaningful concept, because it is a property of the preference ordering, which itself is an explanatory element of the theory.

Digression 17. What Do Preferences and Utility Functions Stand for? The Development of the Modern Concept of Preference Orderings

The view on the concept of utility has gone through substantial changes over the past 100 years or so. What unifies all interpretations is the assumption that individual behavior is somehow related to individual well-being. Initially, economists used the term utility as a proxy for what is called *hedonic* well-being. This position was put forward by *utilitarian* philosophers, like Jeremy Bentham or John Stuart Mill. Mill wrote: "The creed which accepts as the foundation of morals, Utility, or the Greatest-Happiness Principle, holds that actions are right in proportion as they tend to promote happiness, wrong as they tend to produce the reverse of happiness. By happiness is intended pleasure, and the absence of pain; by unhappiness, pain, and the privation of pleasure." Therefore, these philosophers had a specific understanding of what is now called the *theory of mind* and a substantive claim as to what promotes happiness: feeling good. Both the brain and the mind were conceptualized as pleasure- and pain-generating machines and these feelings were considered to be the exclusive motivators for behavior. According to this view, a utility function is a measure for hedonic pleasure (higher utility = more (pleasure minus pain), lower utility = less (pleasure minus pain)) and – together with the assumption that pleasure motivates behavior – is therefore a highly stylized theory of mind. This view of utility was pretty much in line with the leading paradigm of psychology of the time. Psychologists like Gustav Theodor Fechner or Wilhelm Wundt were convinced that mental processes could be measured and compared.

At the turn of the century, however, this view was increasingly scrutinized. The idea that mental phenomena could be measured was mocked as "metaphysical *hocus pocus*," the paradigm in psychology shifted towards what is today called behaviorism and economics followed swiftly. One of the main proponents was Vilfredo Pareto, who wrote in a letter in 1897: "It is an empirical fact that the natural sciences have progressed only when they have taken secondary principles as their point of departure, instead of trying to discover the essence of things. [...] Pure political economy has therefore a great interest in relying as little as possible on the domain of psychology." He replaced the concept of measurable and comparable utility with the concept of an ordinal preference ordering and even went a step further by suggesting that one should not think of a preference ordering as something that summarizes what is going on in the mind or brain, but as a mere *as-if*-device that allows one to explain behavior without giving it a deeper meaning.

However, Pareto kept a minimal theory of mind by assuming that alternatives that individuals rank higher in their preference ordering are better for them (given their own subjective standard). This assumption led to the idea of what is today called *Pareto efficiency* as a normative criterion (see Chap. 5 for the definition).

This concept of preferences and the associated idea that utility functions have no deeper ontological meaning beyond representing preferences led to the development of economic analysis of individual behavior on the basis of indifference curves by Edgeworth and it was perceived as liberating at the time. The enthusiasm can still be sensed in the following quote (Eugen Slutsky (1915/1952)): "[I]f we wish to place economic science upon a solid basis, we must make it completely independent of psychological assumptions [...]." In the wake of this enthusiasm, economics also developed from a rather narrow science of market behavior to a one-size-fits-all tool in an attempt to understand society at large (John Hicks and Douglas Allen 1934): "The methodological implications of [the new] conception of utility [...] are far reaching indeed. By transforming the subjective theory of value into a general logic of choice, they extent its applicability over wide fields of human conduct."

There is one issue remaining before I can move on to applying the theory of preference orderings in order to better understand the market behavior of consumers. Up until this point the assumption has been that preference orderings can be represented by a utility function, but this is far from obvious. In fact, there is a counterexample that is not too far off the mark when it comes to human behavior. Assume that a consumer who has the choice between two goods, x_1 and x_2, has the following preferences: she prefers more of good 1 to less of good 1 and the same for good 2, but for every quantity of good 1 and, irrespective of the quantity of good 2 that she could consume, she prefers more of good 1. These preferences are called *lexicographic*, because the individual orders the quantities of the goods in the same way as a lexicon orders entries: it defines a hierarchy that gives priority of the first letter over the second, the second letter over the third and so on. Only in the event of a tie in the first letter, the second letter becomes relevant and so on. Figure 7.4 illustrates this case.

Here is an example: assume the consumer has the choice between three alternatives $x^1 = (1, 1)$, $x^2 = (1, 100)$, $x^3 = (2, 1)$. With lexicographic preferences, the consumer prefers x^2 to x^1 (more of good 2) and x^3 to x^2 (more of good 1, in which good 2 does not matter as soon as there is more of 1).

Lexicographic preferences may seem rather special and they probably are, but one cannot exclude them from consideration without knowing what people really want. However, the problem with these preferences is that they cannot be represented by a utility function. Understanding the deeper reason for this odd result requires some knowledge in measure theory. Intuitively speaking, there are not

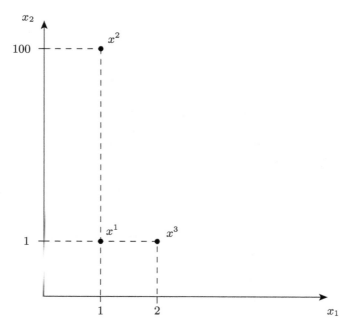

Fig. 7.4 Lexicographic preferences

enough real numbers to represent all the information that is in such a preference ordering. One way to fix the problem is to assume a continuity of preferences, which gives one an explanation for this assumption, and I will henceforth assume that preference orderings are continuous.

7.2   Demand on Competitive Markets

Chapter 4 described several causal factors that explain both individual and market demand on a competitive market. It was argued that demand will most likely depend on the price of the good as well as the prices of other goods, the income of an individual, the individual's tastes and expectations of the future. I am now in a position to replace these intuitive arguments with a sound decision-theoretical analysis using the model of preference or utility maximization introduced before. Remember that economic decision theory comes in two parts: the specification of a choice set and the determination of individual choices from this set for given preferences.

Assume that an individual (Ann) has the choice between two consumption goods, 1 and 2, whose quantities are denoted by x_1 and x_2, both from the set of positive real numbers (including 0). The individual behaves as a price taker and has a budget or income b that she completely spends on the two goods. (The model is very versatile, if one assumes, for example, that x_1 is the consumption today

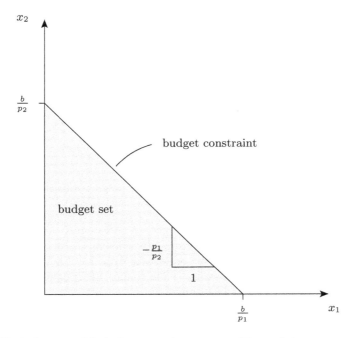

Fig. 7.5 The budget set and the budget constraint on a competitive market

and that x_2 is the consumption tomorrow, it can be interpreted in an intertemporal way to analyze savings behavior.) The prices of the two goods are p_1 and p_2, respectively.

This information can be used to specify Ann's choice set: we know that Ann can spend at most b units of money for the two goods. Expenditures for them are equal to $p_1 \cdot x_1 + p_2 \cdot x_2$. Hence, if expenditures cannot exceed the budget, it must be that:

$$p_1 \cdot x_1 + p_2 \cdot x_2 \leq b.$$

This inequality defines all the pairs x_1, x_2 that Ann can afford to buy, given her income b and prices p_1, p_2. It is her *choice set* that will henceforth also be called her *budget set* and denoted by $B(p_1, p_2, b)$. If Ann completely spends her budget, one will reach a point along the boundary of this set, $p_1 \cdot x_1 + p_2 \cdot x_2 = b$. This equality implicitly defines a function that is called the *budget constraint* or the *budget line*. Figure 7.5 illustrates the budget set.

In this figure, x_1 is drawn along the abscissa and x_2 along the ordinate. Using this convention, one can use the budget constraint to solve for x_2,

$$x_2 = \frac{b}{p_2} - \frac{p_1}{p_2} \cdot x_1.$$

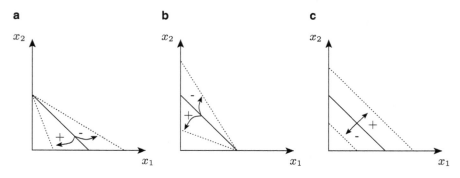

Fig. 7.6 The effects of price and income changes on the budget constraint, **a** Change in p_1, **b** Change in p_2, **c** Change in b

This equation reveals that the budget constraint is a downward sloping straight line that intersects the abscissa at b/p_2, the ordinate at b/p_1 and has a slope $-p_1/p_2$. The set below and to the left of this line is the budget set. It defines the set of all consumption bundles that Ann can afford to buy.

The budget constraint changes with changes in prices or income, as indicated in Fig. 7.6. Note that it shifts outwards (inwards) in a parallel way if the income goes up (down). It rotates outwards (inwards) through the intersection with the ordinate $(0, b/p_2)$ if p_1 goes down (up) and it rotates outwards (inwards) through the intersection with the abscissa $(b/p_1, 0)$ if p_2 goes down (up).

The slope of the budget constraint $-p_1/p_2$ has an important economic interpretation; it measures the rate at which the two goods can be exchanged against each other. Assume that $b = 100$, that $p_1 = 8$ and that $p_2 = 4$. In this example, $-p_1/p_2 = -2$: if one spends one's whole income on the two goods, one has to forfeit two units of good one if one wants to consume an additional unit of good 2, because good 2 is twice as expensive as good 1. The slope $-p_1/p_2$ is, therefore, the *relative price* of good 1 in units of good 2 and measures the opportunity costs of an additional unit of good 2 as defined by market prices.

7.2.1   Graphical Solution

Now one can apply the concept of preference orderings or utility functions in order to analyze choice. The hypotheses that can be derived depend on the assumptions that one makes regarding the structure of the preference ordering. Most of the literature assumes that individual behavior in markets can be described as if individuals would like to maximize a continuous (Assumption 3), monotonic (Assumption 4) and convex (Assumption 5a) or strictly convex (Assumption 5b) preference ordering based on their respective budget sets $B(p_1, p_2, b)$. In order to have an easier diagrammatic representation of the choice problem, it is also assumed that pref-

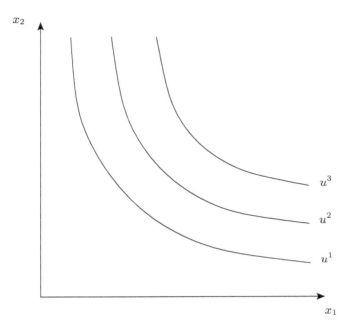

Fig. 7.7 Indifference curves in the context of a competitive market

erences are not only defined on $B(p_1, p_2, b)$ but also on all possible consumption bundles (x_1, x_2), irrespective of whether the individual can afford them or not.

Continuity implies that a preference ordering can be represented by a (utility) function, $u(x_1, x_2)$, and I will henceforth work with this convention. In order to illustrate the choice problem of an individual (Ann) I will assume in the remainder of this subchapter that her preferences are strictly convex and that they can be represented by a continuously differentiable utility function. In that case, her indifference curves for different levels of utility u^j must be inward bending, as illustrated in Fig. 7.7, where I have drawn three indifference curves for utility levels $u^1 < u^2 < u^3$. In order to keep the language simple I will refer to indifference curves that have larger utility indices as "higher" and indifference curves that have smaller utility indices as "lower."

Monotonicity implies that indifference curves that correspond to higher utility levels lie to the upper right of indifference curves that correspond to lower levels of utility. As one can see, the indifference curves provide an ordering of the set of potential consumption bundles. Starting from a given indifference curve, consumption bundles that lie on indifference curves with a larger utility index are preferred and bundles on indifference curves with a smaller utility index make the individual worse off, in comparison.

If one adds the budget set to the picture, one can use the ordering induced by indifference curves to predict behavior.

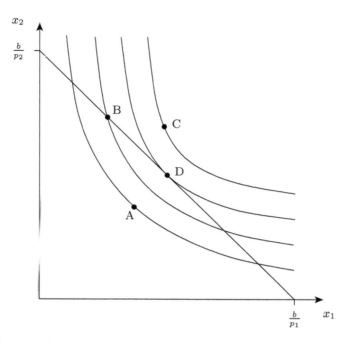

Fig. 7.8 Structure of optimal choices

Figure 7.8 displays a family of indifference curves that is derived from a utility function $u(x_1, x_2)$, and a budget set $B(p_1, p_2, b)$. Qualitatively there are four potential situations that can occur. These are denoted by consumption bundles A, B, C and D. Consumption bundle A is affordable for Ann because it is within her budget set. However, it is not Ann's best choice. If one compares A and B, one can see that B is on a higher indifference curve than A, but still within Ann's budget set. Hence, she would prefer B to A. Is B optimal for Ann? One could argue that C is even better, because it is on an even higher indifference curve. However, note that a consumption bundle like C is outside of Ann's budget set: she would prefer C to B but cannot afford it. Hence, C cannot be her optimal choice either. What one therefore has to do in order to determine Ann's best choice is to look for the highest indifference curve that still belongs to her budget set. Consumption bundle D fulfills this requirement. D is associated with the highest indifference curve that still belongs to budget set $B(p_1, p_2, b)$.

A situation like D has a straightforward economic meaning that is important for understanding the concept of opportunity costs as well as the mechanics of the utility-maximization model. Note that the slope of the budget constraint and the slope of the indifference curve are identical at a point like D. The slope of the budget constraint measures the relative price of the two goods and the ratio at which they can be exchanged on the market. The slope of the indifference curve is the marginal rate of substitution (MRS) and thus the exchange rate between the two goods that makes Ann indifferent between two bundles. At a point like D, both exchange rates coincide and the marginal rate of substitution is equal to the relative price. Why is this condition economically meaningful? Look at the following example: assume that the relative price of good 1 in terms of good 2 is -2 and that the marginal rate of substitution of good 1 in terms of good 2 is -4 at point B. (The budget constraint is less steep than the indifference curve.) Hence, Ann would be willing to give away four units of good 2 for an additional unit of good 1 to stay indifferent. However, given the market rate of exchange, she only has to give away two units. Hence, she can be better off by consuming more of good 1 at the expense of good 2. This logic applies to all consumption bundles for which the "internal" rate of exchange (the MRS) differs from the "external" rate of exchange (the relative price). Hence, only consumption bundles for which the marginal rate of substitution equals the relative price are consistent with the assumption of utility maximization.

The fact that the utility-maximizing consumption bundle is on the budget constraint and not in the interior of the budget set is a consequence of the assumption of monotonicity of preferences. With non-monotonic preferences, it could be that Ann is satiated without fully spending her income. Monotonicity, in this sense, can therefore be thought of as an expression of the underlying assumption of scarcity: with non-monotonic preferences, there could be situations with high incomes b such that all of Ann's desires are fulfilled. This would be the point at which scarcity – at least for Ann – ceases to exist. It is, ultimately, an empirical question as to whether such a point can ever be reached or not. One should, therefore, take the assumption with caution because its unscrutinized acceptance implies that one has implicitly commuted to the idea of quantitative growth in terms of some measure, like gross national product.

In addition, the fact that the MRS equals the relative price of the two goods at the optimum is a consequence of the assumption of the strict convexity of preferences (in fact, indifference curves additionally have to be continuously differentiable in order to guarantee this, otherwise they could have "kinks").

7.2.2 Analytical Solution

The utility-maximization problem can also be studied analytically. In order to do so, one starts by formally stating Ann's choice situation. One is confronted with an optimization problem that has the following structure: p_1, p_2 and b are the explanatory variables of the model, which means that the variables that determine Ann's consumption decisions are x_1 and x_2. They are, therefore, the explained variables. Hence, one needs to determine the functions $x_1(p_1, p_2, b), x_2(p_1, p_2, b)$. How is this possible? By assuming that Ann maximizes her utility function, $u(x_1, x_2)$, under the constraint that she does not spend more on consumption than her income is, $p_1 \cdot x_1 + p_2 \cdot x_2 \leq b$, one assumes that her preferences are monotonic, one knows that Ann will spend her whole income, and one can write:

$$\max_{x_1,x_2} u(x_1,x_2) \quad s.t. \quad p_1 \cdot x_1 + p_2 \cdot x_2 = b.$$

This notation needs some explanation. The $\max_{x_1,x_2}$-term indicates that one is looking for the maximum of the objective (utility) function with respect to the endogenous variables. The term s.t. abbreviates "such that," which indicates that Ann has to respect her budget constraint.

Formally, this is a constrained optimization problem and there are several ways to solve it. As long as one restricts one's attention to two endogenous variables, the solution does not require advanced mathematical techniques; instead one can simply use the constraint to eliminate one of the endogenous variables in the objective function. For more general (and realistic) problems in which Ann can choose between more than two goods, however, one needs a more general procedure that uses so-called *Lagrange* multipliers. I will not get any deeper into the field of optimization problems under constraints, because the results that I can derive with the more elementary approach for the two-goods case gives me sufficient mileage in explaining individual behavior on competitive markets for an introductory course in microeconomics.

In order to solve the problem, one can convert the budget constraint in the same way as I have shown before, $x_2 = b/p_2 - (p_1/p_2) \cdot x_1$, and denote the function that relates x_1 and x_2 by $X_2(x_1) = b/p_2 - (p_1/p_2) \cdot x_1$. This equation can be used to eliminate x_2 in the utility function. One, therefore, ends up with a modified, unconstrained optimization problem:

$$\max_{x_1} u(x_1, X_2(x_1)) = \max_{x_1} u(x_1, b/p_2 - (p_1/p_2) \cdot x_1).$$

In order to illustrate how this problem can be solved, assume that $u(x_1,x_2)$ is twice continuously differentiable and that the underlying preference ordering is strictly convex. If these assumptions are fulfilled, then an interior maximum is characterized by a value of x_1, such that the first derivative is equal to zero (first-order condition):

$$\frac{\partial u}{\partial x_1} + \frac{\partial u}{\partial x_2} \cdot \frac{\partial X_2}{\partial x_1} = \frac{\partial u}{\partial x_1} - \frac{\partial u}{\partial x_2} \cdot \frac{p_1}{p_2} = 0.$$

This condition can be simplified to:

$$\frac{\partial u/\partial x_1}{\partial u/\partial x_2} = \frac{p_1}{p_2},$$

which is the optimality condition for the consumer-choice problem. In order to be able to interpret this condition, one has to understand the term on the left-hand side. In order to do so, one can use the total differential of the utility function

$$du = \frac{\partial u}{\partial x_1} \cdot dx_1 + \frac{\partial u}{\partial x_2} \cdot dx_2.$$

The total differential measures the total effect on utility with a change in the explanatory variable of dx_1 and dx_2, respectively. One is not interested in arbitrary changes but in changes that leave total utility constant, $du = 0$, because this keeps one on the same indifference curve. In other words, the set of all (x_1, x_2) that lead to the same level of utility constitute the marginal rate of substitution:

$$du = \frac{\partial u}{\partial x_1} \cdot dx_1 + \frac{\partial u}{\partial x_2} \cdot dx_2 = 0$$

$$\Leftrightarrow MRS(x_1, x_2) = \frac{dx_1}{dx_2} = \frac{\partial u / \partial x_1}{\partial u / \partial x_2}.$$

However, this is exactly the left-hand side of the optimality condition. One can therefore conclude that a preference- or utility-maximizing individual chooses consumption in a way that the marginal rate of substitution equals the relative price of the goods.

Consumption bundles (x_1^*, x_2^*) that fulfill the first-order condition are the individual's utility-maximizing choices. Formally, they are functions of the explanatory variables $x_1(p_1, p_2, b)$ and $x_2(p_1, p_2, b)$ and are named *Marshallian demand functions* after Alfred Marshall. What is interesting, from the point of view of the structure of individual demand, is whether the Marshallian demand functions have any particular properties that allow one to better understand the structure of individual and, thereby ultimately, market-demand behavior.

However, here comes the challenge: the ultimate test for the usefulness of a theory is – according to critical rationalism (see Chap. 1 for a description of this position in the philosophy of science) – its empirical validity. Hence, one has to formulate a theory in a way that makes it empirically testable. The theory of consumer choice has two building blocks: preferences and choice sets that are determined by prices and income. It is relatively straightforward to empirically measure the latter elements of the theory, but it is not possible to determine individual preferences directly. This is bad news for empirical tests: behavior is determined by both, choice sets and preferences. If one cannot observe preferences, one cannot test the theory. Hence, one can only measure the properties of the theory that are *independent* of the specific preference ordering underlying consumer choices. However, given that any choice of consumption can be rational for some preference ordering, the only hope that one has is that the theory is testable when one looks at *changes* in observable behavior that are caused by *changes* in prices or income. It may be that a change in prices or income induces stable and predictable reactions that can, in principle, be falsified by confronting them with empirical data. In order to be able to do so, however, one has to impose the (dogmatic, see Chap. 1) assumption that preferences remain stable over a period of time. This is why *comparative statics* plays such an important role in economics: if there is any hope for empirically testing the theory, it is because it produces refutable hypotheses regarding the change in Marshallian demand functions when prices or income change. Whether the theory can live up to these standards or not will be the subject of this investigation.

One important property is that Marshallian demand functions are homogeneous of degree zero, i.e. that a proportional change in all prices and income has no influence on individual behavior. Formally, this means that $x_i(p_1, p_2, b) = x_i(\lambda \cdot p_1, \lambda \cdot p_2, \lambda \cdot b)$ for $i = 1, 2$ and $\lambda > 0$. Intuitively this means that it does not matter whether prices and income are measured in Swiss Francs or in Rappen, Euro or Cent; as long as the relative price of both goods and the purchasing power of income remains the same, the individual will not change her behavior. In order to see that this must be the case return to the budget constraint $x_2 = b/p_2 - (p_1/p_2) \cdot x_1$. If all prices and income are multiplied by the same factor λ, one gets

$$x_2 = \frac{\lambda b}{\lambda p_2} - \frac{\lambda p_1}{\lambda p_2} \cdot x_1 = \frac{b}{p_2} - \frac{p_1}{p_2} \cdot x_1.$$

The effect of λ cancels out and, therefore, leaves the location of the budget constraint unaltered. However, with an unaltered budget constraint, the optimal behavior of the individual must be unaltered as well, hence the Marshallian demand functions are homogeneous of degree zero in prices and income.

Digression 18. Money Illusion and the Debate Between Keynesian and Neoclassical Economics

The homogeneity of degree zero in prices and income of the Marshallian demand function may sound like an innocuous mathematical property, but, in fact, it marks a very important watershed in the history of economic thinking. Keynesian and neoclassical economists have profoundly disagreed on the role of economic policy to stabilize the economy. One important field of disagreement is monetary policy. Neoclassical economists are usually sceptic on the role that monetary policy can or should play, with the implication that price stability is usually the primary focus of neoclassical monetary policy. On the contrary, Keynesian economists usually see a much more active role for monetary policy in stimulating and stabilizing the economy (Keynes 1936).

There are several reasons why these schools disagree, but at least one can be traced back to the homogeneity of degree zero of the Marshallian demand functions. If this property holds, the possibility to influence the economy by means of monetary policy are severely limited. Increasing or reducing money supply is like multiplying all prices and income by λ. However, if this is the case and if the model of utility- or preference-maximizing individuals is correct, then the real effects of these changes on the economy are zero: general inflation or deflation is like measuring prices in different currencies without changing the purchasing power of income or the relative prices of goods. This property is sometimes also called the *absence of money illusion*. Without a money illusion, monetary policy has no impact on the economy, because people will not change their behavior and, if people's behavior does not change, then everything remains the same. The only way monetary policy can influence behavior, according to this view, is if inflation or deflation change different prices and incomes differently, hence either changing relative prices, purchasing power or both. This can happen if some prices are nominally fixed, while other prices can adjust to changes in money supply. A Keynesian economist, who sees an active role for monetary policy, therefore, either has to think that some prices or incomes are nominally fixed or that the model of preference or utility maximization is flawed to begin with.

The first-order condition is only a necessary condition for a utility maximum, and one does not know yet if it characterizes a local maximum, a local minimum or a point of inflection. In order to say more, one has to check the second-order condition. The first-order condition is the function:

$$\frac{\partial u(x_1, x_2)}{\partial x_1} - \frac{\partial u(x_1, x_2)}{\partial x_2} \cdot \frac{\partial X_2}{\partial x_1}.$$

It characterizes a local maximum if its derivative, with respect to x_1, is smaller or equal to zero,

$$\frac{\partial^2 u(x_1, x_2)}{\partial x_1^2} + \frac{\partial^2 u(x_1, x_2)}{\partial x_1 \partial x_2} \frac{\partial X_2}{\partial x_1} + \frac{\partial X_2}{\partial x_1} \frac{\partial^2 u(x_1, x_2)}{\partial x_2 \partial x_1} - \left(\frac{\partial X_2}{\partial x_1}\right)^2 \frac{\partial^2 u(x_1, x_2)}{\partial x_2^2} \leq 0.$$

This condition can be simplified, if one remembers that:

$$\frac{\partial X_2}{\partial x_1} = -\frac{p_1}{p_2} = -\frac{\frac{\partial u(x_1, x_2)}{\partial x_1}}{\frac{\partial u(x_1, x_2)}{\partial x_2}}$$

and notes that:

$$\frac{\partial^2 u(x_1, x_2)}{\partial x_1 \partial x_2} = \frac{\partial^2 u(x_1, x_2)}{\partial x_2 \partial x_1}.$$

This leads one to:

$$\left(\frac{\partial u(x_1, x_2)}{\partial x_2}\right)^2 \cdot \frac{\partial^2 u(x_1, x_2)}{\partial x_1^2} + \left(\frac{\partial u(x_1, x_2)}{\partial x_1}\right)^2 \frac{\partial^2 u(x_1, x_2)}{\partial x_2^2}$$
$$-2\frac{\partial u(x_1, x_2)}{\partial x_1} \frac{\partial u(x_1, x_2)}{\partial x_2} \frac{\partial^2 u(x_1, x_2)}{\partial x_1 \partial x_2} \leq 0.$$

If the condition holds at (x_1^*, x_2^*), then the indifference curve is locally convex at that point, hence it characterizes a local maximum. If the condition holds for every (x_1, x_2), then the indifference curve is globally convex. This condition is fulfilled only if the underlying preference ordering is convex. Hence, the assumption that the preference ordering is convex guarantees that the first-order condition characterizes a maximum. If the inequality is strict, then the preference ordering is strictly convex and the solution is unique.

A utility function with this property is called (strictly) *quasi-concave*. Quasi-concavity is weaker than concavity of functions, because it only guarantees that the not-worse-than-x sets (whose boundaries are the respective indifference curves) are convex sets, but makes no assumptions about the concavity of the rest of the function.

7.2.3 Three Examples

There are three utility functions that represent typical preference orderings and that play an important role in a lot of economic applications of the model. This sub-chapter will analyze these examples both graphically and analytically.

7.2.3.1 Homothetic Strictly Convex Preferences

An example of a so-called *homothetic* utility function is given by $u(x_1, x_2) = a \cdot (x_1)^\alpha \cdot (x_2)^\beta$, where a, α, β are positive real numbers. It is an example for a strictly quasi-concave utility function that has the additional property that the MRS is constant for proportional changes of the two goods ($x_1/x_2 = c$, with $c > 0$ being constant).

The following paragraphs focus on a special case in which $a = 1$ and $\alpha = \beta = 1/2$, because it is more convenient to solve mathematically. These assumptions imply that $u(x_1, x_2) = \sqrt{x_1} \cdot \sqrt{x_2}$. Before one derives the Marshallian demand functions, it makes sense to familiarize oneself with the structure of this function. One can, for example, derive the indifference curve for some arbitrary level of utility $\bar{u}$, $u(x_1, x_2) = \sqrt{x_1} \cdot \sqrt{x_2} = \bar{u}$. In order to derive the function, $X_2(x_1)$, that describes the indifference curve, one solves for x_2, $\sqrt{x_2} = \bar{u}/\sqrt{x_1} \Leftrightarrow x_2 = (\bar{u})^2/x_1$. This is a family of hyperbolic functions, one for each value of $\bar{u}$, which implies that the underlying preference ordering is strictly convex.

At this point, one can further illustrate that every monotonic transformation of a utility function represents the same preference ordering. If one squares the utility function (this is a monotonic transformation, because the underlying utility function only has positive values) one gets $v(x_1, x_2) = (u(x_1, x_2))^2 = (\sqrt{x_1} \cdot \sqrt{x_2})^2 = x_1 \cdot x_2$. It follows that the indifference curves of this function are also hyperbolic: $x_2 = \bar{u}/x_1$. The only difference between the two indifference curves is the *absolute* value of utility, but remember that this number has no meaningful interpretation.

With these prerequisites one can move on to analyze Ann's demand if her preferences have a utility representation of $u(x_1, x_2) = \sqrt{x_1} \cdot \sqrt{x_2}$. Figure 7.9 displays a family of indifference curves for different utility levels. Given that they are hyperbolic, the MRS remains constant along a ray through the origin (points A, B, and C).

Figure 7.10a shows the same family of indifference curves and adds different budget constraints for different income levels $b^1 < b^2 < b^3$. The above-mentioned property that the MRS remains constant along a ray through the origin implies that the utility-maximizing choices for different income levels must be on a ray through the origin, as well. This path of optimal choices for different income levels is called

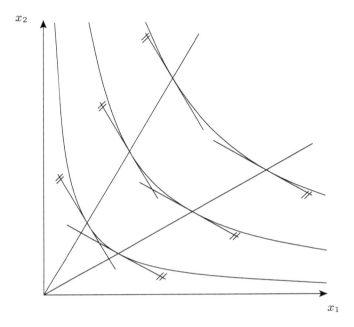

Fig. 7.9 MRS for homothetic strictly convex preferences preferences

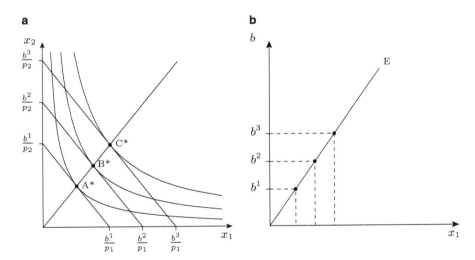

Fig. 7.10 Income-consumption path and Engel curve for homothetic strictly convex preferences

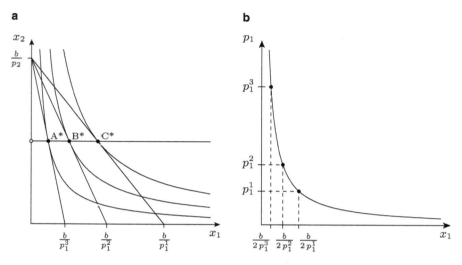

Fig. 7.11 The price-consumption path and demand function for homothetic strictly convex preferences

the *income-consumption path*. This is depicted by A^*, B^*, and C^*. Figure 7.10b displays the demand of one of the two goods (say 1) as a function of income levels. The argument above implies that the relationship between income b and demand x_1^* must be linear. The straight line E reflects this fact. The relationship between income and utility- or preference-maximizing consumption is called an *Engel curve*. In this case, it is upward sloping, which means that the good is normal (see Chapt. 4 for the definition of this term).

Figure 7.11a displays the same family of indifference curves and adds different budget constraints for different price levels of good 1, $p_1^1 < p_1^2 < p_1^3$. An increase in p_1 rotates the budget constraint inwards around the point $0, b/p_2$. The utility-maximizing consumption bundles are, again, depicted by A^*, B^* and C^*. They lie on the horizontal line displayed in the figure that is called the *price-consumption path*. Figure 7.11b displays the demand for good 1 as a function of its price p_1. This is the *individual demand function* that is already known from Chap. 4. It is downward sloping, which means that it is ordinary (see Chap. 4 for the definition of the term).

Alternatively, one can derive Ann's demand function analytically. In order to do so, one can either use the information that MRS has to be equal to the relative price directly, or start with the utility-maximization problem. In order to practice, I will follow the second road in the following paragraphs. Additionally, in order to simplify the mathematics I will use a utility representation $v(x_1, x_2) = x_1 \cdot x_2$ (try the other formulation to see if it leads to the same result):

$$\max_{x_1} \quad (x_1) \cdot (b/p_2 - (p_1/p_2) \cdot x_1).$$

To get the first-order condition one can apply the product rule:

$$(b/p_2 - (p_1/p_2) \cdot x_1) - (p_1/p_2) \cdot x_1 = 0.$$

One can solve this condition for x_1 to get the Marshallian demand function for good 1:

$$x_1(p_1, p_2, b) = \frac{b}{2 \cdot p_1}.$$

Knowing that $x_2 = b/p_2 - (p_1/p_2) \cdot x_1$, one can also derive the Marshallian demand function for good 2:

$$x_2(p_1, p_2, b) = \frac{b}{2 \cdot p_2}.$$

The demand functions have three remarkable properties. First, they are linear in income, which is what one would have expected from the linearity of the Engel curve. Second, they are downward sloping and hyperbolic in their own prices, which is also what one would have expected from the graphic analysis. Third, they do not depend on the price of the other good.

   These utility functions play an important role in economic applications, because of their simplicity. They can be generalized by assuming that the relative importance of the two different goods can be measured by some parameter $\alpha \in [0, 1]$ that yields $u(x_1, x_2) = x_1^\alpha \cdot x_2^{1-\alpha}$. It is, however, not clear if individuals behave as if they maximize preferences of this type. It is more useful as a thought experiment than an empirically supported claim about actual behavior.

7.2.3.2   Perfect Substitutes

The above utility function represents a case in which individuals prefer to consume the goods in relatively balanced bundles, but react to price changes by increasing the relative demand of the good that gets relatively cheaper. This need not be the case. There may be goods for which the individual has more extreme preferences: Ann consumes either one or the other, depending on which one is cheaper. If this is the case, the two goods are called *perfect substitutes* and such a taste can be represented by the following utility function:

$$u(x_1, x_2) = \alpha \cdot x_1 + \beta \cdot x_2,$$

in which α/β measures the relative importance of good 1 compared to good 2. This function is homothetic as well. In the following paragraphs, assume that they are of equal importance to Ann and normalize them to $\alpha = \beta = 1$. In this case, Ann's indifference curves are downward sloping straight lines with a slope of -1. A family of indifference curves is denoted by the dotted lines in Fig. 7.12. I have also drawn a budget constraint in this figure. It is already known that its graph is a straight line with slope $-p_1/p_2$, so the optimal consumption bundle depends on the relative slope of the indifference curves and the budget constraint: if the former is steeper than the latter, then Ann only consumes good 2 (as in the Figure) and *vice*

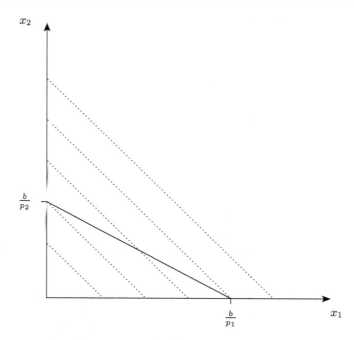

Fig. 7.12 Indifference curves and a budget constraint for perfect substitutes

versa. If both curves have an equal slope, then Ann would be indifferent between both goods.

Given that Ann will only buy the relatively cheaper good, the Engel curves are easy to derive. They are a straight line with slope 0 for the relatively more expensive and a straight line with slope $1/p_i$ for the relatively cheaper good. Price changes can be analyzed as before, but have a somewhat different effect on demand. Figure 7.13a shows a family of indifference curves and – as in the first example of a utility function analyzed before – adds different budget constraints to different price levels for good 1, $p_1^1 < p_1^2 = p_2 < p_1^3$. An increase in p_1 lets the budget constraint rotate inwards around the point $(0, b/p_2)$. The utility-maximizing consumption bundles are, again, depicted by A^*, B^* and C^*. If the price for good 1 is smaller than the price for good 2, then Ann spends all her income on good 1, which is indicated by point A^*. If both prices are equal ($p_1^2 = p_2$), then Ann is indifferent between both goods and we use the convention that she buys equal quantities, in this case. If p_1 rises further, then demand is zero, because Ann prefers the cheaper good. This behavior is illustrated by the demand function in Fig. 7.13b. It is discontinuous at $p_1 = p_2$ and hyperbolic for smaller prices of good 1.

If one remembers the analysis of competitive behavior in Chap. 4, this discontinuity can be problematic, because it may be that there is no intersection between market demand and market supply, in this case. This is why the continuity of both

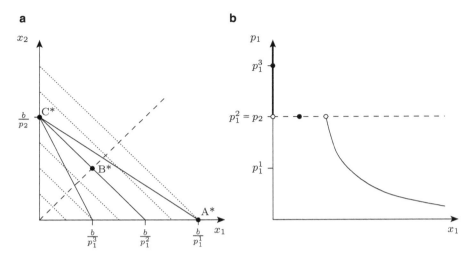

Fig. 7.13 Optimal choices and a demand function for perfect substitutes

demand and supply functions are important to guarantee the existence of a competitive equilibrium and one has now seen a case in which this is not the case. This reveals an advantage of the behavioral foundation of the demand function: it leads to a better understanding of the deeper reasons behind the continuity or discontinuity of demand functions by linking them to individual preferences. As one can see, the potential non-existence of an equilibrium can be a result of preferences that are not entirely absurd. The deeper reason for the discontinuity is that the preference ordering is only convex, not strictly convex. With strictly convex preferences, "small" changes in prices will lead to "small" changes in demand, but this is not the case if goods are, for example, perfect substitutes.

In order to derive the Marshallian demand functions analytically, one has to be careful. Given that both the budget constraint and the indifference curves are linear, the utility maximum cannot be derived from the first-order condition. Fortunately, one has already collected almost all the information that is necessary to determine Marshallian demand. One knows that Ann only buys the cheaper good and one has introduced the convention that she splits her income equally if both goods have the same price. The only step left is to formalize this information:

$$x_1(p_1, p_2, b) = \begin{cases} \frac{b}{p_1}, & p_1 < p_2 \\ \frac{b}{2 \cdot p_1}, & p_1 = p_2 \\ 0, & p_1 > p_2 \end{cases}, \qquad x_2(p_1, p_2, b) = \begin{cases} \frac{b}{p_2}, & p_2 < p_1 \\ \frac{b}{2 \cdot p_2}, & p_2 = p_1 \\ 0, & p_2 > p_1 \end{cases}.$$

7.2.3.3 Perfect Complements
The last example that I discuss is, in a sense, the opposite extreme from the case of perfect substitutes. There are some goods that Ann wants to consume together in fixed proportions, like left and right shoes, printer and toner, or hardware and

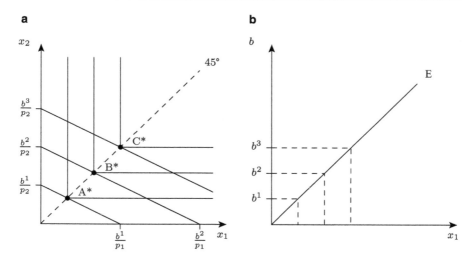

Fig. 7.14 Indifference curves, budget constraints and an Engel curve for perfect complements

software. If this is the case, she wants to spend her income on the two goods in way that makes sure that she buys both goods in fixed proportions. A utility function that expresses such preferences is:

$$u(x_1, x_2) = \min\{\alpha \cdot x_1, \beta \cdot x_2\},$$

in which α/β measures the number of units of good 2 that Ann needs to make use of an additional unit of good 1. This is also a homothetic function. To see this, assume that x_1 is the number of car bodies and x_2 is the number of wheels. It takes four wheels and a car body to assemble a useful car, so if $\alpha = 4$ and $\beta = 1$ one gets $\alpha/\beta = 4$, the number of units of good 2 (wheels) that is needed for one unit of good 1 (car bodies). The following paragraphs will focus on the easiest case in which $\alpha = \beta = 1$, i.e., Ann needs one unit of good 1 together with one unit of good 2, $u(x_1, x_2) = \min\{x_1, x_2\}$.

How do the indifference curves look like? I have drawn a family of them in Fig. 7.14a. They are L-shaped with a kink at the 45-degree line, which is where both goods are consumed in equal quantities. Increasing the quantity of one good while keeping the quantity of the other good constant is useless for Ann, which is why points on the vertical and horizontal lines are on the same indifference curve.

I have also added budget constraints for different income levels $b^1 < b^2 < b^3$. As one can see, the utility-maximizing consumption bundle is always at the kink, which is why they are along the 45-degree line through the origin. This is depicted by A^*, B^* and C^*. It follows immediately that the Engel curve must also be a straight line as in Fig. 7.14b.

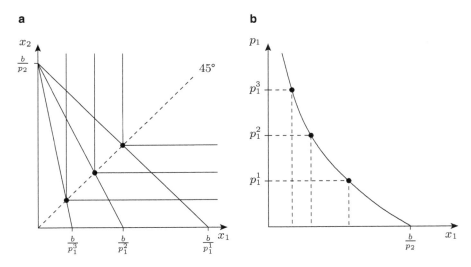

Fig. 7.15 Optimal choices and demand function for perfect complements

How about changes in prices? As before, an increase in the price of good 1 rotates the budget constraint inwards. Once again, I focus on three such prices $p_1^1 < p_1^2 < p_1^3$ and illustrate them in Fig. 7.15a.

One already knows that Ann will always buy both goods in equal quantities, i.e., stay on the 45-degree line. However, this implies that both the demand for good 1 and for good 2 is falling, if the price for good 1 goes up. Hence, the demand function for good 1 is given by the downward sloping graph in Fig. 7.15b. Note that it intersects the abscissa at $x_1 = b/p_2$, because at $p_1 = 0$ Ann can afford b/p_2 units of both goods. It is illustrative to also look at x_2 as a function of p_1, which is done in Fig. 7.16.

One knows that Ann will always buy both goods in equal quantities. However, this implies that the demand of good 2, as a function of p_1, is *identical* to the demand of good 1, as a function of p_1.

In order to derive the Marshallian demand functions analytically, first note that one cannot use first-order conditions in this case either, because the indifference curves have a kink, which implies that they cannot be continuously differentiated. Fortunately, the problem is very intuitive to solve. One knows that Ann is constrained by her budget, $p_1 \cdot x_1 + p_2 \cdot x_2 = b$, and wants to consume both goods in equal quantities, $x_1 = x_2 = x$. This information can be used in the budget constraint to get $p_1 \cdot x + p_2 \cdot x = b$. However, this is a linear function in one endogenous variable, so one can solve it. The solution is:

$$x_1(p_1, p_2, b) = \frac{b}{p_1 + p_2}, \qquad x_2(p_1, p_2, b) = \frac{b}{p_1 + p_2}.$$

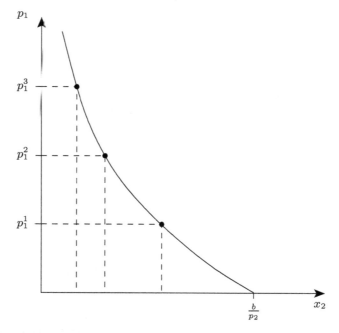

Fig. 7.16 The demand function for good 2 for perfect complements

7.2.4   Comparative Statics and the Structure of Market Demand

The three examples for potential preference orderings with associated utility functions have revealed that there is a stable relationship between the structure of preferences on the one and the structure of market demand on the other hand. In all three cases, one has seen that individual demand is decreasing in the price of the good and (weakly) increasing in income. The cross-price effects, however, seem to be more complex. They do not exist in the strictly convex and homothetic case, are (weakly) positive but extreme in the case of perfect substitutes and negative in the case of perfect complements. One has also seen that the strict convexity of the preference ordering seems to be important in order to guarantee that an equilibrium exists, because individual demand can otherwise be discontinuous.

Now one can find out if these findings can be generalized. Preferences are not directly observable and individuals seem to differ substantially with respect to their tastes. Hence, it would be nice if one did not have to make too many assumptions on the structure of preferences, as every assumption reduces the explanatory power of the theory, because they rule out certain preferences of which one does not know if they accurately describe real-life individuals. Thus, one can see how far one gets if one imposes monotonicity and strict convexity of a preference ordering with respect to the structure of individual demand. In order to do this, one can focus on two comparative-static experiments: a change in income and a change in the price of a good.

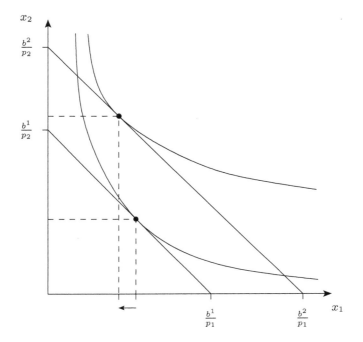

Fig. 7.17 Optimal choices for a change in income: an inferior good

7.2.5   Changes in Income

One has already seen that a change in income leads to a parallel shift of the budget constraint. Furthermore, one has already seen in the above examples that goods can be normal (demand increases if income increases). The remaining question is if this property is an artifact of the specific preference orderings or whether it is a general property of demand functions that are derived from preferences. Figure 7.17 shows that this is unfortunately not the case.

It displays two income levels, b_1 and b_2, and the associated indifference curves that Ann can reach, if she maximizes utility. As can be seen, the demand for good 2 goes up if income goes up, but the demand for good 1 does not. Hence, good 2 is normal and good 1 is inferior for this change in income. (Note that these properties are local and that they can hold for some changes in income, but not for others.) Hence, strict convexity and monotonicity do not rule out the inferiority of one of the goods. Besides, they should not, because there are a lot of goods that are, in fact, inferior, like low-quality products that are replaced by higher-quality substitutes, if the individual gets richer.

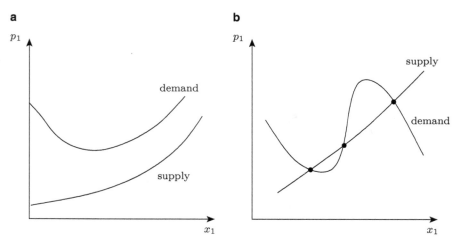

Fig. 7.18 Market equilibrium and non-monotonic demand functions

7.2.6   Changes in Price

In addition to continuity, individual and market demand should be decreasing as the price of the good increases in order to guarantee the existence and uniqueness of a competitive equilibrium. If demand is increasing as its price increases, there may be cases in which an equilibrium does not exist at all or in which multiple equilibria exist. Figures 7.18a and 7.18b illustrate both cases.

Hence, it would be nice if one could show that goods are ordinary if individuals maximize a monotonic and strictly convex preference ordering. Unfortunately, this is not the case. Figure 7.19 gives an example for the so-called *Giffen paradox*, which is a situation in which the demand of a good decreases despite the fact that its price decreases.

Figure 7.19 focuses on a decrease in the price of good 1 from p_1^1 to p_1^2. The utility-maximizing consumption bundle changes from $A = (x_1(p_1^1, p_2, b), x_2(p_1^1, p_2, b))$ to $B = (x_1(p_1^2, p_2, b), x_2(p_1^2, p_2, b))$ and the highest indifference curves that can be reached are denoted by $I(p_1^1, p_2, b)$ and $I(p_1^2, p_2, b)$. As can be seen, the demand for good 1 goes down (and the demand for good 2 goes up) and this follows necessarily from the strict convexity of the preference ordering, because one is moving *along* the indifference curve.

What is going on here? One gets closer to understanding this phenomenon if one focuses on the curvature of the indifference curves. If the change in p_1 did not induce a rotation around $(0, b/p_2)$ but instead induce a rotation *along* the indifference curve (see Fig. 7.20), then the effect of an decrease in the price of good 1 would have the expected negative sign: good 1 gets relatively cheaper compared to good 2 and this isolated effect motivates Ann to buy more of good 1 and less of good 2. However, the reduction of p_1 not only has the effect that good 1 gets relatively

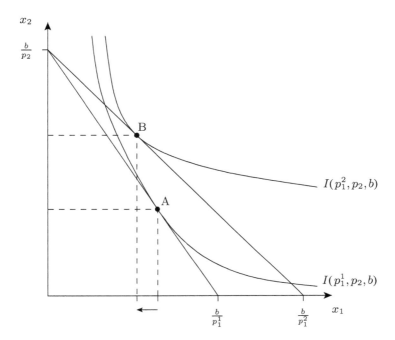

Fig. 7.19 Optimal choices for a change in prices: the Giffen paradox

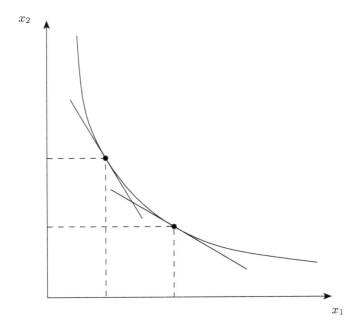

Fig. 7.20 The optimal choice for a compensated change in relative prices

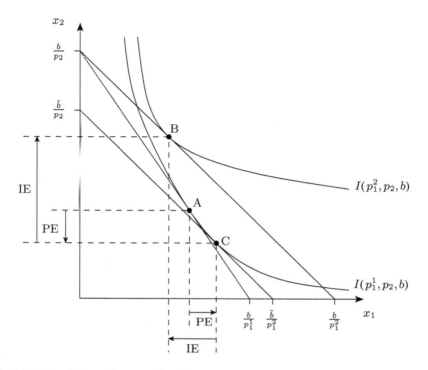

Fig. 7.21 Price- (PE) and Income effect (IE) and the Giffen paradox

cheaper, it also makes Ann richer, because her new budget set contains the old one as a subset. As a matter of fact, it is this latter effect that may cause the Giffen paradox.

In order to understand this, one has to disentangle the two effects in Fig. 7.21.

Figure 7.21 is identical to Fig. 7.19, with the exception that I have introduced an artificial budget constraint for a hypothetical income level $\tilde{b}$. This hypothetical constraint is constructed to allow Ann to reach the same maximum indifference curve as before the price change, $I(p_1^1, p_2, b)$. In order to guarantee this, one has to change her income from b to some hypothetical income level $\tilde{b}$, such that $I(p_1^1, p_2, b) = I(p_1^2, p_2, \tilde{b})$. The utility-maximizing consumption bundle that results from this hypothetical budget constraint $p_1^2 \cdot x_1 + p_2 \cdot x_2 = \tilde{b}$ is denoted by $C = (x_1(p_1^2, p_2, \tilde{b}), x_2(p_1^2, p_2, \tilde{b}))$. One calls it the *compensated demand*. This compensated demand for good 1 is larger than before, $x_1(p_1^1, p_2, b) < x_1(p_1^2, p_2, \tilde{b})$, i.e., the isolated effect of a change in the relative price is negative (a smaller price and a larger quantity). This compensated effect is called the *price effect*, and it brings us from A to C in the figure.

However, the compensation in income from b to $\tilde{b}$ is only the first step in the thought experiment. Therefore, in the next step, one will see what happens if one moves from $(p_1^2, p_2, \tilde{b})$ to (p_1^2, p_2, b). This change holds the relative price constant, but changes Ann's income and one has already seen what can happen. One already knows that this change brings one from C to B, but the additional insight is that this is only possible if the good is *inferior*; comparing C with B reveals that $x_1(p_1^2, p_2, b) < x_1(p_1^2, p_2, \tilde{b})$. This is called the *income effect*.

This thought experiment is important, because it allows one to better understand why individual demand may not fall as its own price falls. Any change in the price of one good has a price as well as an income effect and it is the income effect that may cause the Giffen paradox: if the good is inferior for Ann, then it is possible that her demand is (locally) increasing as the price increases.

This result is perhaps intellectually fascinating and it allows one to understand the mechanics of the preference-maximization model more profoundly, but it is, at the same time, highly unsatisfactory. In the end, the whole exercise to develop a choice-theoretic foundation of market behavior was motivated to better understand the structure of demand functions on competitive markets. These demand functions have to fulfill certain properties, like continuity and ordinary goods, to ensure that a market equilibrium exists. What one can learn from the Giffen paradox and its deeper reasons is that the assumption of preference maximization alone (even with the further restrictive assumptions of strict convexity and monotonicity) is insufficient to guarantee that a unique equilibrium exists. As Chap. 4 described, existence and uniqueness are important for positive economics, because sound economic prognoses depend on them. As one has seen with the three examples above on economic preferences, one can guarantee existence and uniqueness by imposing additional assumptions regarding preferences (like the assumption that all individuals have strictly quasi-concave utility functions), but this comes at the cost of sacrificing generality. Additionally, this cost is substantial indeed, because one does not have epistemic access to individual preferences, so one cannot know if any specific assumptions regarding their structure are empirically justified. However, this is the situation: if one wants a general theory of consumer choice that allows all types of preferences, then one cannot be sure that an equilibrium exists or if it is unique and if one want a unique existing equilibrium, then one has to start from specific assumptions regarding preferences.

References

Hicks, J. R., & Allen, R. G. D. (1934). A Reconsideration of the Theory of Value. Part I. *Economica, 1*(1), 52–76.

Keynes, J. M. (1936). *The General Theory of Employment, Interest and Money*. Palgrave Macmillan.

Slutsky, E. E. (1915/1952). Sulla teoria del bilancio del consumatore. *Giornale degli Economisti, 51*, 1–26. English edition: On the theory of the budget of the consumer (trans Ragusa, O.). In G. J. Stigler, & K. E. Boulding (Eds.), *Readings in Price Theory* (pp. 27–56). Homewood, Ill.: Irwin.

Further Reading

Arrow, K. J. (1989). Economic Theory and the Hypothesis of Rationality, *The New Palgrave: Utility and Probability*.

Becker, G. S. (1976). *The Economic Approach to Human Behavior*. Chicago University Press.

Elster, J. (1989). *Nuts and Bolts for the Social Sciences*. Cambridge University Press.

Mas-Colell, A., Whinston, M. D., & Green, J. R. (1995). *Microeconomic Theory*. Oxford University Press.

Varian, H. R. (1992). *Microeconomic Analysis*. Norton.

Costs

8

This chapter covers ...

- the importance of cost functions and their role in managerial decision making.
- the relationship between a firm's production technology and its cost function.
- different types of costs and their relevance.

8.1   What Are Costs, and why Are They Important?

Two roads diverged in a yellow wood,
And sorry I could not travel both
And be one traveler, long I stood
And looked down one as far as I could
To where it bent in the undergrowth.
    (Robert Frost, The Road Not Taken)

If one goes shopping and buys a new pair of sneakers, the cost for one's sneakers is the price that one pays for them. The monetary price of a good is, however, only part of the story economists tell when they talk about costs. As covered in Chap. 1, scarcity implies that one's decision to go this way makes it impossible to go the other way; that all activities have opportunity costs. The opportunity cost of choosing one alternative is the value that one attaches to the next-best alternative foregone. The implication for the sneakers example is that the total costs of the sneakers are, in general, higher than the price one pays, because one has to invest time and effort to find and buy them. If one could have used one's time otherwise, then one has to take the opportunity costs of time into consideration to get a correct measure of the costs one has to incur to get hold of a new pair of sneakers. To give another example, opportunity costs are the reason why it may be silly to drive the extra mile to refuel your car, only because the gas station is a cent cheaper.

However, the true costs of sneakers are only higher than the monetary costs *in general*. It may be that one actually enjoys going shopping, which implies that the

---

opportunity costs of time are negative, subtracting from the monetary costs. More-over, to make things even more involved, the value that one attaches to the price sticker may depend on one's situation in life and one's expectations. If one assumes that there will be considerable inflation the next day, reducing one's purchasing power substantially, one will most likely do one's best to get rid of one's money that day. Thus, all costs are ultimately opportunity costs, which are psychological and subjective concepts of value that are related, but not identical, to market prices.

The fact that the relevant costs are opportunity costs may be interesting in and of itself, but the real importance of this observation becomes apparent if one considers the implications for decision making. Here is an example: assume one wants to make some extra money parallel to one's studies by offering tutoring services to other students, but one is not sure whether this is a good idea, because one does not fully oversee all the consequences of this decision. In order to get a better idea, one makes a business plan to identify the costs and benefits of one's decision. To keep the analysis simple, assume that one can help one student at a time (class size is one) and that the only things one needs to get one's business going is one's time and a room that one has to rent. Further, assume that one can teach up to 20 hours per month. The monthly rent for the room is CHF 500, and one can charge students CHF 50 for an hour of tutoring. A first back-of-the-envelope calculation reveals that one has to teach for ten hours per month to cover one's monetary costs (this is called the *break-even point*). If one teaches for the entire 20 hours, one ends up with a monetary profit of CHF 500. Given this calculation, the question is if one is willing to enter the tutoring business. Based on the above calculation, one should enter the tutoring business because of the positive monetary profit.

If, however, one does not feel completely happy with starting the business based on this calculation, the reason must be that one puts this number into a different context. What could that context be? For example, the next-best alternative on the job market could be to work as a barista in a café, at an hourly wage of CHF 30 (including tips). Thus, working 20 hours, one could earn CHF 600 per month. Even though the hourly wage is much smaller than the one that one could earn for tutoring, the income exceeds the profits from tuition, because one does not need to pay the rent. Therefore, compared to the barista job, one would *loose* CHF 100 by opening one's business. Hence, one should somehow take these opportunity costs explicitly into consideration.

Now, one could argue that tutoring is a more meaningful way to spend time for one than brewing coffee is. If this is the case, one should also include these psychological rewards and costs into one's calculation. Working may not just be about making money, but also about doing something that one finds meaningful, which implies that there is a difference between costs and expenditures. Assume that one assesses the intrinsic pleasure that one gains from tutoring by CHF 30 and the intrinsic pleasure that one gains from brewing coffee by CHF 20 per hour. In that case, these psychological benefits sum up to opportunity costs of brewing coffee of CHF $20 \cdot$ CHF $30 - 20 \cdot$ CHF $20 =$ CHF 200, which would tip the balance towards opening one's tutorial business.

Table 8.1 Optimal decisions depend on opportunity costs

	Tutor	Barista	Exam
Rental costs	500	0	0
Wages	1,000	600	0
Net	500	**600**	0
Intrinsic pleasure	600	400	500
Net	**1,100**	1,000	500
Future income	0	0	1,000
Net	1,100	1,000	**1,500**

One can elaborate on one more aspect of the problem of getting the business plan straight before summarizing it. Assume that the alternative to opening one's business is not working as a barista, but studying for one's exams. In that case, there are no direct monetary opportunity costs that can be taken into consideration. However, even in this case, one has to figure out how much the additional 20 hours of studying would be worth. These benefits might be completely functional, driven by the effect that one's grades get better and one is, therefore, more likely to qualify for better programs and jobs. On the other hand, they might be purely intrinsic, measuring the pleasure that one derives from learning. Regardless how one evaluates one's own situation, the theory suggests that one should be able to attach some monetary value to these alternatives in order to be able to make the right decision. Table 8.1 gives an overview of the example. It is assumed that one can attach a monetary value of CHF 500 to the intrinsic pleasure of learning and a monetary value of CHF 1, 000 to the better job prospects.

What the above example has illustrated is that costs are a tool that can help one to make smart decisions. However, in order to be able to support your decisions in a rational way, one has to think about costs in terms of opportunity costs. If the costs are calculated incorrectly, then one's decisions will not be smart.

One may wonder if it is always possible to attach a meaningful monetary value to psychological opportunity costs. Numerous psychological studies have shown that, for different reasons, people have trouble specifying their valuations of alternatives in a reasonable way. Section 5.3 discussed some of these reasons. How reliable is the figure that one attaches to the value of 20 hours of additional learning? Will one really use the time to learn? Can one anticipate how much fun it will be to help other students? People are very bad in what is called *affectual forecasting*, i.e. anticipating how they will feel in the future. Is one's perception of the psychological costs and the benefits context-dependent (the anchoring effect from Sect. 5.3)? There is also evidence that people have a tendency to rationalize their gut feelings by developing narratives that selectively focus on aspects that support their "guts." The term *narrative fallacy* describes how flawed stories of the past influence one's perception of the present and future. People have an innate urge to develop a coherent story about the events that shape their lives and simplicity and coherence often more important than accuracy. The mind is a sense-making organ and the narratives it cooks up reduce the anxiety that one would experience if one faced the complexity and unpredictability of life. This may help one in one's life, but it is not the same as descriptive accuracy.

Nevertheless, if one has ample reason to scrutinize the numbers that one assigns to psychological opportunity costs, would it not be better to abandon the idea altogether? This would throw out the baby with the bathwater, because one has to decide somehow and decisions that take all the relevant opportunity costs into consideration are, in expectation, better than decisions that neglect some of the tradeoffs. An awareness of the flaws and biases that exist when one thinks about psychological opportunity costs can help one to put the concept into perspective and to cope with the idiosyncrasies of one's mind.

The following three examples will illustrate how one can proceed in assigning opportunity costs. Assume that a firm produces a good using capital and labor. Profits are revenues minus costs. What are the costs and revenues that are associated with this activity?

- **Case 1, all costs monetary:** The firm borrows capital from capital markets, rents labor from labor markets and sells the good on a goods market. In this case, the revenues of the firm are the market price times the produced and sold quantity of the good (assume revenues are CHF 1,000). The firm's costs are the sum of interest payments for rented capital (CHF 400) and wage payments for hired labor (CHF 500). All relevant costs and revenues are monetary, because they involve market transactions. An accounting system that includes ("takes into account") all three costs and benefits makes the business appear profitable.
- **Case 2, goods not sold:** The firm borrows capital from capital markets, rents labor from labor markets, but the owner of the firm consumes the goods directly. The costs of the firm are, again, the sum of interest payments (CHF 400) for rented capital and wage payments for hired labor (CHF 500). However, it has no monetary revenues. A system of accounting that considers only monetary payments would support the decision to shut down the business, because it would show a deficit of CHF 900. There is no monetary equivalent for the satisfaction or utility of the owner from consuming the goods (again CHF 1,000). Hence, economically meaningful decisions can only be supported by an accounting system that attaches a monetary value to the satisfaction or utility of the owner.
- **Case 3, owner self-employed:** The firm borrows capital from capital markets, sells the good on a goods market, but the owner works himself. In this case, the firm's revenues are, again, the market price times the produced and sold quantity of the good (for example CHF 800 this time). The firm's monetary costs are the interest payments for rented capital (CHF 400). Without incorporating labor costs into the equation, the business appears profitable. However, this calculation would lead to the wrong decision. Assume the owner would make CHF 500, if he worked somewhere else. These opportunity costs should be taken into account to support the right decision. The business now appears deficient and, compared to the next-best alternative, it actually is: if the owner were to shut down the firm, he would earn CHF 500. Staying in business gives a monetary profit of CHF 400, so he actually loses CHF 100 compared to the next best alternative.

What are the consequences of the idea that costs and revenues have to incorporate $\mathcal{B}$
non-monetary opportunity costs? First of all, it can serve as a guideline for the design of managerial accounting systems. One of the primary reasons for the existence
of accounting systems is that they can support decisions. However, as one has seen,
decisions are only accurate according to some objective (profits, in this example)
of the firm, if the accounting system that supports decisions incorporates all opportunity costs. These opportunity costs are sometimes referred to as *imputed interest*
or *calculatory entrepreneur's salary*. *Management accounting*, however, has to be
distinguished from *financial reporting*. The primary purpose of the latter is to communicate a company's financial situation to the outside world. These statements are
subject to legal constraints and regulations that are sometimes incompatible with
the idea of opportunity costs. So-called *imputed costs* are a good example of opportunity costs that are, in general, considered in management accounting, but are
not allowed to be considered in financial statements. It is, for example, possible
to activate interest payments on debt capital, but not imputed interest payments on
equity. Imputed interest payments on equity are opportunity costs, because they are
equal to the interest payments one would have received, if the capital had been lent
to someone else.

Digression 19. Opportunity Costs and Maximization
The idea that rational decisions are based on the correct identification, evaluation and comparison of opportunity costs is closely related to the idea of
maximization. An individual is a maximizer, if she consistently chooses the
best (according to her subjective standard) alternative among the available
alternatives. There is a lot of evidence that people are rarely maximizers in
this sense. One is seldom in a position to know and precisely evaluate all the
alternatives, because of uncertainties regarding the relevant probabilities and
cognitive limitations. Hence, a lot of people are not aiming for the best, but
for a good enough alternative. Think of your decision to meet a friend for dinner. Most people browse their directory and call the first friend with whom
it seems sufficiently interesting to spend the evening. Simon (1957) called
this type of behavior *satisficing*. The idea is that individuals have certain
aspiration levels and choose the first alternative that meets these standards.
Because of that, the resulting choices are, in general, less than optimal. There
may have been friends in your directory with whom you could have spent an
even better evening.
   At first glance, satisficing seems to contradict the idea of maximization
and thereby the concept that one should start by identifying and evaluating
all opportunity costs. However, advocates of the maximization approach
have argued that the opposite is the case: satisficing is optimization where
all opportunity costs, including the costs of processing information and optimization, are considered. Looking for the best friend to spend the evening
with may be so complicated and time consuming that, in the end, one has

dinner alone. It is disputed, however, whether this is a legitimate defense
of the idea of maximization. It brings the whole concept close to a tautol-
ogy, because it comes with the risk of explaining every type of behavior by
identifying arbitrary and non-falsifiable opportunity costs.

What studies with monozygotic and dizygotic twins have shown is that the
tendency to satisfice or to maximize has a strong genetic component and that
people can be categorized into "maximizers" and "satisficers." Interestingly,
maximizers tend to make better decisions than satisficers, but are less happy
with them. One explanation for this apparent paradox is that even maximizers tend to fail to identify the best alternative in complex environments,
but are more aware of the fact that they may have failed to achieve their goals.
Hence, they often feel regretful when they evaluate their choices. Therefore,
in the end, the satisficer goes to the first ok-looking restaurant with the first
ok-looking friend and spends a happy evening, whereas the maximizer continuously questions whether sushi with Sasha would have been better than
pizza with Paul.

8.2   A Systematic Treatment of Costs

One is now in a position to define costs in a systematic way. Costs are the sum of
factor inputs evaluated by their prices (be they monetary or opportunity costs).

In the easiest case, there is only one input, whose quantity is denoted by q and
that can be purchased at a price (per unit) of r. In this case, costs are simply
$C(q, r) = q \cdot r$. If there are $i = 1, \ldots, m$ different inputs with quantities denoted
by q_i and prices by r_i, the *cost equation* can be defined as:

$$C(q_1, \ldots, q_m, r_1, \ldots r_m) = \sum_{i=1}^{m} q_i \cdot r_i.$$

The cost equation is easy to specify, but it is not particularly interesting for economic decision making. What one would like to understand is the relationship
between *output* and *costs* or, more precisely, between output and the *minimum costs*
that are necessary to produce this output. This information is given by the *cost
function*.

▶ **Definition 8.1, Cost function**   A cost function $C(y_i)$ assigns the minimum costs
to the production of y_i units of a good i.

In principle, it would be possible to define certain properties of cost functions
and see what they imply for the behavior of firms in different market contexts.
Economists, however, usually take a detour and establish a causal link between the

cost equation and the cost function, because it allows them to see how the cost function relates to the physical properties of production. This is important for assessing, for example, the effects of technological change on market behavior or market structure, and so on.

Production is, first of all, a physical activity that transforms matter from one state into another, generally more desirable state. The rules of transformation are summarized by the so-called *technology of production*. It is the set of all technologically feasible input-output combinations and is – mathematically speaking – a set. The boundary, or "outer hull," of this set is the subset of all productively efficient input-output combinations because at a point along the outer hull it is (for given quantities of inputs) only possible to increase the production of one good by lowering the production of some other good. This outer hull is called the *production-possibility frontier*. It can – under certain conditions – be represented by a function that one calls the *production function*.

▶ **Definition 8.2, Production function** A production function relates the output of a production process to the necessary inputs. It assigns the productively efficient output to any combination of inputs.

Digression 20. Firms as Production Functions and Firms as Organizations; How Efficient Can One Possibly Be? $\mathcal{B}$

At this point, it is important to scrutinize the basic assumption that a point along the production function can actually be reached. Underlying this assumption is the view that firms are able to organize economic activities within the firm in a perfectly efficient way. Historically, economists were not particularly interested in the management structures of firms and treated the firm as a black box that entered their analysis as a production function. This simplification might be useful, if the primary focus of the analysis is the interaction of supply and demand on markets. As one knows from the short introduction into the philosophy of science, every scientific theory has to make simplifying assumptions; the question is if the simplifications are useful.

The firm-as-production-function view was challenged when economists started to realize that they cannot explain the existence of firms as subsets of transactions that replace decentralized market transactions with more centralized forms of governance (see Sect. 6.2 for more detailed information). Since then, a large body of literature on the internal organization of firms and the boundaries between firms and markets has emerged that allows one to better understand under what conditions and with what kind of organizational structure companies can get to or close to the production function. This issue boils down to understanding if firms can organize economic activities in a way that all interdependencies, which are internal to the firm, are internalized (i.e. no firm-internal externalities exist). The strands of the literature that focus on these problems are called *principal-agent theory*, *contract theory* or merely *theory of the firm*. The important point is that one has to conceptually distinguish between the production function and the relationship between inputs and outputs, which exists given the (possibly imperfect) way economic activities are organized within a firm.

Economists and business economists are usually no experts in the physical laws of production. Nevertheless, they have to be able to communicate with engineers and scientists (who are experts in respect of these laws of production) in order to understand how the production process influences the structure of the cost function. The idea is relatively straightforward and can be exemplified by means of a (hypothetical) production technology that transforms one input, labor (l), into one output, apples (y). The input price is equal to the market wage (w). The production function can then be defined as $y = Y(l)$, and the structure of the function $Y(.)$ summarizes the "laws" for transforming labor into the number of apples picked. A potentially interesting question is by how much the number of apples picked is increased by one additional unit of labor.

▶ **Definition 8.3, Marginal product** The marginal product of a production function measures the change in production y that is caused by an additional unit of an input l.

In order to access the powerful toolbox of Calculus, one has to assume that infinitesimal changes in inputs and outputs are possible and that the production function is continuously differentiable. These assumptions allow one to approximate the marginal product by taking the partial derivatives of the production function. Formally, let dl be a change in labor input and dy the associated change in output. With only one input and a marginal change in l, $dl \rightarrow 0$, the marginal product is given by

$$\frac{dy}{dl} = Y'(l),$$

where $Y'(.)$ is the partial derivative of $Y(.)$. If several inputs $q_1, \ldots, q_m$ are needed for production, the production function can be denoted as $Y(q_1, \ldots, q_m)$, and the marginal product for an infinitesimal change in input i, dq_i, is given by

$$\frac{dy}{dq_i} = \frac{\partial Y(q_1, \ldots, q_m)}{\partial q_i}.$$

The marginal product will be useful later on in the analysis.

Figure 8.1 gives a graphical illustration of a production function for the case of $Y(l) = \sqrt{l}$. The factor input (l) is drawn along the abscissa and the output (y) is drawn along the ordinate. The root function implies that additional labor input increases the output, but at a decreasing rate.

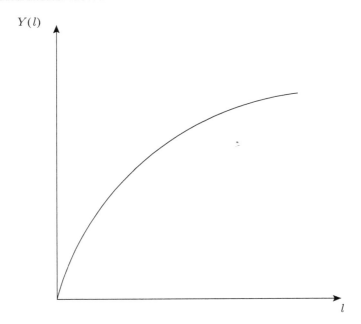

Fig. 8.1 The graph of the production function $Y(l) = \sqrt{l}$

Costs are inputs evaluated by input prices. The production function establishes a link between inputs and outputs. If one had the opposite link between output and input, one would be close to the solution of the problem: if one could associate a level of input with each level of output, the only thing that one would have to do is to multiply the input by the input price to get the cost function. However, the opposite link can be readily established as the *inverse function* of the production function. It gives an answer to the question of how much input one needs for a given output. Multiply this input by the factor price and you have the cost function. More formally, assume the production function is monotonically increasing (i.e. more input generates more output), let $\mathcal{L}$ be the set of all possible inputs and $\mathcal{Y}$ be the set of all possible outputs. The production function is a mapping from $\mathcal{L}$ to $\mathcal{Y}$, $Y : \mathcal{L} \rightarrow \mathcal{Y}$. Denote by $L(y)$ the inverse function of the production function, $L(y) = Y^{-1}(y)$. It is a mapping from $\mathcal{Y}$ to $\mathcal{L}$, $L : \mathcal{Y} \rightarrow \mathcal{L}$. Figure 8.2 illustrates.

This figure displays the relationship between labor input (l, along the abscissa) and output (y, along the ordinate). People are used to following the convention of interpreting the variable on the abscissa as the explanatory variable and the one on the ordinate as the explained variable. This is the interpretation as a production function: how much output can be produced with l units of labor input? Graphically speaking, one looks at the figure from the abscissa to the ordinate, indicated by the stylized eye in Fig. 8.2a. One can, of course, also look at the figure from another angle. In Fig. 8.2b, the stylized eye indicates that one interprets y as the explanatory

a   **b**

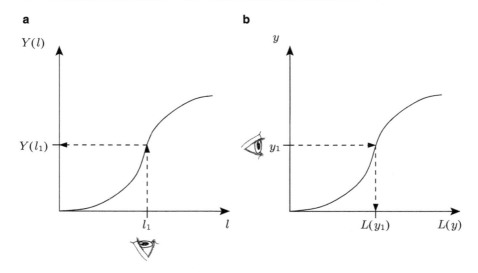

Fig. 8.2 Looking at graphs from two different angles, **a** Production function, **b** Cost function $(w - 1)$

and l as the explained variable. The question that one asks then is how much labor input one needs, if one wants to produce y units of output. The answer to this question is given by the inverse of the production function.

▶ **Definition 8.4, Cost function for one-output-one-input technologies**   The cost function $C(y)$ for a production function $y = Y(l)$ is given by $C(y) = L(y) \cdot w = Y^{-1}(y) \cdot w$.

Figure 8.3 gives a graphical illustration of the inverse production function $L(y) = y^2$ and the cost function $C(y) = y^2 \cdot w$, which results if $Y(l) = \sqrt{l}$. I have used $w = 2$ in the graph.

Now, output (y) is drawn along the abscissa and input (l) along the ordinate. The cost function (see upper graph) is a multiple of the inverse production function (see lower graph).

This link between the production and the cost function allows one to understand how cost functions are related to production technologies. Two qualifying remarks have to be made in order to get the bigger picture.

- First, the assumption that the firm can rent or buy inputs at given input prices reveals the implicit assumption that factor markets are perfectly competitive. If the firm has market power on some input market (for example, if it is the only major employer in the region), then the relationship between costs and technology is no longer so straightforward and is also determined by the power of the firm to set wages as a function of labor input.

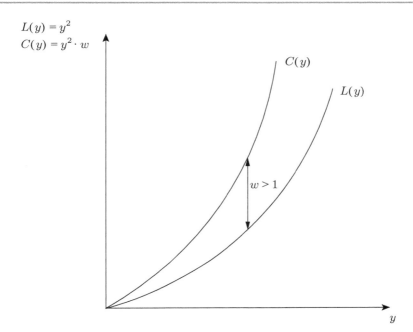

$L(y) = y^2$
$C(y) = y^2 \cdot w$

$C(y)$

$L(y)$

$w > 1$

y

Fig. 8.3 Inverse production function and cost function

- Second, it is, of course, a completely unrealistic assumption that production requires only one input. Some production processes lead to co-production, such
  as when crude oil is separated into its different marketable components. Multi-
  input production gives rise to the more complex question of how to determine
  the optimal mix of inputs. This optimal mix is influenced by the technologically determined degree to which the different inputs can be substituted for each
  other and the input prices. Manufacturing, for example, can be relatively capital-
  intensive or relatively labor-intensive, and the capital-labor ratio depends on the
  relative prices of capital and labor. In order to determine the cost function, in this
  case, one has to solve what is called a *cost-minimization problem.* For understanding this and the following chapters, it is sufficient to work with the insight
  that one gets from the one-input-one-output model.

With this understanding of a cost function, one can now move on and use it as
an explanatory tool for different theories of firm behavior in markets. As one will
see throughout the following chapters, different types of costs will turn out to be
important explanatory factors. Therefore, this subchapter will introduce them now,
filling the toolbox with additional tools that one will use later on.

If one takes total costs and distributes them equally among all the units that one
produces, one gets the average costs of production:

▶ **Definition 8.5, Average costs**   The average costs of production equal the total costs of production divided by the quantity produced, $AC(y) = C(y)/y$.

Some costs vary with, and some are independent of, the quantity produced. Take computer software or cars as examples. Before one can sell a new product, one has to incur development costs. These costs are independent of the number of licenses or vehicles that one produces and sells; they are a prerequisite for their production. The reason why these costs do not vary with the volume produced is technological. They have the property that, from an *ex-ante* perspective (before one makes the investment decision), they are zero, but immediately "jump up," if one decides to develop and sell the new product. One calls these costs *technological fixed costs*.

▶ **Definition 8.6, Technological fixed costs**   The technological fixed costs of production are the costs that occur once a firm starts production and they are independent of the volume of production,

$$TFC(y) = \begin{cases} 0, & y = 0 \\ FC, & y > 0 \end{cases}.$$

Costs are related to prices and prices exist within an institutional framework as part of a contract. This is why costs can also be independent of production volume, because of contractual reasons. Take a long-term rental agreement as an example: the rent has to be paid throughout the duration of the contract, irrespective of the firm's performance on markets. Such costs have the peculiar feature that they are fixed throughout the duration of the contract, but can be variable over longer periods of time or with different contractual arrangements. These costs are called *contractual fixed costs*.

▶ **Definition 8.7, Contractual fixed costs**   The contractual fixed costs of production are the costs that are independent of production: $CFC(y) = FC$ for all $y \geq 0$.

The major differences between contractual and technological fixed costs are, therefore, their causes and their behavior around $y = 0$. Technological fixed costs can become contractual fixed costs, for example, if the engineers who develop a new product have long-term contracts. For a lot of economic applications, it does not matter whether fixed costs are contractual or technological, but there are some important exceptions. If the difference is without relevance, one simply calls them fixed costs; otherwise, one explicitly refers to their cause.

▶ **Definition 8.8, Fixed costs**   The fixed costs of production include the technological and the contractual fixed costs.

Contractual fixed costs are also costs that cannot be recovered: once one has signed the contract, one has to pay them. This is true for all costs that are necessary for production, as well as those that occurred in the past. The fact that the monthly payments fixed in a long-term rental agreement will occur in the future is irrelevant for this, because the legally relevant act, the signature of the contract, happened in the past.

▶ **Definition 8.9, Sunk costs**   Sunk costs are costs that have already been incurred at a given point in time and thus cannot be recovered.

It can be argued that sunk costs are, necessarily, fixed costs: from today's perspective, they cannot be influenced and are, therefore, independent of the volume of production. However, if all costs are opportunity costs, then sunk costs are not "proper" costs at all, because they refer to events in the past and are, therefore, irrelevant for one's decisions. They are like gravity: one can complain that it exists and that it makes flying difficult, but the only thing one can do is to cope with it.

The costs that vary with production are called variable costs. If one is in the fruit-picking business and one hires fruit pickers every morning, then labor costs are variable on a daily basis.

▶ **Definition 8.10, Variable costs**   The variable costs of production are the costs that vary with the quantity produced: $VC(y) = C(y) - FC$.

One can now look for averages in fixed and variable costs, which motivates the following two definitions:

▶ **Definition 8.11, Average fixed costs**   The average fixed costs of production equal the fixed costs of production divided by the quantity produced: $AFC(y) = FC/y$.

The concept of average fixed costs can be applied to both contractual and technological fixed costs.

▶ **Definition 8.12, Average variable costs**   The average variable costs of production equal the variable costs of production divided by the quantity produced: $AVC(y) = VC(y)/y$.

Last, but not least, one might be interested in the costs that result if one produces an additional unit of output:

▶ **Definition 8.13, Marginal costs**   The marginal costs of production are the costs that result from the production of an additional unit of output: $MC(y) = dC(y)/dy$.

Marginal costs $MC(y)$ are approximately equal to the partial derivative of the cost function $C'(y)$, if one allows for infinitesimal changes in inputs and outputs. Marginal costs are a key concept in mainstream economics, because they play a prominent role in determining the behavior of firms, which seek to maximize their profits.

The above office-rental and fruit-picker examples suggest that the relationship between fixed and variable costs is a matter of time frame and contract structure. As soon as one has signed a contract with a specific duration, the costs that emerge from this contract are sunk and, therefore, fixed during the term of the contract, but variable for larger time spans. Thus, the question of which inputs contribute to fixed costs and which to variable costs depends on the contract structure and the contract structure may be culture-specific. In countries with extensive employment-protection laws, it is difficult to fire employees on short notice, so one gets a certain downward rigidity. In countries with hire-and-fire cultures, it is much easier to adjust one's workforce on short notice. The same is true for capital, where one needs to understand the contract structure in order to know whether rental agreements, etc., contribute to sunk and fixed or variable costs. In the long run, however, all costs are either variable costs or technological fixed costs. (For nerds: strictly speaking, technological fixed costs are a special case of variable costs, because they depend on production, even if only at $y = 0$. Classifying them as such, however, makes it harder to build an understanding about the underlying economic phenomena, which is why one summarizes them in a specific category.)

Digression 21. The Profits for the Apple Watch Are ...
According to the news agency Reuters (May 01, 2015), Apple Inc's Watch has the lowest ratio of hardware costs to retail price of any Apple phone. The hardware costs of the Sport edition was about 24% of the suggested retail price (29–38% for Apple's other products). The suggested US retail price for the watch is $349, and the hardware costs (including manufacturing costs of $2.50) amount to $83.70.

One could argue that this is a nice profit margin and a proof of Apple's strong monopoly position in the industry. Nevertheless, be careful! It may be safe to say that Apple will make a nice profit with its new product line, but to conclude from the above report that it amounts to $265.30 per watch would be premature.

Based on the costs analysis, one knows that total costs consist of variable plus fixed costs and that hardware and manufacturing costs are usually variable costs. Hence, $265.30 is closer to what we have called producer surplus.

Fixed costs are usually substantial for products like the Apple Watch, because they include costs for research and development and also for marketing. What makes the above numbers almost impossible to interpret is the fact that one does not know to which extent these costs are already figured into the hardware costs.

Here is an example why this is complicated. Apple buys the processor from Samsung. If Samsung did all the research and development, these costs must already be figured into the price Samsung charges Apple for this part. The same is true for all the other components that make up an Apple Watch. Therefore, if Apple mainly assembles components developed by other firms, a substantial part of the technological fixed costs are not contractual fixed costs from the perspective of Apple. In this case, the per-unit profit is closer to $265.30. But wait! Maybe the contract between Apple and Samsung specifies a so-called two-part tariff where Apple makes a fixed, upfront payment and pays a lower price for each processor. Such a contract would bring the price structure more closely in line with Samsung's cost structure, which may be beneficial from an efficiency point of view. Nevertheless, if this were the case, the upfront payment should not count as part of the variable hardware costs ... unless Apple divides them among the expected sales, which again would make them appear in the hardware costs.

To make a long story short: one should be very cautious when reading news like the above, because the information is usually much too crude to draw any reliable conclusions about profits.

References

Simon, H. A. (1957). *Models of Man: Social and Rational.* New York: John Wiley.

Further Reading
Allen, W. B. (2009). *Managerial Economics Theory, Applications, and Cases.* Norton.
Ferguson, C. E. (1969). *The Neoclassical Theory of Production and Distribution.* Cambridge University Press.
Pindyck, R., & Rubinfeld, D. (2001). *Microeconomics.* Prentice-Hall.
Png, I., & Lehman, D. (2007). *Managerial Economics.* Wiley.

Part IV
Firm Behavior and Industrial Organization

A Second Look at Firm Behavior Under Perfect Competition

9

This chapter covers ...

- how profit-maximizing firms behave in competitive markets (behavioral foundation of the supply function).
- how the supply function is related to marginal and average cost functions and what this says about the informational demands and effective organization of firms.
- the technological prerequisites for the functioning of competitive markets.
- how competition drives profits to zero and why this is not bad.

9.1 Introduction

> The natural price or the price of free competition ... is the lowest which can be taken. [It] is the lowest which the sellers can commonly afford to take, and at the same time continue their business. (Adam Smith, The Wealth of Nations (1776[1991]), Book I, Chapter VII)

This chapter will take a closer look at the supply decision of a firm that sells in a market with perfect competition. To make the problem manageable, one has to specify the objective of the firm and say a few words about its ownership structure as well as its internal organization.

The standard assumption in the literature is that firms seek to *maximize profits*. If p is the price of some good produced by the firm, y is the quantity produced and $C(y)$ are the costs of production, then the profits are $\pi(y) = p \cdot y - C(y) = R(y) - C(y)$, where $R(y)$ stands for revenues. One way to think about this objective function is in terms of the interests of the owners. Assume that a single person owns the firm and uses it as a vehicle to maximize her income. What objective would she try to give the firm in order to pursue her goal? Obviously, the increase in income that the owner can extract from the firm is equal to the firm's profit: the owner deploys capital and labor, which costs her $C(y)$ for y units of output, and

© Springer International Publishing AG 2017

M. Kolmar, *Principles of Microeconomics*, Springer Texts in Business and Economics, DOI 10.1007/978-3-319-57589-6_9 she gets the revenues of the firm, $p \cdot y$. Therefore, the surplus or the increase in income is equal to the firm's profit. Hence, if income-maximizing owners invest in firms, it is in their best interest that the firms maximize profits. If owners invest in firms because they want to reach something else, then the imputed objective may be different. Nevertheless, it is a good starting point to conjecture that most shareholders invest in corporations because they want to make money.

$\mathcal{B}$

Digression 22. The Limits of Profit Maximization: Information, Contracts, and the Organization of Firms

The idea that income-maximizing owners would like to make sure that the managers of the firm maximize profits is simple and powerful. However, it is the source of a lot of controversy for both normative and positive reasons. From the positive point of view, it is sometimes argued that firms do not, in fact, maximize their profits. Deviations from this objective may have several reasons. They can be a result of imperfect information about costs and revenues. Limited information is definitely a relevant problem and it may lead to decisions that are apparently not in line with profit maximization. Nevertheless, it does not falsify the objective *per se*. As previous chapters have shown, it is the purpose of managerial accounting to provide information to support decisions. If the information is bad, the decisions are bad, and the first impulse should be to develop a better accounting system, not to abandon profit maximization.

Another important reason for deviations from profit maximization results from the fact that firms are usually complex networks of individuals with their own objectives. The key question then becomes whether it is possible to align the interests of the owners with the interests of the workers. Take the CEO of a firm that is not managed by the owner and assume further that both, the owner and the manager, want to maximize their incomes. The income of the manager depends on the contract, so it becomes a problem of contract design whether the owner's and the manager's income maximizations coincide. (One can think of such a contract as an incentive mechanism. An optimal contract is one that creates no externalities between manager and owner.) The key question is, therefore, how such a contract has to be designed to make sure that the manager internalizes the interests of the owner. If the contract is ill-designed, the manager will use her discretionary power to maximize her own income, which is not compatible with the owner's income maximization and, therefore, is in conflict with profit maximization. An example might be a contract with bonus payments that incentivizes short-term profits, despite the fact that they are in conflict with the long-term interests of the firm.

To simplify things, one assumes that owners want to maximize income, that contracts perfectly align owners' and managers' interests, and that the accounting system is sufficiently precise to allow for a realistic view of costs and revenues.

Hence, firms maximize profits. This case acts like a benchmark. If one understands the benchmark, one can get a better understanding of the effects of deviations from it.

It makes sense to state the objective function explicitly. The firm maximizes profits by the choice of the quantity of the good produced. Formally, it is expressed as follows:

$$\max_{y} p \cdot y - C(y).$$

The assumption of perfect competition enters the above choice problem, because the price is treated as a parameter, which means that the firm takes it as given. The above formulation also assumes both that the quantity produced and the quantity sold is identical and that firms do not produce for or sell from stocks.

The concept of marginal revenues will be helpful in understanding optimal firm decisions.

▶ **Definition 9.1, Marginal revenues** The marginal revenue of production is the revenue the firm makes by an additional unit of production: $MR(y) = dR(y)/dy$.

What are the implications of the assumption of profit maximization for the supply decision of the firm? The following thought experiment allows one to gain a better understanding. Assume that the firm produces a quantity such that marginal revenue is larger than marginal cost and marginal costs are strictly positive. What would be the effect on profits of an increase in production by one unit be? Given that profit is revenues minus costs, profit must increase, if marginal revenues exceed marginal costs, so it would be rational to increase production, because the firm would make more money with the additional unit than it would cost. Next, assume that the firm produces a quantity such that the marginal revenues are smaller than the marginal costs. In this case, profit would go down, because the next unit of production costs more than the firm would get for it on the market. Therefore, it would be rational to reduce production. These two observations pin down the profit-maximizing behavior of a firm: the optimal quantity is the one where marginal revenues are equal to marginal costs.

This condition can also be derived analytically, by setting the first derivative of the profit function equal to zero. Given that the price is fixed for a firm on a competitive market (the firm is a price-taker), the marginal revenues are equal to the price of the good, $MR(y) = p$, and one gets:

$$\pi'(y) = p - C'(y) = p - MC(y) = 0.$$

Denote the quantity of the good that fulfills this condition by y^*. This result is a very important finding in the theory of firm behavior: a profit-maximizing firm on a competitive market produces according to the "price-equals-marginal-costs" rule, because marginal revenue is equal to the price under perfect competition. This rule has several implications that the following paragraphs will discuss. Its applicability also depends on several factors that one has to make explicit for an in-depth understanding of its role in the theory of firm behavior. I will start with the implications.

The first and most important implication, for an economist, is the link between the cost and supply functions. The condition $p = MC(y)$ formally establishes a relationship between price p and quantity y. The supply function of a firm establishes the same type of relationship with $y = y(p)$. It maps each price onto a quantity produced by the firm. If one looks at the inverse of the marginal-cost function, one gets $y = MC^{-1}(p)$. But this mapping has prices as domain and quantities as codomain. This mapping described by the supply function has quantities as its domain and prices as its codomain. The implication of this is that these two mappings are inverse to each other and that a competitive firm's supply function is identical to its inverse marginal-cost function. When one observes a firm's market behavior, one can "look through" the supply decision and get information about the firm's marginal costs.

This finding also allows for a more in-depth understanding of the willingness-to-sell concept, which Chap. 5 introduced: I have argued that one can interpret a point along the supply function as the minimum price the producer has to get in order to be willing to sell an additional unit of the good. This price, as we have seen, is equal to the marginal costs of producing this unit, which makes perfect sense: marginal costs measure how much it costs to produce an additional unit of the good. If one is paid more than that, one makes a profit with this unit, if one is paid less, one takes a loss. Therefore, one is indifferent between selling and not selling, if one gets exactly one's marginal costs.

$\mathcal{B}$ 　　The "price-equals-marginal-costs" rule also has important managerial implications: In order to be able to behave in accordance with this rule, a manager needs information about the market price and the marginal-cost function. This has implications for the organization of her company: in addition to the factory that produces the goods that the firm sells, the firm needs a controlling department that collects information about costs as well as current (and maybe also expected future) market prices. The organization of a competitive firm is not very complicated. However, getting the controlling correct is crucial for the firm, because the quality of decisions depends on the accuracy of the information about marginal costs.

Unfortunately, life on competitive markets, as either a manager or an economist, is not as simple as the above rule suggests. Next, one has to put the "price-equals-marginal-costs" rule into perspective. I will discuss three aspects in the following subchapter: technological conditions under which perfect competition works, short- versus long-run decisions, and the relationship between firm and market supply.

9.2   The Relationship Between Production Technology and Market Structure

As this chapter has already shown, the profit-maximizing production decision of a firm can be characterized by the condition:

$$\pi'(y^*) = p - C'(y^*) = p - MC(y^*) = 0.$$

I will scrutinize this approach from a purely technical point of view and discuss the economic implications thereafter. This approach illustrates how a back-and-forth between economic thinking and mathematical reasoning can improve one's understanding of the economy.

One may remember, from one's mathematics classes in high school, that the so-called first-order conditions are necessary, but not sufficient, for the characterization of a maximum. The only thing that a first derivative of zero guarantees is that the function has a "flat" point, which can be a maximum, a minimum or a point of inflection. In order to make sure that one characterizes a maximum, one has to check the so-called second-order conditions, i.e. one has to check if

$$\pi''(y) = -C''(y)$$

fulfills a certain property at the potential optimum. The second-order condition says something about the curvature of the function. In order to make sure that the first-order condition characterizes a maximum, one has to make sure that the profit function is "hump shaped" or, more technically, strictly concave. (In addition, one has to make sure that there is an interior optimum. A technical condition that guarantees this is $p > MC(0)$ and $p < \lim_{y \to \infty} MC(y)$.) This is guaranteed if the second derivative of the profit function is negative,

$$\pi''(y) = -C''(y) < 0 \Leftrightarrow C''(y) > 0.$$

Figure 9.1 illustrates this case.

The upper part of Fig. 9.1 shows the profit of a firm for all possible production levels. It is inversely u-shaped. The first-order condition identifies the quantity where the slope of the function is zero, which characterizes the maximum profit. The lower part of Fig. 9.1 disentangles the profit into revenues and costs. The straight line represents revenues as a function of y. Revenues are linear in production, because it is simply the product of an exogenous price and the endogenous quantity of the good. Costs increase disproportionally. Profit in Fig. 9.1 is equal to the *vertical* difference between the revenue and the cost curve. What the firm tries to do is to identify the output where this vertical difference is at its maximum. This point is at the place where both functions have the same slope. The slope of the revenue function is p and the slope of the cost function is $MC(y)$.

In concluding that purely technical argument, is there anything one can learn from this condition as an economist? First of all, one can see that the condition restricts the class of admissible cost functions to those that increase overproportionally in production. Hence, the marginal costs of production increase in the quantity produced. Here is an example: assume that one invests time to study for the final exam. The more time one spends, the better one's expected grade becomes. It is relatively easy to pass, but it becomes more and more difficult to get the best possible grade. This property is nicely reflected by an application of the so-called 80-20 rule (also called the Pareto principle), which states that, for many events, 80% of the effects come from 20% of the causes. Applied to the example, it would say that one

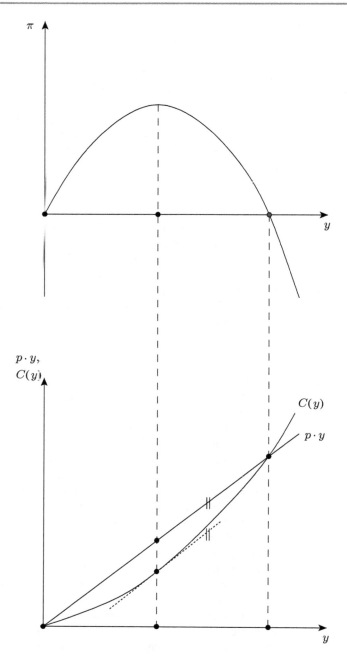

Fig. 9.1 Profits as a function of output

gets 80% of the output in 20% of the time, and the additional 20% of output in the remaining 80% of the time. This principle applies to a large number of production processes, and intellectual or physical skills are only one example. If one exploits natural resources, it is usually relatively easy at the beginning and gets increasingly difficult, when the source becomes depleted. Increasing the crop on a given piece of land is relatively easy at the beginning but, the larger the crop, the more difficult it gets to further increase it, and so on.

The above arguments used technological explanations for increasing marginal costs and this is exactly one of the reasons why I have linked costs with production functions. Given that input markets are competitive, the structure of the cost function is determined by the structure of the production function, because they are, in the one-factor example (up to a scaling factor, which is determined by input prices), inverse to each other. Increasing marginal costs exist, if the marginal product decreases, i.e. if it gets more difficult to increase production the more one is already producing.

Assume that marginal costs are decreasing. In this case, the first-order condition characterizes a minimum. What are the economic implications of this finding? Figure 9.2 illustrates this case.

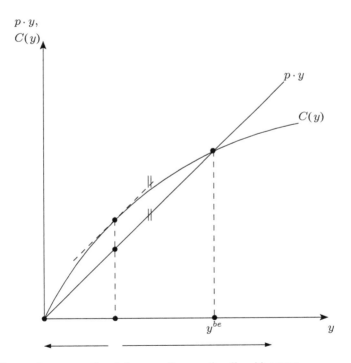

Fig. 9.2 Costs and revenues, if costs increase disproportionally with output

Figure 9.2 shows revenues (straight line) and costs (curved line) as functions of output. What one can see is that the output level, where price equals marginal costs, now characterizes the profit minimum and the marginal-cost curve is downward-sloping. At this point, the firm basically has two strategies. It can leave the market and make zero profits (arrow to the left) at $y = 0$, or it can try to grow as large as possible (arrow to the right). If the firm is successful in growing beyond the point y^{be}, it starts making profits. However, the firm should not stop here; the figure reveals that the difference between revenues and costs get larger the more the firm produces. Therefore, it is the best strategy for the firm to grow as large as possible. However, this strategy is incompatible with the assumption that the firm takes prices as given because of its smallness relative to the rest of the market. If a firm gets so large that it can serve the whole market, then it is able to influence the market price. The assumptions of perfect competition and decreasing marginal costs are logically incompatible.

The implication of this inconsistency is that perfectly competitive markets are no one-size-fits-all institution that can be used to organize economic activities. Such markets can only sustain themselves if the industry produces with the "right" type of technology, and an important number of industries do not fit into this picture.

The intermediate case, of constant marginal costs, deserves some attention, as well. Figure 9.3 shows revenues and costs as functions of output.

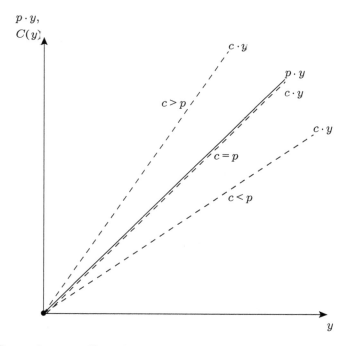

Fig. 9.3 Costs and revenues, if costs increase proportionally with output

Constant marginal costs imply that the cost function is linear. One can denote it as $C(y) = c \cdot y$ with $c > 0$. There are three possible cases: the cost function is steeper than the revenue function, $c > p$; flatter, $c < p$; or both have the same slope. The economic implications of these scenarios are straightforward: the best the firm can do, if $c > p$, is to shut down its business and to leave the market. If $c < p$, however, the opposite is the case. The firm should grow indefinitely, which is in conflict with the assumption of perfect competition. Thus, there is only one case left, where perfect competition is compatible with constant marginal costs: $c = p$. In this case, the firm is indifferent between all production levels, because it makes zero profits irrespective of how much it produces. (At some other point, I explain in detail why zero profits do not imply zero gains from trade. Zero profits imply that equity owners cannot expect a rate of return that exceeds the market interest rate for a similar investment. Hence, one can assume that the firm continues to produce, even with zero profits.)

9.3 The Short Versus the Long Run

I have argued in Chap. 8 that, depending on the time frame and the term structure of contracts, some costs of the firm can be contractually fixed and some are variable. The "price-equals-marginal-costs" rule made the implicit assumption that the firm is active on the market. However, it can always decide to leave the market and this may be a wise decision, if the losses that occur when leaving the market are smaller than the losses would be, if it stays. This statement may sound dubious at first, so one has to dig a little deeper to understand what exactly is meant by it.

Assume that a firm produces with technological fixed costs $FC > 0$ and variable costs:

$$C(y) = \begin{cases} 0, & \text{for } y = 0 \\ VC(y) + FC, & \text{for } y > 0 \end{cases},$$

and further assume that marginal costs are increasing. This situation is depicted in Fig. 9.4 with the example of a variable cost function being $VC(y) = 0.5 \cdot y^2$. Please note that this function implies that the marginal costs $MC(y) = y$ for all $y > 0$.

The horizontal line (p) is the market price and the linear monotonic line represents the marginal costs. If the firm decides to stay in the market, it will choose the quantity that equals price and marginal costs, indicated by y^* in the figure. Total revenues for an output of y^* are $p \cdot y^*$, and can be represented by the rectangular area $0pAy^*0$. Given that marginal costs are the first derivative of the variable-cost function, the triangular area $0Ay^*0$, under the marginal-cost curve, represents the variable costs. Hence, the producer surplus, $PS(y^*)$, is given by the triangular area $0pA0$ and is equal to revenues minus variable costs.

This is a general property. One may have wondered why the measure for the gains from trade of a firm is called producer surplus instead of profit. The reason is that the profit includes fixed costs, whereas producer surplus does not. One can

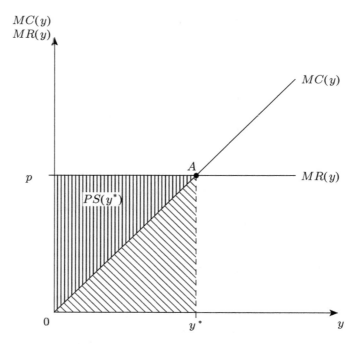

Fig. 9.4 Revenues and variable costs

establish the following relationship between profit and producer surplus:

$$\pi(y) = PS(y) - FC.$$

It follows that producer surplus and profits coincide, if fixed costs are zero, $FC = 0$. In this case, the area $0pA0$ is also the profit of the firm. However, if production requires upfront investments, then the profit is lower than the producer surplus.

The easiest way to see how technological fixed costs (remember that these costs can be avoided by producing a quantity of 0) influence profits is by adding the average-cost curve to the picture. It is equal to

$$AC(y) = \frac{VC(y)}{y} + \frac{FC}{y} = 0.5 \cdot y + \frac{FC}{y},$$

if $y > 0$. A closer look at the expression reveals that it is u-shaped: the first term is a linear, increasing function, whereas the second term is hyperbolic. Hence, the sum must be u-shaped. Is there anything else that one can say about the average-cost curve? Yes: it intersects with the marginal-cost curve at the minimum of the average costs. To understand this intuitively, think of the range over which the average-cost curve is declining. Within this range, marginal costs must be smaller than average costs: if the average is declining at a given point, then the cost of the last unit needs to be below the average costs up to this point. By the same token, if the average-cost curve is increasing, then the costs of the last unit must be higher than the average costs at any given point, because otherwise marginal costs would not have sufficient "leverage" to bring up average costs.

To see this algebraically, note that, applying the quotient rule, a necessary condition for the minimum of the average-cost curve is

$$AC'(y) = 0 \Leftrightarrow \frac{C'(y) \cdot y - C(y)}{y^2} = 0.$$

For $y > 0$, this condition can be simplified to

$$C'(y) \cdot y - C(y) = 0,$$

if one multiplies by y^2 (which is possible because $y > 0$). Dividing by y and rearranging terms gives the desired result:

$$C'(y) = \frac{C(y)}{y} \Leftrightarrow MC(y) = AC(y).$$

(The same calculation can be carried out for average variable costs.) To illustrate, Fig. 9.5 shows the marginal-cost curve and the average-cost curve for $FC = 10$.

Note that $AC(y) = C(y)/y$, or $C(y) = AC(y) \cdot y$, which means that total costs for any output y can be measured by the rectangular area $0ABy0$.

With this prerequisite, one can return to the relationship between producer surplus and profit, or the role of technological fixed costs in firm behavior. Different levels of technological fixed costs give rise to a family of average-cost curves, where higher curves correspond to higher technological fixed costs.

Figure 9.6 shows the marginal-cost curve and the family of average-cost curves for different values of FC.

There is one average-cost curve that is of particular interest: the one that intersects with the marginal-cost curve exactly where price equals marginal costs. Call this level of fixed costs FC'. This situation is represented in Fig. 9.7.

One already knows that total revenues are equal to $0pAy^*0$ at y^* and that producer surplus is equal to $0pA0$. What one knows, in addition, is that total costs are $0pAy^*0$ at y^*, so total profits are equal to zero. What is happening here is that the producer surplus is sufficient to cover the technological fixed costs. If fixed costs are lower, the firm stays in business and makes a profit. What happens, however, if the technological fixed costs are higher? In this case, the firm would end up with a negative profit or loss. Is there anything the firm can do about this loss? Yes, given that one is talking about technological fixed costs, it can *ex-ante* (before it starts the development process), anticipate that the producer surplus will be insufficient to cover the technological fixed costs, at the expected market price, and stay out of business. At this *ex-ante* stage, the total profit from staying out of business is 0, which is better than the loss that results from entering the market. This finding

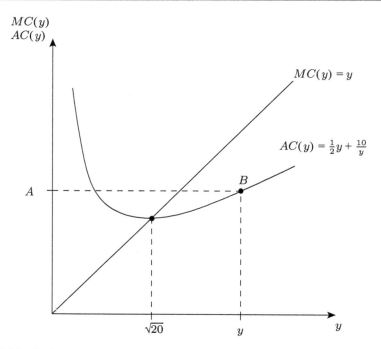

Fig. 9.5 Marginal and average costs

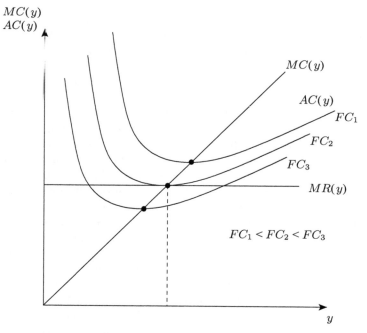

Fig. 9.6 A family of average-cost curves

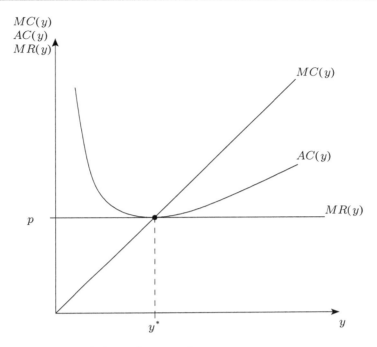

Fig. 9.7 Average costs such that profits are equal to zero

leads to an important modification of the optimal supply decision. In the long run, when all costs can be avoided by not entering the market, the optimal strategy of a competitive firm is to determine the optimal quantity, according to the price-equals-marginal-costs-rule, if the market price is (weakly) above the average costs, and to stay out of the market otherwise. The individual supply function is, therefore, identical to the inverse of the marginal-cost function, if the price is (weakly) larger than average costs.

One can now turn to a slightly modified case. Contrary to the above example, assume that the firm has already entered a contractual arrangement that turns the fixed into sunk costs, i.e. "costs" that cannot be avoided by shutting down the business. In this case, the same analysis as in Fig. 9.7 applies, but the economic consequences are different. For all levels of fixed costs $FC > FC'$ the firm makes a loss, but this loss cannot be avoided by going out of business. Hence, the best the firm can do is to minimize losses, which means sticking to the price-equals-marginal-costs rule. This rule will lead to losses in the end, but they are smaller than the loss that would occur with any other strategy, including going out of business. Figure 9.8 compares the two scenarios.

The upward-sloping function is the marginal-cost curve and the u-shaped function is the average-cost curve. The supply curve equals the section of the marginal-cost curve above the average-cost curve, if the technological fixed costs are not yet

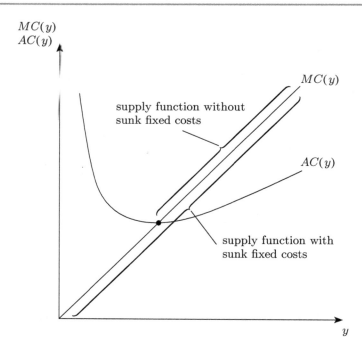

Fig. 9.8 Supply functions with and without sunk fixed costs

sunk, and it equals the complete marginal-cost function, if the technological fixed costs are sunk.

$\mathcal{L}, \mathcal{B}$ The above example carved out the implications of the difference between technological and contractual fixed costs. Contractual fixed costs can, however, also exist in situations where technological fixed costs are zero. Assume, for example, that a farmer has a cherry orchard with a given number of trees. The only additional input that he needs at harvest time is labor. Assume also that the quantity of cherries picked is increasing in the number of hours fruits are picked, but that the increase is declining (it gets harder and harder to pick additional cherries). At a given market wage, this "picking technology" creates increasing marginal costs of fruit picking. If the farmer can hire fruit pickers on a daily spot market, this assumption turns wage payments into variable costs. Assume, on the contrary, that a union of fruit pickers negotiated a three months dismissal protection. In this case, labor costs become fixed and sunk once the employment contract is signed. The analysis of situations like this is qualitatively identical to the analysis above and the basic understanding is simple: *costs that the firm cannot influence have no significance for the optimal behavior of the firm.* Alternatively, to put it shortly: *sunk costs are sunk.*

    At this point, a remark is in order about the role that sunk costs play in standard economics. It is a generally accepted view that rational decision makers ignore sunk costs in their decisions: if one cannot influence them, they should be irrelevant for one's decisions. One calls this the *sunk-cost principle*. Generally, it is a wise and important normative principle: one should not care about the past, if one wants to make rational decisions (but one should have a look at the digression below). However, it is less clear that its predictive power in positive theory is very high. In a number of cases, people care about sunk costs, even if they should not, according to the sunk-cost principle.

An example is the empirical phenomenon of *mental accounting* that describes the tendency of individuals to keep different financial titles in different "mental accounts," and to evaluate the performance of the different titles separately, despite of the fact that a rational decision maker should aggregate them and evaluate the performance of the whole portfolio. For example, assume that someone made equal investments in two stocks. If he sold them today, stock A would have gained CHF 5,000 and stock B would have lost CHF 5,000. Assume that he has to sell a stock because he needs some extra liquidity. A rational person would take the past performance of stocks into consideration, only if he thinks that past performance is correlated with future performance, such that one can learn from the past. Otherwise, past gains and losses should be irrelevant for one's decision to sell stock A or stock B. However, empirical evidence shows that most people have a preference for selling the winning stock A, which could only be rationalized, if the good past performance is an indicator for bad future performance. Much more likely is the explanation that they hold both stocks in different mental accounts and react emotionally to realized losses and gains. Capitalizing the gains from selling stock A gives one pleasure and, at the same time, it allows one to avoid the confrontation with the pain of realizing the losses of stock B. These emotional predispositions may influence one's behavior and make it incompatible with the sunk-cost principle. The tendency to invest additional resources into losing accounts or to "throw good money after bad" is sometimes also called the *sunk-cost fallacy*.

Digression 23. Evolution, Emotions, and Sunk Costs: When Caring about Sunk Costs Can Be Beneficial
It appears that deviations from the sunk-cost principle are always bad. However, if this were the case, one may wonder why humans' brains evolved in a way that make us vulnerable to the sunk-cost fallacy. Recent research in evolutionary biology challenges the theory that such behavior is necessarily bad. Take the so-called ultimatum game as an example. In this game, two players have to decide how to divide a sum of money. The first player can propose how to split the sum between the two players and the other player can then accept or reject the offer. If she accepts, the money will be split according to the proposal; if she rejects, neither player receives anything. According to the sunk-cost principle, the second player should accept any positive amount, because the proposal of the first player is in the past and cannot be influenced. Nevertheless, with this logic, a selfish player 1 should offer the minimum

amount possible. Hence, the sunk-cost principle guarantees that player 2 gets almost nothing.

This prediction has been consistently tested and falsified in the laboratory. It turns out that subjects in the role of player 2 very often reject small offers, because they find them unfair or even outrageous. However, rejecting positive offers violates the sunk-cost principle. In the end one walks home without any money when one could at lease have had some. From an evolutionary point of view, however, the apparently dysfunctional emotions of anger, frustration or rage that lead one to turn down flimsy offers may play a very functional role. Within a community, reputation takes on a vital role in human interaction, because it is not unlikely for one to be in a position to do business with the same person more than once or with people who have heard about one's previous business dealings. Thus, player 2 would like to commit to a strategy that turns down bad offers because, if player 1 knows that bad offers will be turned down, he has an incentive to make better ones. The problem is, of course, how to make such an announcement credible. An important role emotions seem to be playing in regulating human interactions is exactly this: to make credible commitments possible. Assume player 2 reacts with anger and frustration to bad offers, so that he happily rejects them and player 1 knows this (either by introspection or because he knows player 2). This knowledge would motivate player 1 to make a better offer, with the consequence that the resulting allocation is more egalitarian, which gives an "emotional" player 2 a fitness advantage over a purely "rational" player 2. What this example shows is that one's behavioral dispositions and emotional reactions evolved over a long period of time and that they are usually functional adaptions to certain environments. In different environments, however, they may become dysfunctional. This is why it would be completely premature to classify the sunk-cost principle as the only rational way to make decisions; it all depends on the context.

9.4   Firm and Market Supply

This chapter has, up until this point, concentrated on the behavior of an individual firm. It has also shown that one can interpret the individual supply function as the inverse of the marginal-cost function, if the market price exceeds a certain benchmark, which is defined by the relevant average costs of the firm. It has neglected, thus far, to cover the relationship between individual and market supply. The assumption that firms seek to maximize profits allows one to say a little bit more than one already knows from Sect. 4.2.

Section 4.2 worked under the assumption that there is a given number of firms l in each market i. In this case, market supply for a good j was defined as

$$y_i(p_i, r, w) = \sum_{j=1}^{l} y_i^j(p_i, r, w).$$

Depending on market prices, fixed and sunk costs, the optimal policy of a firm can lead to positive profits, zero profits or even losses (during a period of time when contractual obligations restrain firms). Assume now that, for the given number of firms l, profits are strictly positive. Under certain conditions, such a situation is unsustainable in the long run, if firms maximize profits. Positive profits in a market attract additional firms to enter the market.

A good example is Lidl and Aldi, two German grocery stores that entered the $\mathcal{L}$ Swiss market a few years ago. The old situation, in which two major incumbents, Coop and Migros, divided the lion's share of the Swiss market, was no longer sustainable after Switzerland signed the Bilateral Agreements II with the European Union, which became relevant for the food industry in 2004. These treaties opened the Swiss market for new entrants from the European Union and the relatively high profit margins, in fact, encouraged Aldi (in 2005) and Lidl (in 2009) to enter.

As the example shows, there may be legal impediments to entering a market, but there may be technological ones, as well: for example, if one has to invest in an infrastructure for the distribution of one's products, whose value depreciates if one leaves the market again. The loss in value is like a barrier to entering the market, because it defines a minimum producer surplus below which market entry is not profitable.

There may not only be barriers to enter, but also barriers to exit a market. Most of them are related to unfinished contractual obligations, which create financial liabilities even after leaving the market, as seen in the sunk-costs example above. However, there may also be technological closure costs, like shipping costs of equipment. With positive exit costs, a firm might be forced to stay in the market because the costs of leaving are higher than the operative loss.

I will focus on the extreme case, where entry and exit costs are zero, because it allows one to derive a very strong conclusion about the effects of competition. If the number of firms is fixed, profits can be positive, if the price exceeds average costs. Therefore, without market entry, it may be possible that profits are positive. Without entry and exit costs, these profits will encourage other firms to enter the market. This process will continue until they drive profits down to zero. Any other solution would be incompatible with the assumption of profit maximization. However, this is only possible if the market price is equal to the average costs of the firm, which is the situation that Fig. 9.7 illustrates.

This equilibrium is bad news for the owners of firms and good news for the general public. It is bad news for owners, because any expectation about positive profits will ultimately be discouraged, because market forces drive them down to zero. This finding illustrates Adam Smith's quote from the beginning of this chapter.

Zero profits does not mean that being in business is meaningless, as one can see from the cost equation. I will focus on two factors, capital K and labor L, for simplicity. Zero profits means that revenues $p \cdot y$ equal costs $C(K, L, r, w) = r \cdot K + w \cdot L$. Assume the owner provides all the capital and works himself. In this case, zero profits means that he cannot expect a compensation for his capital and labor that exceeds the market interest rate r and the market wage w, because all the revenues of the firm are completely used for factor payments, whose opportunity costs are evaluated at the input prices. Therefore, the owner is indifferent between investing in her own firm, renting out the capital at an interest rate r and working for his own firm or working for someone else for a wage w. Zero profit, in other words, does not mean that there are no gains from trade; it only implies that the owners of a company do not get rents larger than the current market rates.

From the point of view of the general public, zero profits are good news: they imply that production takes place at minimum average costs, because marginal costs intersect with average costs at the minimum of the average costs. As long as profits are positive or negative, the average costs of production are not at their minimum. The allocation is efficient, given the number of firms in the market, but the number of firms (or factories) is not yet optimal. Free entry and exit implies that, in the long run, even the number of firms adjusts such that goods are produced in the cheapest possible way.

$\mathcal{L}, \mathcal{B}$ **Digression 24. The Ethics of Profit Maximization**

> Profit is useful if it serves as a means towards an end that provides a sense both of how to produce it and how to make good use of it. Once profit becomes the exclusive goal, if it is produced by improper means and without the common good as its ultimate end, it risks destroying wealth and creating poverty. (Benedict XVI (2009), Caritas in Veritate)

One of the most intensely scrutinized assumptions of mainstream economics is profit maximization. Most people find it unethical, or even morally offensive, and claim that profit maximization is a major source of the problems of capitalist societies. The idea of *corporate social responsibility* (CSR) is seen as an alternative to profit maximization, which helps firms to better align their behavior with society's interests.

The debate about ethical and moral standards in business is probably as old as business itself. One of the oldest deciphered writings of significant length in the world, the Code of Hammurabi (1700s B.C.), lays down the rules of commerce and prescribes prices and tariffs, as well as penalties for noncompliance.

According to the 2001 Greenbook by the European Union, CSR is a "concept whereby companies integrate social and environmental concerns in their business operations and in their interaction with their stakeholders on a voluntary basis." In addition, since 2011 the European Union defines CSR as "the responsibility of enterprises for their impacts on society." This concept goes far beyond the narrow idea of profit maximization, which was put forward by Milton Friedman (1970): "In [a free economy] there is one and only one social responsibility of business – to use its resources and engage in activities designed to increase its profits so long as it stays within the rules of the game." This quote nicely expresses the mainstream view that normative concerns should, and can be, addressed at the level of the foundational institutions of society: the "rules of the game." One has seen examples for this approach in the preceding chapters: externalities should be internalized by the design of property rights, contract law, taxes, regulations and so on, but not by appealing to firms to voluntarily internalize them by non-profit-maximizing business practices.

Given these opposing views, is it possible to bridge them? For starters, one gets a lot of support for the so-called *Friedman doctrine* from the model of firm behavior under perfect competition, which was developed in this chapter. First, note that the existence of a complete set of competitive markets implies an ideal institutional framework which, in the language of Milton Friedman, could be understood as the perfect rules of the game. This is expressed in the First Theorem of Welfare Economics. Second, with free entry and exit, competition has the tendency to drive profits to zero. However, if profits are zero at the maximum, firms do not have much choice but to maximize them. Paying higher wages to employees or selling at lower prices simply drives firms out of business. The only exceptions to this rule are short-run profits, or a situation where entry and exit are restricted, such that profits are positive even in the long-run. However, in this case, advocates of free markets would argue that one should first try to reduce the entry- and exit barriers to the largest extent possible, in this case. Like it or not, under perfect competition, there is not much room for anything but profit maximization.

A lot of firms voluntarily choose ethically sound business practices. One has to be careful to judge these practices correctly, though. Their existence does not necessarily imply that firms incorporate other objectives than simply profits into their business models. There are a number of cases where a more comprehensive understanding of the factors that influence the adoption of these practices is necessary. For example, there are apparently cases in which ethical practices are profit maximizing in a long time horizon. Paying decent wages may motivate employees to work harder and to be loyal to the firm, thereby increasing profits. Sustainability standards may lead to higher prices, if consumers have a willingness to pay for sustainable production, and so on. In fact, a lot of proponents of CSR reduce the concept to this "enlightened self-interest" of the firm, the argument being that a lot of potential conflicts of interest between the owners and managers of firms ("shareholders") and other groups in society ("stakeholders") result from a too-narrow perspective of the shareholders. This view implicitly accepts the profit motive, but aligns it with social interests by declaring them compatible. The approach could also be called *responsible profit maximization*.

However, this is not the end of the story. One has seen that perfect competition depends on technological prerequisites, which are not always fulfilled, and that externalities may make an equilibrium inefficient. In situations like these, there is room for discussion about the adequate way to address inefficiencies and problems of sustainability, as well as distributive justice on the firm level. Here is an example: one of the major challenges of globalization is exactly the lack of a consistent global regulatory framework – the rules of the game – that create a perfectly level playing field, and institutions like the WTO or OECD are too weak to fill the holes and gaps in the playground. (Nevertheless, CSR goes beyond the problems imposed by globalization.) Nation-states even enter into race-to-the-bottom types of international competition, where they reduce taxes and social standards to attract internationally mobile capital. This type of competition can, in principle, be beneficial, if it is primarily utilized as a disciplinary device for nation states to provide public services more efficiently, but it often drives standards below the efficient level. Especially large, multinational corporations can profit from these developments and, for the foreseeable future, there is no other institutional actor able to address the ethical issues that result from these developments other than the corporations themselves. Again, like it or not, if they do not care, no one will.

References

Friedman, M. (1970). The Social Responsibility of Business is to Increase Its Profits. *The New York Times Magazine*.

Pope Benedict XVI (2009). *Caritas in Veritate*. Veritas Publications.

Smith, A. (1776)[1991]. *An Inquiry into the Nature and Causes of the Wealth of Nations*. Everyman's Library.

Further Reading

Mas-Colell, A., Whinston, M. D., & Green, J. R. (1995). *Microeconomic Theory*. Oxford University Press.

Varian, H. R. (1992). *Microeconomic Analysis*. Norton.

Firm Behavior in Monopolistic Markets

10

This chapter covers . . .

- cognitive, technological, and regulative prerequisites for the existence of monopolies.
- how firms can use their monopoly to develop basic and sophisticated pricing strategies.
- the role of price discrimination in markets with imperfect information about the willingness to pay of the customers, and why the findings help to better understand pricing behavior in, for example, airline, software and hardware markets.
- the role of price discrimination between market segments and why the findings help to understand the debate about international price differences.
- how the informational demand for optimal pricing strategies is related to the optimal organization of firms.
- the economic-policy consequences of the above pricing models.

10.1 Introduction

> Like many businessmen of genius he learned that free competition was wasteful, monopoly efficient. And so he simply set about achieving that efficient monopoly. (Mario Puzo (1969), The Godfather)

The model of firm behavior under perfect competition has shown how a firm's supply is determined, if it takes prices as given. If it produces a positive quantity, a firm's optimal policy is generally determined by the condition "marginal revenues equal marginal costs," which simplifies to "price equals marginal costs," because marginal revenues and prices are identical under perfect competition. Therefore, the firm's supply function is the inverse of its marginal-costs function. The fact that it is willing to sell at marginal costs is also a prerequisite for the Pareto-efficiency of a competitive market.

However, the model has also shown that not all goods can be traded under conditions of perfect competition. One prerequisite is that there are many perfect

© Springer International Publishing AG 2017
M. Kolmar, *Principles of Microeconomics*, Springer Texts in Business and Economics,
DOI 10.1007/978-3-319-57589-6_10

substitutes for the good a firm produces. Another prerequisite is that the firm's production technology has to guarantee that the (long-run) marginal costs are non-decreasing. For different reasons, both conditions cannot be taken for granted. Consequently, one has to ask how markets function, if there is no perfect competition and keep one's focus on imperfect competition on the supply side. A similar logic applies to imperfect competition on the demand side as is, for example, frequently observed in regional labor markets, when there are only a few firms and labor has a low mobility. Another example is public procurement with firms that specialize in public projects. However, because imperfect competition on the supply side is the more commonly analyzed case, I will derive the implications of a supply-side monopoly for the functioning of markets.

One can start by analyzing a situation in which a firm has a monopoly for the supply of some good. The definition of a monopoly as a market with only one supplier of a good seems pretty obvious. However, this definition is not very operational, because it is unclear what exactly is meant by the idea that there is only one supplier. Hence, one first has to get to grips with a more operational understanding of what it takes for a firm to have a monopoly. Then one studies the optimal policy for a monopolist and analyze what this implies for the functioning of markets.

10.2 Conditions for the Existence of a Monopoly

Assume that one wants to bake a cake and needs flour. If one compares different retailers, one will find that each of them has different brands. In this sense, for example Migros is a monopolist for flour sold as "M-Budget Haushaltsmehl" (M-Budget flour), because Migros is the only supplier of this brand in the world. However, does this mean that Migros has a monopoly on "M-Budget Haushaltsmehl" in any meaningful economic sense? This question cannot be answered without further information. The reason is that two conditions have to be met in order to leverage the unique characteristics of a brand onto a monopoly position.

1. The customers have to be able to differentiate the product from other products and this ability to differentiate is reflected in the fact that alternative goods are not perceived as perfect substitutes. If consumers of flour are aware that there are different brands, but if this fact does not influence their decision which one to buy (because, for example, all they care for is the price of the flour), then flour is a homogeneous good sold by different suppliers, irrespective of the different brands. The fact that no other firm sells "M-Budget Haushaltsmehl" does not translate into the ability of Migros to raise prices above those of its competitors. However, if the customers consider the different brands to stand for non-homogeneous goods, then firms can use this willingness to differentiate between brands to charge brand-specific prices and optimal pricing becomes an integral aspect of the optimal firm policy. Indeed, flour seems to be a homogenous good to most customers and one may thus conjecture that the market for flour is, in fact, competitive. However, it is important to note that homogeneity

of goods has nothing to do with the good's *physical* characteristics or the brand name *per se*. It is the customers' willingness and ability to differentiate between goods of different suppliers that is a necessary prerequisite for a monopoly.

The willingness to differentiate can be assessed empirically by estimating the price and cross-price elasticities of demand. Intuitively, elasticities measure the percentage change of a variable that is caused by a one percent change in some other variable. If demand is ordinary and very price elastic, then it reacts strongly to price changes and there is no leeway to set prices actively. Similarly, the cross-price elasticity describes how demand changes, if the price of another good changes. If this elasticity is very large (in absolute terms) and the goods are close substitutes, then there is, again, little scope for price setting. An introduction to the concept of elasticities can be found in the mathematical appendix in Chap. 14.

If the existence of a monopoly position depends on the customers' ability and willingness to differentiate between products, then it must be an integral element of corporate communications and marketing to define and communicate relevant differences to other firms' products or to create them in the first place. From this point of view, even an ordinary product like flour becomes interesting: in recent years, the market has displayed increasingly differentiated products. For example, wheat flour has been differentiated by cultivation method (organic vs. conventional), origin (local vs. from somewhere else), etc. This differentiation has the purpose of transforming a formerly homogeneous product into a set of heterogeneous products for which – if the efforts are successful – differences in the willingness to pay exist that can be exploited by the firms. Two other examples are denim jeans and coffee.

- Denim jeans do not fundamentally differ in their functionality: they protect $\mathcal{B}$ from weather, have pockets to store and carry small items, and so on. The physical characteristics of jeans seem to suggest that they are a fairly homogeneous product, which is sold on competitive markets. However, this reasoning does not take into account that producers of jeans can use advertising campaigns in an effort to create a specific brand image that adds additional "cultural" content to the product, from which customers can benefit: jeans do not only protect from weather, but customers send a specific social message by wearing a specific brand. The brand's image is transferred onto the customer, allowing the customer to perform a specific societal role; to belong to a specific group whose values are implicitly communicated by the brand name. In our societies, jeans and many other products are sophisticated mediums of communication and the communicative function often dominates, or even replaces, the primary, utilitarian one (think of intentionally ripped jeans). This is why firms often produce cultural narratives in which their products play an important role. If successful, there are many differentiated products with their own differentiated markets, like the markets for Levi's jeans, Diesel jeans, Wrangler jeans, and so on, and the firms have more or less extensive leeway to set prices. One can, for example, buy a pair of H&M jeans for $19.95 and a pair of Tom Ford Jeans for $990 (spring 2015).

- For many firms in the food industry, the wine market is *the* reference point for the development of marketing strategies. Because, for large parts of the population, it is fashionable to be a "wine connoisseur," a plethora of differentiated products exist, such that producers are, to some degree, able to exploit the customers' ability and willingness to differentiate by charging higher prices (which, *ceteris paribus*, translates into larger profits).

  Currently, coffee is probably one of the more interesting products in the food industry, because many producers – inspired by the market for wine – try to escape the dead end of the homogeneous-good market by "educating" the customer to distinguish different types of coffee. Historically, the lion's share of the market sold homogenous quality and customers were very price-sensitive, which implied a high degree of competition between suppliers. The central elements of the "third wave of coffee culture," as it is called, are referencing the origin of the coffee all the way back to the farm where it was cropped: an accentuation of the varieties of tastes of coffee and differences in cultivation methods, from the coffee cherry to the final beverage. This includes the introduction of quality standards, like the "Cup of Excellence" seal, and the training of customers with respect to the flavors and brewing methods. On top of this, the emergence of the third wave as a subcultural phenomenon in Portland, Chicago and San Francisco gives third-wave coffee specific cultural overtones, which makes it an attractive symbol for certain groups of customers (as in the jeans example) and which is important for the evolution of a niche product into the (profitable) mainstream. If this project succeeds, the "third wave" has the potential to change the logic of coffee markets.

2. If the demand for some product is not perfectly elastic (the market demand curve is not flat), then the potential for a firm to create a monopoly exists. However, a somewhat inelastic demand function is not sufficient, but merely necessary for a monopoly. In addition, it has to be impossible for other firms to undercut the privileged position by simply imitating the first firm's products. There are different reasons for why imitation might not be a possibility.

- The producer has exclusive control over some necessary resource. For example, the "De Beers diamond monopoly" existed because De Beers controlled a large share of raw diamond mines.
- The producer is a technology leader, such that other firms are not able to imitate the product, because of a lack of skills. A good example is the US-American telecommunications company AT&T. It became the first long-distance telephone network in the USA and made huge investments in research and development that allowed it to acquire crucial inventions. As a result, the company obtained near-monopoly power on long-distance phone services.
- The state regulates market entry by creating public monopolies or by patents and trademarks. An example for high profit margins that are protected by

patents and trademarks was the Nespresso capsule system (most patents have expired by now). A slightly awkward example for a public monopoly was the German monopoly for safety matches (Zündwarenmonopol) from 1930, which granted exclusive rights to distribute safety matches within the borders of the German Empire to a monopolist (Deutsche Zündwaren-Monopolgesellschaft). It ended in 1983. Historically, most countries had public monopolies in the post and telecommunications industries, as well as for rail transportation. A lot of these industries were, at least partially, privatized and opened up for competition in the 1980ies and 1990ies.

- One has already seen that competitive markets cannot function, if either or both marginal and average costs are decreasing. In industries with such technologies, firms with larger market shares have an advantage, because they can produce at lower marginal and average costs and, hence, can sell at lower prices compared to their smaller competitors. Therefore, a large market share protects firms to some extend against competition. I have already discussed a special case of such a technology in Chapter 6, where I argued that club goods are sometimes called natural monopolies, because they imply decreasing average costs by increasing the number of users.

- Some products are characterized by the fact that the utility and, therefore, the customers' willingness to pay is increasing with the number of customers. This phenomenon is called a positive "network externality." Examples include telephones, word processing software and social media like Facebook and Twitter. Positive network externalities benefit firms with large market shares. Market share protects, to some degree, against market entries and competition, because a competitor with a smaller market share offers a *ceteris paribus* less attractive product.

If one or more of the above conditions holds, then the customers' willingness to differentiate translates into a (possibly temporary) monopoly position. The next subchapter analyzes how firms can make use of such a position.

10.3   Profit Maximization in Monopolistic Markets

The problem of a monopolistic firm is quite complex. As the previous subchapter explained, it has to decide about brand management and product development, needs to develop pricing strategies and has to take into account the political environment to protect and further its interests. In the following subchapter, I will reduce this complexity by focusing on the pricing aspect. Thus, the following analysis will assume that a firm has a monopoly in an already existing market with an established product. To understand optimal pricing policies, different cases have to be distinguished:

- The firm is not able to discriminate prices between customers; each customer buys at the same price. This is the standard model of monopoly theory. It will allow one to understand the important elements of an optimal pricing strategy better. However, it is not quite realistic, because firms will usually try to differentiate prices between customers. Hence, the standard model alone cannot give one an appropriate picture of monopolistic behavior. The fact that price discrimination is an important tool of a firm's policy will become clear during the analysis.

- The firm is able to discriminate prices.

  - Perfect *(first degree) price discrimination* is possible. This is a theoretical benchmark, where the firm is able to set individualistic prices for each customer and can, in addition, discriminate by the quantity demanded. This model helps one to understand the consequences of price discrimination by bringing it to its extreme. It is, however, not particularly realistic, because it assumes that firms have all the relevant information about their customers and that a legal environment exists that allows them to use this information to charge different customers different prices. The availability of "big data" and the development of sophisticated algorithms that analyze the behavior of individuals on the internet may, however, allow them to move closer in the direction of perfect price discrimination in the future.

  - Price discrimination according to the quantity, quality or time demanded *(second degree price discrimination)*. A firm often knows that there are different "types" of customers, who differ in their willingness to pay. From its market research, it may also know the different demand functions of these types. However, it does not know the willingness to pay of a specific customer. Therefore, the firm cannot condition the price on the customer's type directly, but needs to find alternative, indirect ways to skim off the different types' willingnesses to pay by appropriately designing products and prices. Examples are economy- and business-class airline tickets. Flexibility regarding the altering of the booking, leg space and service on board are important quality dimensions and airlines have an incentive to play with these variables to optimize profits.

  - Price discrimination according to specific customer attributes or customer segments *(third degree price discrimination)*. The monopolist is able to discriminate between groups of customers, but cannot discriminate within groups. An example is price discrimination of multinational firms between countries.

The following subchapters will discuss these four cases.

10.4   The Optimal Production and Pricing Decision of a Monopoly Without Price Discrimination

The most studied case is the non-price-discriminating monopolist. Before discussing the circumstances under which it may be optimal to dispense with price discrimination, one should analyze how such a market functions. Assume the monopolist acts as a profit maximizer and produces with a technology that leads to a cost function $C(y)$.

A firm's market-research department estimates a market-demand function $x = X(p)$ for one of the firm's products. This function shows the amount of the good or service that can be sold at a given price p. In perfectly competitive markets, the perceived demand function is perfectly price elastic at the market price. This implies that the only information necessary to determine the optimal output is the existing or expected market price. However, this is no longer the case for a monopoly, where the firm needs to estimate the market-demand function with as much precision as possible. The organization of the firm is hence more complex: while in perfectly competitive markets, firms only need managerial accounting to determine marginal and average costs, but a firm in a monopoly market also needs a market research unit to estimate the demand function, because it is no longer a price taker. $\mathcal{B}$

Digression 25. Measuring Willingness to Pay $\mathcal{B}$
This chapter's analysis shows that an understanding of the likely responses of potential buyers to price changes is of considerable importance for firms. Despite this fact and despite the advances in pricing research, many firms price and develop products without an adequate knowledge of their customers' willingness to pay. Research has shown that only 8% to 15% of all companies use pricing strategies that are based on empirical assessments of buyer responses (Monroe and Cox, 2001), despite the fact that there is empirical evidence that even minor changes in prices can have important effects on profits (Marn et al., 2003). A lot of firms would rather use a strategy that may be dubbed "intuitive" pricing.

Marketing research offers a large variety of different techniques for measuring the willingness to pay. Broadly speaking, they fall into two different categories: "revealed" and "stated" preference models. Revealed-preference techniques infer a customer's willingness to pay from observed data. This can be market data, data that is generated while browsing the internet or data generated in experiments. Stated-preference techniques are based on surveys that are designed to elicit information about the willingness to pay. Examples of these include, among others, expert or customer surveys, and a technique that is called conjoint analysis. Conjoint analysis is a statistical technique where a product is partitioned into different attributes that together generate value for the customer (in the case of a car, these attributes might be mobility, versatility, status, etc.). Customers are then asked to rank or rate different bundles of these attributes. The results are used for the design and pricing of future products.

The different techniques to measure one's willingness to pay have their own strengths and weaknesses and it depends on the specific product and the available budget of the market-research department which method is applied.

At this point, cne has to make a decision. One can assume that the monopolist sets a price and passively adjusts the produced quantity, $X(p)$, or one can alternatively assume that the monopolist decides on the quantity and demand determines the price at which the market clears. Both approaches lead to the same result but, since the second is somewhat simpler, it is the one that is usually applied. In order to do so, however, one has to infer the so called "inverse demand function" from $X(p)$, knowing that, for any price (p), the quantity (x) that can be sold is given by $x = X(p)$. Taking the inverse function of this demand function yields $p = X^{-1}(x)$, which determines the price that clears the market for any quantity offered by the firm. The convention from the previous chapters is to denote demand by x and supply by y. Given that one is analyzing the problem from the position of the monopolist who decides how much to supply, it makes sense, therefore, to replace x by y in the inverse-demand function that one denotes as $p = P(y)$.

If π denotes the firm's profit, then one can use this information to express it as revenues minus costs:

$$\pi(y) = P(y) \cdot y - C(y).$$

The problem faced by the firm's manager is to determine the quantity that maximizes profits. This quantity is implicitly defined by the necessary ("first-order") condition $\pi'(y) = 0$. (Assume in the rest of the book that this condition characterizes the global profit maximum. This is guaranteed, for example, if the second derivative of the profit function is globally negative, has a positive slope at $y = 0$ and has a negative slope for $y \to \infty$.) This yields:

$$P'(y) \cdot y + P(y) \cdot 1 - C'(y) = 0.$$

This condition has a straightforward economic interpretation: the first two terms represent the marginal revenues of an additional unit of the good, which can be decomposed into a price effect (first term) and a quantity effect (second term). The quantity effect is known from perfectly competitive markets. It measures by how much revenues increase, if an additional (marginal) unit is sold and the price stays constant. For infinitesimal changes, it is equal to the good's price. The price effect is new, however, and measures the loss in revenues of the firm, if it wants to sell another unit. To be able to sell another unit, the firm has to lower its price a bit to gain more customers. Since price discrimination is not possible, the firm also has to lower the price for those customers who would have paid a higher price. This "loss" can be interpreted as an opportunity cost and is measured by the price effect. The third term represents marginal costs and hence the "marginal revenues = marginal costs" rule holds.

$\mathcal{B}$ The first-order condition can be transformed into an easy rule of thumb, which is of great relevance in the management and pricing literature. Simple manipulations of the first-order condition show:

$$P'(y) \cdot y + P(y) = C'(y)$$

$$\Leftrightarrow P(y) \left(P'(y) \cdot \frac{y}{P(y)} + 1 \right) = C'(y)$$

$$\Leftrightarrow p\left(\frac{1}{X'(p) \cdot \frac{p}{X(p)}} + 1\right) = C'(y)$$

$$\Leftrightarrow p\left(\frac{1}{\epsilon_p^x(p)} + 1\right) = C'(y).$$

The transformation from the second to the third row follows from the definition of the inverse-demand function and the fact that demand is denoted by x. The manipulation between the third and fourth rows follows from the definition of the price elasticity of demand: $\epsilon_p^x(p)$.

One can start by making a short plausibility check: when the market-demand function has a negative slope, the price elasticity of demand is negative. Hence, the expression in brackets is smaller than one. Therefore, the condition can only hold if the price exceeds marginal costs. If there are perfect substitutes for the good, then the price elasticity converges to $-\infty$, such that the expression in brackets converges to 1, which leads to an intuitive conclusion: to comply with the above condition the price has to be equal to marginal costs, which is, of course, the case in perfect competition.

If the price elasticity is finite, however, then the optimal price exceeds marginal costs. The difference between price and marginal costs is called the "markup" and this rule of thumb is called "cost-plus pricing." Generally, the optimal markup is the higher, the less elastic market demand reacts to price changes. Thus, a manager who wants to set the optimal price needs information about two things: the marginal costs and the price elasticity of demand.

To further illustrate the optimal pricing decision of a monopolist, I derive the solution for the special case of a linear demand function $p(y) = a - b \cdot y$ and constant marginal costs $MC(y) = c$. In this case, revenues are equal to $R(y) = a \cdot y - b \cdot y^2$, with marginal revenues being $MR(y) = a - 2 \cdot b \cdot y$. Equating marginal revenues and marginal costs and solving for y yields the optimal solution $y^* = (a - c)/(2 \cdot b)$ and a price of $p^* = (a + c)/2$.

The demand function $x(p) = a/b - (1/b) \cdot p$ has a price elasticity of demand:

$$\epsilon_p^x(p) = -\frac{1}{b} \cdot \frac{p}{a/b - (1/b) \cdot p} = -\frac{p}{a - p}.$$

Thus, the absolute value of the price elasticity of demand is equal to:

$$|\epsilon_p^x(p^*)| = \frac{p^*}{a - p^*} = \frac{a + c}{a - c} \geq 1.$$

The fact that the optimal quantity and price are in the elastic part of the demand function is no coincidence: if demand is inelastic, then the monopolist can increase revenues by reducing output, because a one percent decrease in output increases the price by more than one percent. However, in this case, the initial level of output could not have been profit maximizing, since reducing output also reduces costs.

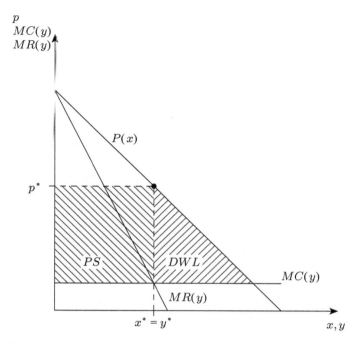

Fig. 10.1 The supply of a monopolist without price discrimination

Figure 10.1 shows the graphical solution to the profit-maximization problem of the monopolist. The optimum is given at the quantity where the marginal-revenue and the marginal-cost curves intersect. The associated price is defined by the value of the demand function for that quantity.

The solution to the linear model looks abstract, but is, in fact, rather intuitive. Assume for the moment that $c = 0$. In this case, profit maximization boils down to revenue maximization and revenues $p(y) \cdot y$ are the rectangular area under the demand function for any output y. Hence, the monopolist maximizes the size of this area. The optimal output is, therefore, given at $y = a/(2 \cdot b)$. If costs are positive, one has to restrict attention to the area where gains from trade are positive. $(a - c)$ are the maximum gains from trade of the customer with the highest willingness to pay, in this case. Thus, what one effectively gets is a "truncated" demand function $\bar{p}(y) = \bar{a} - b \cdot y$ with $\bar{a} = a - c$. The same argument as was used in the case of zero marginal costs also applies to this truncated function.

Digression 26. What Factors Determine Price Elasticities?
There are two factors that determine the price elasticity of demand: the customer's purchasing power (when the good becomes more expensive, customers can *ceteris paribus* afford less of it) and the willingness and possibility to substitute the good with another one. This second determinant makes the model applicable to markets with close but imperfect substitutes, e.g. different brands of jeans. The model's implication for markets with close substitutes is that markups have to be relatively moderate. The markup rule also gives one a first clue about how firms should invest in advertising and public relations: cost-plus pricing indicates that the markup in a market is negatively related to the price elasticity of demand. A marketing campaign should hence aim at making demand less elastic. In order to determine the optimal advertising budget, the firm needs to know the marginal revenues and marginal costs of advertising. The marginal revenues are determined by the change of the price elasticity of demand that is induced by another unit of advertising.

The monopolist's optimal price policy has interesting implications for economic policy: since the optimal price exceeds marginal costs, unexploited gains from trade will remain. There are still customers who are willing to pay a price that exceeds the firm's marginal costs, but at which the monopolist is not willing to sell. Consequently, the sum of consumer and producer surplus is below its maximum and the market is inefficient, a situation that is also called "market failure." The reason for this market failure is easy to grasp: if the monopolist wants to sell another unit of the good, he needs to lower the price by a bit. However, because price differentiation is not possible, the price needs to be lowered for all: not only for the marginal customers, but also for those who would buy the product at higher prices. If the firm wants to sell at marginal costs, the decrease in revenues due to the price effect exceeds the increase due to the quantity effect; therefore, the monopolist prefers to constrain the quantity to keep the price high. The inefficiency is given by the triangular area DWL in Fig. 10.1. DWL stands for *deadweight-loss* and is a measure for the inefficiency of the allocation. Before one can discuss the implications for economic policy one needs to have a better understanding of the causes of this inefficiency. Hence, the next analysis is a monopolist's optimal pricing policy, if price discrimination is possible.

10.5   Price Discrimination

Reasonable charges
Plus some little extras on the side!
Charge 'em for the lice, extra for the mice
Two percent for looking in the mirror twice
Here a little slice, there a little cut
Three percent for sleeping with the window shut
When it comes to fixing prices
There are a lot of tricks he knows
How it all increases, all them bits and pieces
Jesus! It's amazing how it grows!
   (Alain Boublil (2013), Les Miserables (based on the novel by Victor Hugo))

10.5.1   First-Degree Price Discrimination

This subchapter covers the problem of a monopolist who is able to discriminate prices perfectly. Although this is not a very realistic assumption, as firms are usually unable to get all the relevant information, it is a useful theoretical benchmark and allows one to better understand the reasons for the above-mentioned inefficiency as well as current trends in firms' pricing strategies.

Perfect price discrimination is easy to analyze. In order to be able to pursue this strategy, the monopolist needs to know the willingness to pay of each individual customer. If this information is available, the firm will charge individualized prices for each customer, which equal that customer's willingness to pay. (It may be necessary to lower the price a bit to induce customers to actually buy the product. The remainder of this book will assume that indifferent customers behave in the interest of the firm. This assumption is innocuous with respect to its implications and simplifies the analysis.) Hence, in such a market, there is no uniform price, but a price function that is exactly equal to the inverse-demand function.

What is the minimal price at which the monopolist will supply the good? His profit increases as long as the price of the last unit exceeds the marginal costs of that unit. Hence, he will expand his supply up until the point where price equals marginal costs. This brings about a surprising result: the resulting market equilibrium is Pareto-efficient and the sum of consumer and producer surpluses are maximized. However, contrary to the case of perfect competition, gains from trade are not shared between the producer and the customers. Instead, the monopolist is able to skim off all the surplus in the market (see Fig. 10.2).

> **Digression 27. Price Discrimination in the Digital Age**
> Compared to second- and third-degree price discrimination, first-degree price discrimination has long been seen as a theoretical benchmark without much practical relevance, because the need for customer-specific information that

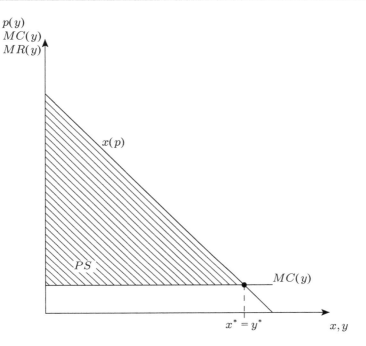

Fig. 10.2 Supply of the efficient quantity with first-degree price discrimination

is necessary to charge personalized prices was considered too extensive. On the other hand, moving in the direction of perfect price discrimination is extremely tempting for firms, because of its obvious consequences for profits. It should therefore not come as a surprise that e-commerce sites experiment heavily with pricing strategies that are based on the tracks people leave while browsing the internet.

A 2015 report in the hands of the President of the United States concludes that "the combination of differential pricing and big data raises concerns that some customers can be made worse off, and have very little knowledge why. [...] [M]any companies already use big data for targeted marketing, and some are experimenting with personalized pricing, though examples of personalized pricing remain fairly limited. [...] [P]roviding consumers with increased transparency into how companies use and trade their data would promote more competition and better informed consumer choice."

Hannak et al. (2014) analyzed the search results of 300 people who visited 16 online retailers and travel agencies from the US. They found that customers were shown different prices or different results for the same searches on nine of these 16 sites. For example, the online-travel company Expedia discriminates prices according to the browsing history stored on the customers' computers. It is unclear, however, which type of browsing history triggers high prices. Another travel-agency, Travelocity, offered hotel rooms that were $15 a night cheaper if viewed from an iPhone or iPad. Home Depot displays higher prices and pricier products for smartphone users than for customers using desktops. In 2012, Wall Street Journal found that Staples discriminated prices according to the location of the device, and Orbitz discriminated prices between Mac and PC users, because data analysis revealed that Mac users are willing to pay higher prices for hotels.

These attempts to discriminate prices are still relatively crude, but the availability of more information and better algorithms may soon change the picture. Calo (2013) concludes that big data and better algorithms will enable companies to profile customers and deliver advertisements in a much more personalized way, also making use of the limited rationality of individuals. For example, Apple and Microsoft have filed patents for so-called "mood-based advertising" and Amazon is developing algorithms that tell them what the customers are likely to want before they place an order. This information is crucial for price discrimination, because it allows them to adjust prices or tweak choices while the customer is still searching. Google, for example, is filing a patent for an algorithm that can decide if a customer is likely to buy something and then to display a high price, while lowering the price for customers who have a low likelihood of buying.

Shiller (2014) studies the effects of including more information into pricing strategies on profits in the case of Netflix. He found that, compared to standard second-degree price discrimination, using the full set of information about web-browsing behavior increases variable profits by 1.39%, compared to 0.15% if pricing strategies are based on demographics alone. This may not sound like much but, compared to net profit margins of 2.34% in the US online retail industry, it makes a big difference.

What are the consequences of this discovery for economic policy and the regulation of monopoly markets? If one compares the case of a non-price-discriminating monopolist with that of a perfectly price-discriminating one, one can see that the monopolist will always choose to discriminate prices, if he can. Hence, the inefficiency in the market with a non-price-discriminating monopolist is caused by the inability to discriminate prices. There are three reasons why this instrument may be infeasible:

𝓛 1. Price discrimination is illegal. The monopolist is then forced to charge the same price for any customer. In this case, it is the regulation of the market that causes the inefficiency. Market failure is not a result of some inherent tendency of the monopolist to be inefficient, but by a failed regulation of the monopoly (if the objective of regulation is to achieve efficiency).

An example for efforts to impede price discrimination by legal action is the "Robinson Patman Act," specifically Title 15, Chapter 1 §13 of the United States Code titled "Discrimination in price, services, or facilities". It is worthwhile studying the first paragraph of the act. "It shall be unlawful for any person engaged in commerce, in the course of such commerce, either directly or indirectly, to discriminate in price between different purchasers of commodities of like grade and quality, where either or any of the purchases involved in such discrimination are in commerce, where such commodities are sold for use, consumption, or resale within the United States or any Territory thereof or the District of Columbia or any insular possession or other place under the jurisdiction of the United States, and where the effect of such discrimination may be substantially to lessen competition or tend to create a monopoly in any line of commerce, or to injure, destroy, or prevent competition with any person who either grants or knowingly receives the benefit of such discrimination, or with customers of either of them: Provided, That nothing herein contained shall prevent differentials which make only due allowance for differences in the cost of manufacture, sale, or delivery resulting from the differing methods or quantities in which such commodities are to such purchasers sold or delivered: Provided, however, That the Federal Trade Commission may, after due investigation and hearing to all interested parties, fix and establish quantity limits, and revise the same as it finds necessary, as to particular commodities or classes of commodities, where it finds that available purchasers in greater quantities are so few as to render differentials on account thereof unjustly discriminatory or promotive of monopoly in any line of commerce; and the foregoing shall then not be construed to permit differentials based on differences in quantities greater than those so fixed and established: And provided further, That nothing herein contained shall prevent persons engaged in selling goods, wares, or merchandise in commerce from selecting their own customers in bona fide transactions and not in restraint of trade: And provided further, That nothing herein contained shall prevent price changes from time to time where in response to changing conditions affecting the market for or the marketability of the goods concerned, such as but not limited to actual or imminent deterioration of perishable goods, obsolescence of seasonal goods, distress sales under court process, or sales in good faith in discontinuance of business in the goods concerned."

2. The monopolist cannot prevent the resale of his products. In this case, resale markets evolve and so-called arbitrageurs specialize in buying and reselling the monopolist's products. For example, if there are two customers and one of them can buy at a high price only while the other can buy at a low price, it is worthwhile for both to trade at a price that is somewhere in between the two monopoly prices. Under ideal conditions, this process continues until only a uniform price prevails in the market.

Why should a monopolist be unable to prevent the emergence of resale markets?   To answer this question one needs to take a closer look at the types of contracts a monopolist can use because, from a legal perspective, the sale of a product or service is a transfer of a bundle of rights that is (explicitly or implicitly) specified in the underlying contract. If the monopolist can freely choose and constrain these rights, he can prohibit the resale of his products. He grants his customers the right of usage, but not the right of resale. The formation of resale markets can be precluded, if such contracts are legal and enforceable. However, in reality, it is often the case that courts do not enforce such contracts. They are sometimes legal in insurance markets, where insurance policies cannot be traded freely whereas, in traditional, consumption-goods markets, such contracts are usually illegal (for example in the European Union). However, the picture is more complicated for digital products, where complicated arrangements exist that regulate user rights. If resale were possible, then one would have to conclude that the source of market failure is, once again, an inefficient regulation of the market.

3. The monopolist does not have the information that is necessary to discriminate prices. The next chapter will cover the implications for the monopolist's profit maximization in more detail.

The preliminary conclusion that one can draw at this stage is somewhat surprising, because one cannot make a case against monopolies that is based on efficiency arguments. In light of the two models that have already been covered, one has to conclude that market inefficiencies are a result of an insufficient regulation of the market, not of the monopoly as such. This conclusion is, however, at odds with the intuitive feeling that most people and also most economists share, which states that monopolists are inherently inefficient. There are two ways to align this idea with the realizations discussed above:

First, it is, indeed, possible that a monopolist, who can perfectly discriminate prices, can lead to an efficient market outcome. Still, society might have goals that go further than efficiency. For example, distributive justice is a goal that many societies pursue. However, since monopolies are owned by individuals, who are also customers, customers will, in the end, receive the monopolist's profits. Therefore, one cannot judge the distributive properties of monopoly markets without further knowledge of the distribution of property shares among the population. However, there is empirical data about asset ownership in different countries. The demand for products and services is usually widely scattered, whereas property is concentrated in the hands of relatively few, rich individuals such that, from a more egalitarian point of view, a tradeoff between efficiency and distributive justice can exist. This may explain why some inefficiency is seen as the necessary price for a more egalitarian society. However, then the question arises of why the problem of distributive justice is not addressed more directly, for example by redistributive taxation, which would be more in line with the idea of the Second Theorem of Welfare Economics.

Digression 28. Pricing and Bounded Rationality
Finding ways to more effectively discriminate prices is a key topic in many industries. Strategies to discriminate go under names such as "dynamic pric-

ing," "power pricing," or "yield management." The basic problem behind all of these strategies is the same: how can a firm segment its customers into groups, which differ in their willingness to pay, and charge group-specific prices? Such strategies can actually lead to win-win situations between firms and customers, if there are close substitutes to the offered products (customers do not have to accept the offer, which is why they have to be better off if they accept it) and customers economize on search costs (e.g. finding an appropriate hotel for the planned trip to Vienna).

A related problem has to do with irrational or boundedly rational behavior. Based on findings from behavioral economics, some legal scholars criticize pricing strategies that systematically exploit customers' behavioral biases. Research in this field is still in its infancy.

Here is an example: assume a health club or gym uses a two-part tariff with an upfront-payment of L and a per-visit charge of p. If p equals marginal costs and L contributes to the financing of the club's fixed costs, then the contract is efficient. There is a lot of evidence, however, that customers overestimate the number of times that they will go to the club. This form of irrationality can be used by the club by charging p below its marginal costs and increasing L, which widens the gap between the surplus that the customer expects to receive from accepting the contract and what she actually receives. The customer finds this contract more appealing, but may end up with a negative consumer surplus.

Another example is a pricing strategy that is based on the *anchoring effect* discussed in Chap. 5. The rule of thumb on how to sell a good, for which customers have an unclear willingness to pay, is to place it right next to a similar, but much more expensive good. Williams-Sonoma added a $429 breadmaker next to their $279 model. The consequence was that sales of the cheaper model doubled, even though practically nobody bought the $429 machine (Ariely 2008): in this case, the expensive option acted as a price anchor.

A similar effect occurred in a study on purchasing patterns for beer (Poundstone 2011). In the first test, subjects had the choice between a regular beer for $1.80 and a premium beer for $2.50; 80% chose the premium beer. In the next test, a smaller and cheaper ($1.60) option was added. No one chose the cheap option, but orders for the premium beer dropped to 20%. In the final test, the cheap option was replaced by a large, expensive ($3.40) option. In this case, orders for the premium beer rocketed to 85%. This experiment shows that customers react to the pricing brackets in which products are displayed. Most people go for the "middle" option, which gives firms a lot of leeway in manipulating choices by developing adequate contexts for their products.

It might also be the case that the intuitive problems many people have with monopolies are not adequately grasped by the model. It is possible that the reason for a lack of efficiency of a monopoly is inherently dynamic, for example, because an ironclad monopoly position decreases the incentives to innovate. Such an argument, however, suggests that a completely different model is necessary to tackle the problem.

10.5.2 Second-Degree Price Discrimination

One central problem a monopolist faces when trying to discriminate prices is his lack of information about customers. There are two ways to solve this problem: investing in better information or using the given information to discriminate prices with maximum possible effectiveness. This subchapter will analyze the latter strategy.

In order to keep the problem simple and manageable, assume that there are two groups of customers, which can be differentiated by their willingness to pay. The firm knows each group's willingness to pay and also the respective group sizes, but cannot identify a customer as a member of one group or the other. An example for this situation is an airline that offers a flight from Zurich to Frankfurt, which is frequented by both business and leisure travelers. Business travelers have a higher willingness to pay for the flight and, in particular, for altering bookings flexibly. The airline knows the respective willingness to pay, as well as the groups' relative sizes, but cannot distinguish between individuals at the ticket counter (or on their homepage for that matter).

If the firm had all the relevant information, it would charge each customer according to his or her willingness to pay, such that both groups would receive their respective optimal offers, as in the case of first-degree price discrimination. From the point of view of the firm, the problem with asymmetric information is that a customer with a high willingness to pay may prefer the offer that is being provided for customers with a low willingness to pay. Their "own" offer gives the customer zero consumer surplus, whereas the offer provided for the other group not only differs in the quantity or quality of the good, but is also sold at a different price (both lower). Hence, buying a lower quality or quantity at a lower price might be profitable for the customer, if the lower price compensates for the loss in quality or quantity. In that case, all customers choose the offer that was designed for the group with the lower willingness to pay and the other offer remains a shelf warmer. This observation begs the question of what a firm's optimal reaction should be.

In order to answer the question, one should give the problem a more formal structure and analyze it graphically. In the following figures, one can see the quantity or quality of a good along the abscissa and the customers' willingness to pay along the ordinate. "Quantity" or "quality" can thereby be interpreted as an attribute, for which there are differences in the willingness to pay. In the airline example, quality can be interpreted as the flexibility to alter a booking, how much leg space there is or the level of service provided. If the good is a printer, quantity could refer to the

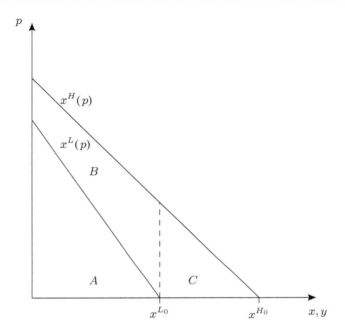

Fig. 10.3 Two types of customers, L (dashed line) and H (solid line)

number of pages the printer can print per minute. Depending on the specific context, it may be hard to distinguish between the quantity and the quality of a product. I will use the convention to talk about quantity in the following example.

In Fig. 10.3, one can see the inverse demand functions $p^H(x)$ and $p^L(x)$ of an individual with a high (H-type, solid line) and low (L-type, dashed line) willingness to pay. In order to simplify matters, assume that there are as many H-types as there are L-types and that there is only one individual of each type. The monopolist has marginal costs of zero, such that the efficient quantities supplied are equal to the maximum demand levels x_o^H and x_o^L. An individual's aggregate willingness to pay for a quantity x is equal to the area under the demand function, $P^H(x) = \int_0^x p^H(x)dx$, $P^L(x) = \int_0^x p^L(x)dx$. Her respective willingness to pay for the efficient quantity is, therefore, given by the areas $P^L(x_o^L) = A$ and $P^H(x_o^H) = A + B + C$.

In the preceding chapters, one has implicitly assumed that a firm sets a price per unit of the good and the customers choose how much to buy. For effective price discrimination with asymmetric information, the firm needs to restrict the customers' sovereignty by offering pre-specified quantity-price bundles. For example, $\{y, P^H(y)\}$ is a possible offer where the monopolist offers quantity y at the maximum price the H-type is willing to pay. An arbitrary pair $\{y, P\}$ is also called a "contract." P is the price for y units, not for one unit of the good, as it was before.

It is immediately clear that, with perfect information and, therefore, price discrimination, the monopolist will offer the efficient contracts, $\{x^L, P^L(x^L)\} = \{x_o^L, A\}$, $\{x^H, P^H(x^H)\} = \{x_o^H, A + B + C\}$. (In the following analysis, assume that a customer is willing to purchase a contract, if he is indifferent between buying and not buying, and that he is also willing to purchase the contract designed for him in case that he is indifferent between two contracts. This assumption simplifies the analysis and is without relevance for the qualitative results.) The monopolist's profit is then $2A + B + C$, and the consumer surplus is zero, $CS^H(x_o^H, P^H(x_o^H)) = CS^L(x_o^L, P^L(x_o^L)) = 0$.

However, with asymmetric information, these contracts are not enforceable. A H-type individual would prefer to buy the contract of the L-type, because it leads to a higher consumer surplus of B. The L-type would never buy the H-contract, because she has no willingness to pay for the additional quantity and, therefore, has no willingness to pay the higher price.

How will the firm react to this problem? In order to answer this question, one first needs to understand whether it is possible to change the contracts in a way that increases profits. If nobody buys the H-contract, because the H-types prefer the L-contract, then the firm's profits are $2A$. In order to induce the H-type to buy "his" contract, the firm can decrease the price of the H-contract until the H-type is indifferent between the two. Because his consumer surplus is B, this is achieved when the H-contract is $\{x_o^H, A + C\}$ (see Fig. 10.3). Because altering the contract in that way increases profits to $2A + C$, it is always profitable. The profit is smaller than it would be with perfect price discrimination, but larger than $2A$ is.

Is this the profit-maximizing pair of contracts? The answer is no, because the firm has another policy parameter that it can use to increase the effectiveness of price discrimination, namely the quantity of the product. If it is reduced in the L-contract, then both types' willingness to pay for this contract decreases. Therefore, the firm has to complement this change with a decrease in the price for this contract in order to be able to sell it. This seems like a bad idea, because it decreases the profit from the L-contract. The fact that a change of the contract in this direction can increase profits can be seen once one takes into account that the H-type has a higher marginal willingness to pay for additional quantity; he is willing to pay more for the last unit than the L-type is. This fact has the following consequence: the reduction of the quantity that is offered in the L-contract does not only decrease the profit from selling to the L-type, but can also be used to increase the price for the H-contract. The H-types' implicit "threat" to choose the L-contract becomes weaker given that $\{x^H, P^H(x^H)\} = \{x_o^H, A + C\}$. The L-contract becomes less attractive for both, but this effect is stronger for the H-type, whose willingness to pay for additional quality is higher than the L type's is. Thus, this quantity reduction can be used as an instrument for type selection. In the limit, as the L-contract's quantity goes to zero at a price of zero, it becomes possible to increase the H-contract's price to $A + B + C$ again.

Contracts that make the H-type indifferent between both contracts fulfill the so-called *self-selection constraint*. Figure 10.4 shows the possible and necessary price adjustments accompanying a change of the quantity of the L-contract from x_o^L to $x_o^L - dx_o^L$.

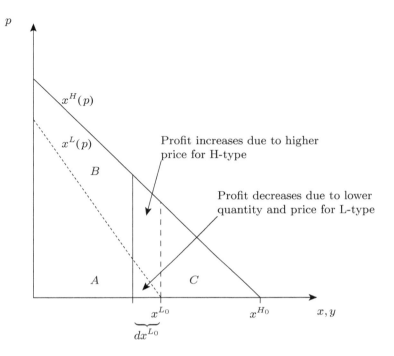

Fig. 10.4 Effect of a reduction in L-quantity

The adjustment of the contract stops when the marginal increase in profits, due to the increase in the price for the H-contract, equals the marginal loss in profits caused by the reduction of the price for the L-contract. Graphically, this means that the line segment between the H-type's and the L-type's demand functions have to be of equal length as the line segment between the L-type's demand function and the abscissa. This situation is depicted in Fig. 10.5.

(In Fig. 10.5, both contracts offer positive amounts of the good. This finding results from the assumption that both types have equal frequencies in the population and the specific demand functions. If there are either very few L-types, or their willingness to pay differs significantly from the H-type's, then it can be the case that the monopolist prefers not to sell to the L-types at all.)

The previous analysis has revealed some general characteristics of optimal price discrimination with asymmetric information.

- The H-type always consumes his optimal quantity, unlike the L-type. This property is also called *no distortion at the top*, because it is a general property of models with asymmetric information to not distort the allocation of the "best" type.
- With such contracts, the H-type always receives positive consumer surplus, if the L-quantity is positive. The L-type always gets a surplus of zero.

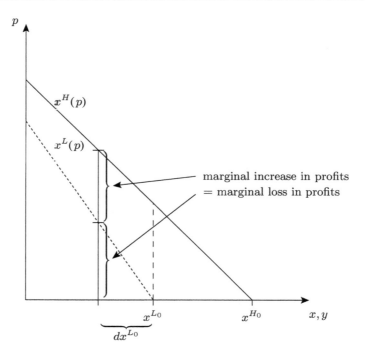

Fig. 10.5 In the optimum, the marginal decrease equals the marginal increase in profits

While these characteristics of optimal contracts sound quite abstract, they are very useful for understanding real-world pricing decisions. In the aforementioned example of airline pricing, business travelers usually have a higher willingness to pay for flexibility than leisure travelers do. In order to apply the model, one can interpret x as a variable that measures the flexibility of a ticket. The results of the model can then be interpreted in the following way: the airline should discriminate between the two groups of customers by offering economy and business class tickets. Business class tickets offer the optimal flexibility and comfort to business travelers, but economy tickets come with less flexibility and comfort than economy customers would like (and are willing to pay for). This reduced flexibility and comfort of economy tickets is the reason that business travelers choose 'their' higher-priced tickets, because they get a larger surplus from doing so. This can lead to strange incentives for airlines that may, for example, make economy seating purposely uncomfortable.

This logic can be applied to many other markets, for example ones in which customers can be grouped as "professional" and "private" users (software, computer hardware, ...). The strategy to play around with product quantity and quality in a way that makes sure that market segments are kept separate, in order to prevent demand spillover from high-priced segments to low-priced segments, is also called *price fencing.* Price fences are very important for effective market segmentation and, therefore, for profit maximization. As in the above example, they are designed such that customers who can afford and are willing to pay higher prices are not tempted by the lower-priced versions.

Looking at the pricing of the MacBook Pro with Retina display (US prices, 2013), there were three versions of the computer, priced $1,299, $1,499 and $1,799. The only difference between the $1,299 and $1,499 versions was the RAM and flash capacity. Assume that one prefers the cheapest version, but would like to have more flash capacity. One option would be to customize the cheapest version by increasing the flash capacity. However, this option was not available. The only way to get more flash capacity was to buy the $1,499.00 version. The only option offered was to increase RAM from 4 GB to 8 GB for $100 more. It could be argued that the lack of such an option resulted from a technical problem, but it is not because the option to upgrade flash capacity was offered for the non-Retina version of the MacBook Pro. Hence, constraining options for customers must be seen as a purposeful strategy to "fence them in."

Sometimes, firms even have an incentive to incur costs to make products worse. Manufacturers of printers, for example, standardly equip their printers with soft- and hardware that is designed for fast printing speed and then equip a series of these printers with (costly) additional hard- and software to slow them down and sell them at a lower price. Thus, if a private user wonders at times why firms do not offer the perfect products for her needs, this may be the answer: from the point of view of a firm, the marginal costs of production are not the only ones it has to take into consideration. Additionally, there are opportunity costs that exist because customers from a different group may buy a version of the product that is not designed for them. These opportunity costs are relevant for the firm, but not for society, which is why there are externalities in the resulting equilibrium.

A related strategy for profit maximizing firms, which should briefly be mentioned, is called *bundling*. The underlying problem for a firm is to decide on the number of characteristics or features of a product. A car manufacturer can, for example, include features like driver alert, adaptive cruise control or other safety features in the standard package, or sell them separately. A flower shop can sell bouquets or separate flowers, a computer can come with word processor, spreadsheet and presentation program, or the soft- and hardware can be sold separately, and so on.

If one thinks of different features of a complex product as separate, simpler products, then the problem of bundling is to determine which products should be included in a bundle and which ones should be sold separately. There are several rationales for bundling. There could, for example, be complementarities between products, which is why shoes are usually sold as pairs. Alternatively, bundling might economize on costs, because bundles are more efficient to produce or distribute.

A subtler reason for bundling results from the fact that it allows the producer to skim off the willingness to pay in situations where it would otherwise be impossible

Table 10.1 An example for product bundling

Customer type	WP	SS	Sum
N	120	100	220
A	100	120	220

to do so. This is why it makes sense to discuss bundling in the context of asymmetric information and second-degree price discrimination.

Here is an example. Assume there are two types of customers in the market, who are interested in two different products: word-processing (WP) and spreadsheet (SS) software. One type of customer has a high willingness to pay for WP and a low willingness to pay for SS (say, a novelist, N), and the other has a high willingness to pay for SS and a low willingness to pay for WP (an accountant, A). Table 10.1 gives an example for the two types of customers' willingness to pay for the two different products. Type N is willing to pay up to CHF 120 for WP and CHF 100 for SS, and type A is willing to pay up to CHF 100 for WP and CHF 120 for SS. Hence, the total willingness to pay for both products is CHF 220 for both types.

Assume that the firm that sells the software knows that these two types of customers exist (one of each type), but that it cannot verify the identity of a customer when she buys the software. Assume further that the marginal costs of an additional software license are zero. Now, consider two pricing strategies for the firm: the unbundled and bundled selling of the two products.

What happens if the firm sells both products separately? I start with WP and denote its price by p^{WP}. If $p^{WP} \leq 100$, the firm can sell two licenses. If $p^{WP} \in (100, 120]$, it can sell one license and, for all prices $p^{WP} > 120$, the number of licenses sold drops to zero. Hence, the profit-maximizing price is $p^{WP} = 100$, yielding a profit of $\pi = 200$. The same calculation applies for SS. Therefore, total firm profits with unbundled selling are $\pi = 400$.

What happens if the firm decides to bundle the products? Denote the price of a bundle by p^B. Demand for each bundle is two, if $p^B \leq 220$, and drops to zero for higher prices. Hence, the profit-maximizing price for each bundle is equal to $p^B = 220$, yielding profits of $\pi = 440$.

Compared to the unbundled selling, bundling increases profits by $40 = 440 - 400$. What has happened? The underlying rationale is that differences in the willingness to pay "average out" by bundling. If the licenses are sold separately, the minimum willingness to pay becomes decisive in the example. This effect cancels out, if the two products are bundled and sold as a "package."

This result is robust and especially relevant for digital products that are produced at almost zero marginal costs: bundling large numbers of unrelated goods makes it easier to predict the customers' valuations for a bundle than their valuation for an individual good does when it is sold separately. This "predictive value of bundling" makes it possible to increase sales and profits. Examples are cable television, an internet site's content (e.g. the New York Times), or copyrighted music (for example Spotify).

To conclude this subchapter, it is important to note that the above findings also $\mathcal{B}$
have implications for the optimal organization of firms. As seen, there are different contracts for different groups of customers. The two dimensions of the optimal
contracts, which are price and – depending on the specific interpretation – quantity or quality, are not independent from each other, but can only be understood in
combination. Hence, to take these important interdependencies into account, the
responsibility for the different customer groups should not be given to different,
independent product managers whose responsibility is to maximize profits for their
departments (profit centers). This system ignores the fact that modifications of the
contracts cause externalities in the other departments thereby leads to a situation
where each manager maximizes the profits of his profit center, but not the total firm
profits.

10.5.3   Third-Degree Price Discrimination

The last case is third-degree price discrimination. This variant is characterized
by the firm's ability to discriminate between different segments of customers, but
not within each segment. A prominent example is price discrimination between
national markets, which is often practiced by internationally operating firms. Especially the pharmaceutical industry repeatedly makes it into the headlines for selling
the same active ingredients at higher prices in Switzerland than in, for example,
the European Union. However, prices for ordinary consumption goods are also discriminated in this way and Swiss customers quite often pay more for a good than
others do. Apple Inc., for example, makes extensive use of international price discrimination. The average price for a song in the iTunes store was $1.29 in the US
and CHF 2.30 in Switzerland in 2011 (the average exchange rate in 2011 was about
US$ 1.10 per 1 CHF). However, there are many other forms of third-degree price
discrimination, for example according to age group or status (student or senior discounts, discounts for military members in the US) or according to gender ("Ladies'
night" in nightclubs or at the dry cleaners, which typically charge higher prices for
women's clothes).

Assume that a firm produces a given product at a given production facility (say
in China) and sells it to two countries, Switzerland (country 1) and France (country
2). The respective quantities are y^1 and y^2, and the production and logistics costs
depend on the total quantity produced, $C(y^1 + y^2)$. The market-research department estimates the demand functions in the two countries as $P^1(y^1)$ and $P^2(y^2)$.
Consequently, total profits are given by:

$$\pi(y^1, y^2) = P^1(y^1) \cdot y^1 + P^2(y^2) \cdot y^2 - C(y^1 + y^2).$$

From the manager's point of view, the problem is to choose the quantities supplied
to the different markets in order to maximize profits. The optimal decision is char-

acterized by the following necessary conditions:

$$\frac{\partial \pi(y^1, y^2)}{\partial y^1} = \frac{\partial P^1(y^1)}{\partial y^1} \cdot y^1 + P^1(y^1) \cdot 1 - \frac{\partial C(y^1 + y^2)}{\partial y^1} = 0,$$

$$\frac{\partial \pi(y^1, y^2)}{\partial y^2} = \frac{\partial P^2(y^2)}{\partial y^2} \cdot y^2 + P^2(y^2) \cdot 1 - \frac{\partial C(y^1 + y^2)}{\partial y^2} = 0.$$

If one looks at the two conditions in isolation, the result is not very surprising: as in the model without price discrimination, the firm chooses the quantity that equalizes marginal revenues with marginal costs for each market. Only if one takes into account the fact that marginal costs are identical irrespective of the market where the products are sold (production takes place in the same factory) can one learn something new. Then one can establish the following relationship between the two markets:

$$\frac{\partial P^1(y^1)}{\partial y^1} \cdot y^1 + P^1(y^1) = \frac{\partial P^2(y^2)}{\partial y^2} \cdot y^2 + P^2(y^2).$$

Thus far, the above condition only states that marginal revenues are equal in both markets. However, rewriting the equation to transform it into the rule of thumb that was developed before, one gets:

$$p^1 \left(\frac{1}{\epsilon_{p^1}^{y^1}(p^1)} + 1 \right) = p^2 \left(\frac{1}{\epsilon_{p^2}^{y^2}(p^2)} + 1 \right).$$

Further assuming that demand is falling with respect to price in both markets (which implies that the elasticities are negative), one ends up with:

$$p^1 \left(1 - \frac{1}{|\epsilon_{p^1}^{y^1}(p^1)|} \right) = p^2 \left(1 - \frac{1}{|\epsilon_{p^2}^{y^2}(p^2)|} \right).$$

In order to understand the economic reasoning underlying this condition, assume that the price elasticity in market 1 (Switzerland) is lower than the price elasticity in market 2 (France), $|\epsilon_{p^1}^{y^1}(p^1)| < |\epsilon_{p^2}^{y^2}(p^2)|$ (Swiss demand is less elastic). This implies that the expression in brackets is smaller in market 1 than in market 2. Hence, the condition can only be fulfilled if $p^1 > p^2$: the good is sold at a lower price in the market with the higher price elasticity.

To further illustrate this condition, assume that demand in both markets is linear, $p^i(y^i) = a^i - b^i \cdot y^i, i = 1, 2$, and that marginal costs are constant and equal to $c > 0$. In this case, one knows from above that:

$$y^{i*} = \frac{a^i - c}{2 \cdot b^i}, \quad p^{i*} = \frac{a^i + c}{2}, \quad \epsilon_{p^i}^{y^i}(p^{i*}) = \frac{a^i + c}{a^i - c}, \quad i = 1, 2.$$

Comparing the elasticities between both markets reveals that:

$$\epsilon^{y^1}_{p^1}(p^{1*}) > \epsilon^{y^2}_{p^2}(p^{2*}) \Leftrightarrow \frac{a^1+c}{a^1-c} > \frac{a^2+c}{a^2-c} \Leftrightarrow a^1 > a^2.$$

However, this is the case if and only if $p^{1*} = \frac{a^1+c}{2} > \frac{a^2+c}{2} = p^{2*}$: the price on the less elastic (in equilibrium) market is higher.

This result gives an important hint as to why prices in Switzerland are generally higher than abroad. The willingness to pay (as reflected by a low price elasticity) is higher in Switzerland than elsewhere, which implies that firms sell their products at higher prices. One can use this theoretical result to test the theory empirically. All one needs to do is to estimate the price elasticities in different markets and compare them with prices. If the hypothesis of the model cannot be rejected, then one has a valid explanation for an important empirical phenomenon.

Without going into the analytical details, it is time to contemplate the consequences of a regulation that forbids price discrimination between markets. Such a regulation might, for example, prevent price discrimination directly, or it might allow the emergence of resale markets that make profit out of price arbitrage between markets. Such a "single-price philosophy" can, for example, be found in the European Union with its "Single European Market," which is enforced by the European Commission.

Taking the theoretical results from above as a point of departure, the monopolist, who is no longer in a position to discriminate prices, needs to determine the new, aggregate demand function for the joint market. This new demand function results from adding up the individual market demand functions, $X(p) = x^1(p) + x^2(p)$, and it follows that the new inverse demand function is $P(x) = X^{-1}(x)$. The resulting problem is equivalent to the problem of a monopolist who cannot discriminate prices. Hence, even without a formal analysis of the situation, one can determine the differences between the new and the old situation. There are three different constellations possible:

1. The demand structure is similar in both countries and markets are of approximately the same size. The monopolist will sell to customers in both markets and the new price will be in between the prices that would be charged with price discrimination. The redistributive consequences between the customers in both markets are easy to determine: customers in the previously high-price market profit, because they can buy at lower prices, and more customers buy, while those in the previously low-price market lose, because they are paying higher prices and fewer customers buy.

2. The country with less elastic demand is relatively large or the price difference between countries is large, or both. In this case, it can happen that the monopolist will not serve the smaller market anymore. The reason is that, in order to sell to the smaller market, he must lower the price in the larger market to an extent that makes it rational to not serve the more elastic, or smaller market, at all. The effect is that nothing changes for the large country, but the situation in the

smaller market deteriorates, because its customers are excluded from the consumption of the product. Therefore, prohibiting price discrimination in different markets may lead to inefficiency and exclude customers from consumption.

3. The country with higher prices is relatively small or the price difference is relatively small, or both. In this case, the price in the more expensive country might decrease almost to the level of the other country. There is no change for the formerly low-price country, whereas customers in the formerly high-price country benefit.

This qualitative analysis shows that no clear prediction about the consequences of market integration can be made without further information about the relative size of the markets and the relative willingness to pay in each market. Only if this kind of information is available, is it possible to make a reliable prognosis about the effects of such a policy change.

$\mathcal{L}$ **Digression 29. Parallel Imports in the European Union (EU)**

Firms that sell in different markets have an incentive to discriminate prices according to the market-specific elasticities of demand. However, the creation of a common market within the EU has made it possible for parallel imports to move freely across the EU. Parallel imports are sales by authorized or unauthorized distributors to another country without the permission of the initial property owner.

Parallel importers use price differences between markets to make a profit out of price arbitrage. This puts pressure on high prices and, thereby, creates a tendency towards uniform prices within a common market. The only industry-specific exemption from the general competition principles is the automotive industry. The purpose of the so-called "block exemptions" is to restrict competition between car dealers. Nevertheless, even with these special agreements, EU rules require the car dealers to sell their products to any EU citizen regardless of where they live.

This regulation is, of course, a thorn in the side of car manufacturers, who try to find ways to limit competition due to parallel imports. The European Commission fined Volkswagen an amount of €102 million (later reduced to €90 million) for preventing Austrian and German customers from buying cars in Italy. It also fined PSA Peugeot Citroen €49.5 million. Peugeot Netherlands tried, for example, to incentivize its franchise dealers to restrict sales to other countries by withholding bonus payments and limiting the supply of Peugeot cars.

10.6 Monopolistic Competition

I have argued that the models of monopolistic behavior apply to situations where a firm faces a demand function that is not perfectly elastic. This situation allows the firm to charge prices that are above marginal costs. The associated producer surplus is higher than in a situation of perfect competition. This has two consequences.

First, if production involves fixed costs, the firm can stay profitable in situations where firms in competitive markets would have to leave the market, as long as prices are above average costs. Such a situation is illustrated in Fig. 10.6, where it is assumed that the monopolist can not discriminate prices.

In this figure, average costs are above marginal costs, but below the price of a non-price-discriminating monopolist. This leads to a situation where a competitive firm makes a loss equal to area A, whereas a monopolist makes a profit that is equal to area B.

Second, positive profits in a monopolistic market, as in Fig. 10.6, make it attractive for other firms to develop similar products. Even if these firms are legally or otherwise prevented from simply imitating the profitable product, they can try to develop and sell similar ones. Such a situation is called *monopolistic competition* and there are several examples for such industries:

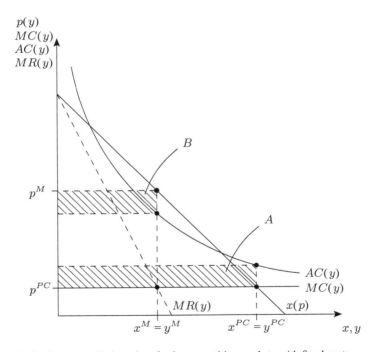

Fig. 10.6 Profits in monopolistic and perfectly competitive markets with fixed costs

- Cars of a given type from different manufacturers (like SUVs from Audi, Mercedes, BMW, Volkswagen, etc.).
- Books or music that are variations of the same topic (romantic novels or textbooks in economics) or style (Jazz, Pop, Classical Music).
- Smartphones, tablets or notebooks from different manufacturers.
- Pubs and restaurants in a city.

The above list illustrates that monopolistic competition is a very prominent market structure, especially in an economy where brands are important for customers (which is the same as saying that they are willing to pay for a specific brand). This is why it makes sense to understand the functioning of this type of a market. The different varieties of similar goods are called *differentiated products*.

The main question is about the number of similar products that exist in such an industry. When one compares the total number of different SUVs with the total number of different romantic novels, one sees that there are huge differences. Are there any patterns that allow one to explain why some industries produce a relatively small number of variants, whereas others produce far more?

The basic idea for answering this question is to blend the analysis of a single monopoly with the idea of market entry: assume there is free market entry and exit and that a monopolist makes a profit with a product, say an SUV from a given manufacturer A. Profits exist, if the price exceeds the average costs, $p^A > AC^A(y^A)$.

These profits encourage another firm, B, to enter the market and to sell a similar product. The availability of this additional product increases the choice of the customers. They still consider the products to be different, but the existence of another model in the SUV market makes the first one less exclusive and manufacturers A and B have to somehow share the market. The effect is that the demand for SUV A is likely to shift leftward and to become more price elastic, which reduces profits. With free entry and exit, additional firms will enter the market as long as profits are still positive.

By the same token, if the number of different products is so large that the firms (for example i) are making losses, $p^i < AC^i(y^i)$, some of them will have to leave the market. The effect is that the number of products from which the customers can choose decreases. This effect likely shifts the demand rightward and makes it less price elastic.

The long-run equilibrium must, therefore, be a situation where the prices of the products equal average costs, $p^i = AC^i(y^i)$, because at this point firms make zero profits. This situation is illustrated in Fig. 10.7.

This is the situation where no further competitor is willing to sell another similar product and no existing competitor is willing to leave the market. The figure reveals two properties of the long-run equilibrium in such a market.

First, if firms can develop similar products, there is no escape from the zero-profit equilibrium in the long run. As in the case of perfect competition, competitors will react to positive profits by entering the market with differentiated products.

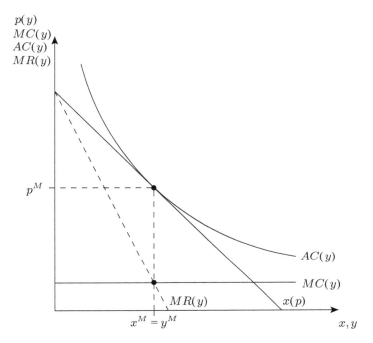

Fig. 10.7 Long-run equilibrium in a market with monopolistic competition with free entry and exit

Second, even though long-run profits are zero, the resulting allocation is not Pareto-efficient. The single firm is still facing a downward-sloping demand function for its product, so prices in this industry will be above marginal costs.

In addition, one can say something about the number of differentiated products that can survive in such an industry. Figure 10.6 shows a situation where the monopolist has profits π_M. Profits depend on the relationship between the demand and the average-cost curve. The bigger the difference is, the higher the profits are. Now, assume that this firm is the first one to sell a new type of product (the first SUV), such that Fig. 10.6 refers to a situation where no other firm has entered the market with a similar product.

Profits depend on fixed costs. An increase in fixed costs shifts the average-cost curve upwards, which implies that profits decline. In the extreme case, profits are equal to zero without any competition from differentiated products. If a second firm enters the market with a similar product, any leftward shift of the demand function that is caused by market entry implies losses for the firm. If profits had been positive but relatively small, the shift in the demand function reduces them, but not necessarily to zero, which would imply that there is room for a second firm selling a similar product. The pattern that becomes visible here is a general one:

the more fixed-cost-intensive the production of the product is, the fewer firms can succeed in the market; the number of differentiated products is inversely related to the fixed costs of production.

The above argument was pretty loose, so it makes sense to develop it formally. In order to do so, assume that there are n differentiated products in a market and that demand for a single product is given by the demand function $y^i = Y(1/n - b(p^i - \bar{p}))$, where $\bar{p} = \sum_{j=1}^{n} p^j / n$ is the average price level and $Y = \sum_{j=1}^{n} y^j$ is the total output in the industry. Here, b represents the responsiveness of a firm's output to its price. The demand function implies that the n different firms share the market equally, if they charge identical prices: $p^i = \bar{p}$ implies $y^i = Y/n$. When we solve the model we will see that this function implies that individual and market demand are not absolutely fixed so that we have a degree of freedom in determining the equilibrium. Assume further that all firms produce with identical cost functions $C(y^i) = c \cdot y^i + FC$ and maximize profits. These assumptions may not be particularly realistic, but they simplify the analysis considerably without changing the qualitative insights.

From the point of view of a single firm, the inverse demand function is given as $P(y^i) = 1/(b \cdot n) + \bar{p} - y^i/(b \cdot Y)$, which leads to profits as a function of output:

$$\pi(y^i) = P(y^i) \cdot y^i - c \cdot y^i - FC = \left(\frac{1}{b \cdot n} + \bar{p} - \frac{y^i}{b \cdot Y} \right) \cdot y^i - c \cdot y^i - FC.$$

The profit-maximizing output of firm i is, again, characterized by the first-order condition of the profit function. (In general, Y and $\bar{p}$ are functions of y^i. Assume that the firm neglects this effect.) If one does the math, one ends up with the following price and output of product variant i:

$$y^{i*} = \frac{bY \left(1/(b \cdot n) + \bar{p} - c \right)}{2}, \qquad p^{i*} = \frac{1/(b \cdot n) + \bar{p} + c}{2}.$$

The solution has intuitive economic properties: the profit-maximizing output and price of variant i is decreasing as the number of variants n increases. This property illustrates the effect of competition on the market for product variant i: the larger the number of similar products, the fewer the number of products of a given variant that can be sold and the lower the price level for this variant.

What is even more interesting is if one can say something about the number of differentiated products that can be supplied in this market. In order to gain insight into this question, assume that all firms charge equal prices in equilibrium, $p^{i*} = \bar{p} = p^*$. If one uses this assumption, one can solve for the equilibrium price level in this industry for a given number of differentiated products n:

$$p^* = \frac{1/(b \cdot n) + p^* + c}{2} \quad \Leftrightarrow \quad p^* = \frac{1}{b \cdot n} + c.$$

This finding nicely illustrates that the markup rule still applies: the firm is able to sell the variant above marginal costs c, but the markup is the lower, the higher the competitive pressure is that results from the number of similar products n.

If we use the information that all prices in the industry are the same, we get back to the above-mentioned property that individual and market demand are not fixed in absolute terms such that we have a degree of freedom that we can use to fix one of these variables and solve the rest of the model relative to it. We use $y^{i*} = 1$ as normalization because it is easy to solve, but any other convention would work as well:

$$y^{i*} = \frac{bY\left(1/(b \cdot n) + 1/(b \cdot n) + c - c\right)}{2} = \frac{Y}{n} = 1.$$

Hence, the total supply of variant i is equal to 1, and industry output is, therefore, equal to n, the number of firms in the industry. One last step is missing to determine how many differentiated products exist in the long run. Given the equilibrium outputs and prices, profits of an arbitrary firm i are equal to

$$\pi^{i*} = \left(\frac{1}{b \cdot n} + c\right) \cdot 1 - c \cdot 1 - FC.$$

This equation, again, shows the effects of competition: equilibrium profits are decreasing with an increasing number of differentiated products.

One knows that free entry and exit into this industry drive profits down to zero,

$$\left(\frac{1}{b \cdot n} + c\right) \cdot 1 - c \cdot 1 - FC = 0.$$

This information can be used to finally determine the long-run number of differentiated products:

$$\left(\frac{1}{b \cdot n} + c\right) \cdot 1 - c \cdot 1 - FC = 0 \quad \Leftrightarrow \quad n = \frac{1}{b \cdot FC}.$$

This result confirms the intuitive conclusion that I have discussed before: there is a negative relationship between the fixed costs of an industry and the number of differentiated products that exist in a long-run equilibrium. This result sheds light on the question of why there are fewer SUVs than romantic novels on the market: product categories differ with respect to fixed costs. Writing a beach novel is far less expensive than developing a new car. (Both, the costs of writing the novel and the development costs, are part of the technological fixed costs, because they occur independently of the number of copies or cars sold.)

The result also contains an important message for managers in an industry with differentiated products, because profits for a given number of competitors and the long-run number of competitors allows estimations as to whether market entry is still profitable and how many variants a product can survive in the long run.

$\mathcal{B}$

References

Ariely, D. (2008). *Predictably Irrational*. Harper Collins.
Boublil, A., & Michel-Schonberg, C. (2013). *Les Misérables*. Wise Publication.
Calo, R. (2013). Digital Market Manipulation. *George Washington Law Review* 2013(27), 995–1051.
Executive Office President of the United States (2015). Big Data and Differential Pricing. Report.
Hannak, A. (2014). Measuring Price Discrimination and Steering on E-Commerce Web Sites. In *Proceedings of the 2014 Conference on Internet Measurement Conference*.
Marn, M., Roegner, E., & Zawada, C. (2003). Pricing New Products. *The McKinsey Quartely, 3*, 40–49.
Monroe, K., & Cox, J. (2001). Pricing Practices that Endanger Profits. *Marketing Management, 10*(3), 42–46.
Poundstone, W. (2011). *Priceless: The Myth of Fair Value (and How to Take Advantage of It)*. Simon and Schuster.
Puzo, M. (1969). *The Godfather*. Signet.
Shiller, B. (2014). *First-Degree Price Discrimination Using Big Data*. Working Paper.

Further Reading

Allen, W. B. (2009). *Managerial Economics Theory, Applications, and Cases*. Norton.
Frank, R. H. (2008). *Microeconomics and Behavior*. McGraw-Hill.
Haisley, E., Mostafa, R., & Loewenstein, G. (2008). Subjective Relative Income and Lottery Ticket Purchases. *Journal of Behavioral Decision Making, 21*, 283–295.
Pindyck, R., & Rubinfeld, D. (2001). *Microeconomics*. Prentice-Hall.
Tirole, J. (1988). *The Theory of Industrial Organization*. MIT Press.
Varian, H. R. (1992). *Microeconomic Analysis*. Norton.

Principles of Game Theory

11

This chapter covers ...

- game theory as a mathematical method for analyzing situations of strategic inter-dependence.
- the basic definitions and solution concepts of games.
- how games can be used to analyze complex social interactions.
- how games can be used to help one understand real-world problems, like the decision of firms to enter a market, the economic mechanisms underlying climate change, the political incentives to engage in tax competition, etc.

11.1 Introduction

> I am willing to take life as a game of chess in which the first rules are not open to discussion. No one asks why the knight is allowed his eccentric hop, why the castle may only go straight and the bishop obliquely. These things are to be accepted, and with these rules the game must be played: it is foolish to complain of them. (W. Somerset Maugham (1949))

Game theory addresses the analysis of strategic interdependencies between the actions of different decision makers. Many of the early contributions to game theory dealt with the analysis of parlor games. Two of the most important works on game theory of the early 20th century were Ernst Zermelo's "Über eine Anwendung der Mengenlehre auf die Theorie des Schachspiels" (On an Application of Set Theory to the Theory of the Game of Chess, 1913) and John von Neumann's "Zur Theorie der Gesellschaftsspiele" (On the Theory of Parlor Games, 1928). The term *game* obtained a totally different meaning, referring to all kinds of situations where individuals interact. Game theory is, by now, an indispensable tool in many scientific disciplines, apart from economics, mostly in political science and finance, but also in biology, law and philosophy. The goals of game theory are various and reach from explaining societal phenomena, to predicting individual decision making, to providing consultation.

© Springer International Publishing AG 2017
M. Kolmar, *Principles of Microeconomics*, Springer Texts in Business and Economics,
DOI 10.1007/978-3-319-57589-6_11

$\mathcal{B}$ Many economists would even state that it was the development of game theory that made economics a scientific discipline of its own. Traditional economic reasoning has been based on more or less informal theories or on models, which were frequently adapted from physics. The analysis of imperfect competition on markets, bargaining, conflicts between individuals and groups, and competition between states are only some of the fields in which techniques from game theory are successfully applied. In the course of the appearance of game theory, economists developed a subfield called "market design." This discipline has strong similarities to engineering, because it applies scientific theories to design market mechanisms that help to facilitate structured transactions and improve efficiency. Market design became known to a wider audience because of the auctioning of UMTS telecommunications licenses, for which many countries used auction formats that had been developed by economists using methods from game theory. Another example for market design is the development of algorithms facilitating organ donations and labor markets. For instance, in some countries, they are used to facilitate kidney exchanges or to allocate doctors to hospitals. Game theory is also the backbone of behavioral economics, a field of research in which economists study the structure of cooperative behavior and the limitations of rational decision making.

A classic field of study within economics, in which game theory is often used, is the analysis of oligopoly markets. The central characteristic of those markets is that firms have some control over prices, similar to a monopolist, but the existence of competitors restricts this control and makes the optimal price and quantity decision dependent on each firm's expectations about the behavior of the competitors and, hence, the decisions are interdependent. In order to study these interdependencies and, thus, be able to make predictions about these market types (as well as many other societal phenomena), I give a short introduction to game theory in this chapter.

11.2   What Is a Game?

A "game" describes a situation of strategic interdependence. The involved decision makers, for example individuals and firms, are called "players." Strategic interdependence in the game-theoretic sense is given, if the actions of players potentially influence one another. Here is a simple example that illustrates the problem. Assume there were no rules about on which side of the road one should drive. Two cars, moving in opposite directions, meet. The driver of each car wants to go on driving without an accident. If both stay on their right or left lane, then no harm is done. However, if one goes left and the other one goes right, then the result is an accident. Probably, everybody knows similar situations from crowded market places and sidewalks, where people go in opposite directions and try to avoid bumping into each other. A good way to navigate through the crowds depends on the actions of all the others and, hence, this is a situation of strategic interdependence.

Another illustrative example is the "rock-paper-scissors" (RPS) game. Two players face each other and have to choose one of the following gestures: 'rock', 'paper' or 'scissors.' 'Rock' beats 'scissors', 'scissors' beats 'paper', and 'paper' beats 'rock'. If both players choose the same gesture, then nobody wins and the game ends in a draw. If a player wants to win the game, then his optimal gesture depends on the gesture of his opponent. If the other player chooses 'rock', then one optimally chooses 'paper'; if the other chooses 'scissors,' then 'rock' would be the optimal gesture.

11.3 Elements of Game Theory

In order to analyze a situation of strategic interdependence using game theory, it is necessary to (1) systematically describe a game, (2) to form hypotheses about players' behavior and (3) to apply a so-called solution concept.

The description of a game Γ usually starts with listing the involved decision makers, the *players*. The set $N = \{1, 2, \ldots, n\}$ denotes all $n \geq 1$ players involved in a game. In the example of RPS, this set is $N^{\text{RPS}} = \{1, 2\}$, i.e. the set listing players 1 and 2.

Next, one has to specify what the players can do. An action of a player is called a *strategy*. The set of all, m_i, possible strategies of a player, $i \in N$, is denoted by $S_i = \{s_i^1, s_i^2, \ldots, s_i^{m_i}\}$, and s_i^j, $j \in \{1, 2, \ldots, m_i\}$ denotes one specific strategy from this set. In the example of RPS, both players have the same strategy sets: $S_1^{\text{RPS}} = S_2^{\text{RPS}} = \{\text{rock}, \text{paper}, \text{scissors}\}$.

A *strategy profile* assigns a strategy to each player and is denoted by $s \in S = S_1 \times S_2 \times \cdots \times S_n$ (the mathematical operator '×' refers to the Cartesian product of the sets S_i, $i = 1, 2, \ldots, n$). S is the set of all possible strategy profiles. In RPS, it is equal to the set of all combinations of the form $(s_1 \in S_1^{\text{RPS}}, s_2 \in S_2^{\text{RPS}})$, which is the set with the elements (scissors, scissors), (rock, scissors), (paper, scissors), (scissors, rock), (rock, rock), (paper, rock), (scissors, paper), (rock, paper), and (paper, paper).

Each strategy profile is a possible course of the game. Starting from the strategy profiles $s \in S$, one can determine the possible outcomes of a game induced by the different profiles. The outcome function, $f : S \to E$, assigns to each strategy profile $s \in S$ an outcome e, from the set of potential outcomes E. In the example of RPS, the set of possible outcomes is $E^{\text{RPS}} = \{\text{player 1 wins, player 2 wins, draw}\}$. The function $f(s)$ determines an outcome $e \in E$ for every $s \in S$. For example if, in RPS, the strategy profile is $s = (\text{scissors}, \text{rock})$, then the function $f(s)$ determines the outcome '$e = $ player 2 wins.' If $s = (\text{paper}, \text{paper})$, then the outcome is '$e = $ draw.'

Finally, in order to be able to determine what the players will do, one has to connect the outcome of a game with the players' evaluations of this outcome. The functions $u_i(e)$ assign an evaluation for each player and for each possible outcome, namely $u_i : E \to \mathbf{R}$ for player i. Economists use the convention that larger numbers are assigned to preferred outcomes. This convention suggests that one calls u_i player i's *utility*. In RPS, if one assumes that players prefer winning to having a draw, and having a draw to losing, then any assignment of numbers to outcomes with the following property is consistent with this evaluation:

$$u_i(\text{player } i \text{ wins}) > u_i(\text{draw}) > u_i(\text{player } i \text{ loses}).$$

For example, each player could assign 1 to the outcome 'win,' 0 to the outcome 'draw,' and -1 to the outcome 'lose.'

The above elements describe a game and can be summarized in the following way:

$$\Gamma = \{N, S, f, \{u_i\}_{i=1,\dots,N}\}.$$

It is often quite useful to sidestep the somewhat lengthy definition by directly conditioning the players' utilities on strategy profiles instead of outcomes. This is possible, because a strategy profile determines an outcome, which in turn determines utilities: $S \to E \to \mathbf{R}$. One can therefore skip the step in the middle and assign utilities directly to strategy profiles: $u_i : S \to \mathbf{R}$. This shortens the description of a game and one gets:

$$\Gamma' = \{N, S, \{u_i\}_{i=1,\dots,N}\}.$$

This representation of a game will be used in the remainder of this chapter. However, it loses some of the societal content of the situation that is being analyzed: one no longer knows why players prefer this strategy over that strategy. This is not relevant from a technical point of view, but it may be important for understanding the social context that is represented by the game. In RPS, one only knows that a player prefers (rock, scissors) to (scissors, rock), if one specifies Γ'. The more lengthy specification Γ allows one to answer why this is the case: because the player wins with the first strategy profile and loses with the second.

In order to be able to make predictions about the way players play the game, one needs a hypothesis about the players' behavior and the way this behavior is coordinated. Usually, economists work with the so-called (expected) utility-maximization hypothesis, which states that each player chooses a strategy to maximize her (expected) utility. If the players are competing firms and if utility can be identified with profits, then the already familiar profit-maximization hypothesis is an example. However, altruistic or even malevolent motives can also be taken into account, if one uses the more general concept of utility. For example an altruistic player prefers a distribution of profits (5,5) to a distribution of profits (10,0), whereas a profit maximizer always prefers (10,0), irrespective of the other player's profits.

Knowing this, one can assign the optimal reaction of a player to the strategies of the other players. This information is contained in the so-called *reaction function*. Let $s_i \in S_i$ denote a strategy of each player, $i \in N$, and let the strategy profile of all players except i be denoted by $s_{-i} \in S_{-i}$ ('$-i$' refers to the set of all players except i). Player i's best responses to the other players' strategy profile s_{-i} specifies the subset of strategies that maximize player i's utility, given strategy profile s_{-i}. The reaction function of player i collects this player's best responses to all possible strategy profiles of the other players. The idea can again be exemplified by using

RPS. If player 2 chooses 'scissors', then the best response of player 1 is to choose 'rock'. If player 2 chooses 'rock', then player 1's best response is 'paper'.

▶ **Definition 11.1, Reaction function** A strategy, $s_i^* \in S_i$, that maximizes a player's utility, $u_i(s_i, s_{-i})$, given the strategies of all other players, $s_{-i} \in S_{-i}$, is called his or her *best response* to s_{-i}:

$$u_i(s_i^*, s_{-i}) \geq u_i(s_i, s_{-i}) \text{ for all } s_i \in S_i.$$

A function that specifies a best response for all possible strategy profiles of all other players is called the *reaction function* of player i.

The concepts 'best response' and 'reaction function' are convenient for solving a game. A particular kind of best response is called a *dominant strategy*, which means that a strategy is a best response to all the other players' possible strategy profiles:

▶ **Definition 11.2, Dominant strategy** A strategy, $s_i^d \in S_i$, is called a *dominant strategy*, if it is a best response to all possible strategy profiles, $s_{-i} \in S_{-i}$:

$$u_i(s_i^d, s_{-i}) \geq u_i(s_i, s_{-i}) \text{ for all } s_i \in S_i \text{ and for all } s_{-i} \in S_{-i}.$$

If a player has a dominant strategy, then her best response is the same for all s_{-i}. Therefore, a dominant strategy is a borderline case of strategic interdependence, because the strategies of all other players, s_{-i}, may influence the utility of player i, but do not impact which strategy she optimally chooses. However, dominant strategies often do not exist, as in the example of RPS.

11.4 Normal-Form Games

In a so-called normal-form game, all players choose their strategies simultaneously and are not allowed to alter them during the course of the game. If there are only two players with only a few strategies, then a normal-form game can be represented in matrix form, see Table 11.1 for an example. The m_1 strategies of player 1 are represented by the different rows of the matrix, and the m_2 strategies of player 2 are represented by the different columns. Each field of the matrix represents a strategy profile and displays the corresponding utility levels. For example, $u_2(s_1^2, s_2^{m_2})$ is player 2's utility level from the strategy profile $(s_1^2, s_2^{m_2})$, which is implemented, if player 1 chooses strategy 2 and player 2 chooses strategy m_2.

The best-response function tells us what each player is expected to do when confronted with the other players' strategy profiles. What is not known, at this point, is how these best responses are coordinated. In order to be able to make predictions about the way people are playing games, one has to make an assumption about how they coordinate their behavior. Such an assumption is called an *equilibrium*

Table 11.1 Matrix representation of a game

	s_2^1	...	$s_2^{m_2}$
s_1^1	$u_1(s_1^1, s_2^1), u_2(s_1^1, s_2^1)$	...	$u_1(s_1^1, s_2^{m_2}), u_2(s_1^1, s_2^{m_2})$
s_1^2	$u_1(s_1^2, s_2^1), u_2(s_1^2, s_2^1)$	...	$u_1(s_1^2, s_2^{m_2}), u_2(s_1^2, s_2^{m_2})$
$\vdots$	...	...	...
$s_1^{m_1}$	$u_1(s_1^{m_1}, s_2^1), u_2(s_1^{m_1}, s_2^1)$	...	$u_1(s_1^{m_1}, s_2^{m_2}), u_2(s_1^{m_1}, s_2^{m_2})$

concept. The most important equilibrium concept for normal-form games is called a Nash equilibrium, which is named after the US mathematician John F. Nash. A Nash equilibrium is defined in the following way:

▶ **Definition 11.3: Nash equilibrium** A strategy profile, $s^{ne} = \{s_1^{ne}, \ldots, s_n^{ne}\}$, is called Nash equilibrium, if the strategies of all the players are best responses to the equilibrium strategies of all the other players:

$$u_i(s_i^{ne}, s_{-i}^{ne}) \geq u_i(s_i, s_{-i}^{ne}) \text{ for all } s_i \in S_i \text{ and for all } i \in N.$$

The idea behind a Nash equilibrium is relatively easy to grasp. Assume there are two players. Player 1 has two strategies, going to the movies or going to a bar, and player 2 has two strategies, as well: going to the movies or going to a bar. Each player i assumes that the other player will stick to his strategy no matter what player i does. This allows i to determine reaction functions (in which they treat the other players' strategies as parameters).

However, the players do not only have to figure out what they, but also what the other player will do. Assume that it is the best response of player 1 to go to the movies, if player 2 goes to the movies, and to go to a bar, if player 2 goes to a bar (he wants to meet the other player). What should player 1 do? In order to answer this question, he has to get into the head of player 2. Assume that player 2 will go to the movies no matter what 1 is doing. Then, player 1 knows that he should go to the movies if player 2 does so, and that player 2 will go to the movies no matter what: thus, the best responses are mutually consistent. The conjecture that player 2 will go to the movies induces player 1 to go there as well and it is a best response of player 2 to stick to his plan. This mutual consistency is the missing link between individual reaction functions and the outcome of the game. A Nash equilibrium is nothing more than such a consistency condition. To see why, focus on the other possible conjecture that player 1 could make, namely that player 2 will go to a bar. In that case, the best response would be to go to a bar, as well, to which player 2 reacts by going to the movies, which is not consistent with the conjecture that player 2 will go to the bar.

The above argument shows that the players have to be able to figure out the planned equilibrium strategies of the other players and that they have to believe that deviations in their own strategy will not cause deviations by any other player (which is why they can treat their strategies as parameters). However, this is not all. At this point a player can figure out his or her best strategy for some strategy profile of the other players and also the best strategies of the other players for some given strategy profile. What is missing is that the players know that the other players will use this logic to solve the game and furthermore that the players know that the other players know that they will use this logic, and so on. The term *common knowledge* refers to a situation where the players have this special kind of knowledge about the beliefs of the other agents.

There is common knowledge of some state, z, in a group of players, N, if all players in N know z, they all know that they know z, they all know that they all know that they know z, and so on *ad infinitum*. The next digression illustrates why common knowledge is important.

Digression 30. A Tale about the Importance of Common Knowledge
On an island in the South Seas, there live 100 blue-eyed persons. The rest have a different eye color. They are perfect logicians and never talk about eye color. An old custom, to which all citizens adhere, demands that, as soon as a citizen knows that he or she has blue eyes, he or she will leave the island during the subsequent night. However, because the citizens never talk about their eye color and because there is no reflecting surface on the island, no one knows his or her eye color. Consequently, no one ever leaves the island.

One day, an outsider comes to the island. He is allowed to stay and soon acquires a reputation for being completely trustworthy. After a while, a ship lands and the outsider leaves the island again. At the time of his departure, all citizens gather at the harbor and the last thing the outsider tells the citizens is: "By the way, there is at least one blue-eyed person on the island!"

What happens during the subsequent nights? Additionally, what does all this have to do with the concept of *common knowledge*? The answer is that, during the 100th night after the announcement, all the blue-eyed people will leave the island.

Why does the announcement of the outsider make a difference? Before his announcement, each islander knew that there are blue-eyed persons on the island, but she did not know that the other islanders knew it as well, knew that she knows it, etc. Thus, the knowledge that there are blue-eyed islanders was not common knowledge. This changed with the announcement by the outsider. From that moment on, the existence of blue-eyed persons became common knowledge.

Why does it make a difference? To see this, one can use an inductive argument. If there is exactly one person with blue eyes, that person knows that there is no other person with blue eyes on the island. Before the announcement of the outsider it was a possibility that there is no one with blue eyes on the island, so there was no need to leave. However, given the information by the outsider, the blue-eyed person learns that she must have blue eyes, so she leaves at night one.

Next, assume that there are two persons with blue eyes. There is no need for any of them to leave during the first night, because there is a possibility that there is only one person with blue eyes and that it is the other person. Thus, both will still be around the next day. However, given that both are still around the next morning, they have to realize that both of them must have seen another person with blue eyes. Given that there is no one else around, it must be herself. Therefore, both will leave during night two.

The same argument holds if there are n blue-eyed persons: induction states that no one will leave during the first $n - 1$ nights. However, given that everyone is still around after night $n - 1$, each blue-eyed person has to conclude that there are n persons with blue eyes in total, one of them being him- or herself.

Thus, the rather innocuous-sounding announcement by the outsider allows the islanders to eventually figure out the color of their eyes.

To further illustrate, take the game represented in Table 11.2 as an example. In this game, two players $i = 1, 2$ have two strategies each. The game has one Nash equilibrium, (U, L). First, one has to show that this strategy profile is, in fact, an equilibrium. Suppose player 2 chooses 'L.' Player 1's best response is then to choose 'U' because $4 > 2$. Hence, 'U' is a best response to 'L.' If it is a Nash equilibrium, then 'L' must also be a best response to 'U,' which is indeed the case, because $3 > 2$. Thus, no player has an incentive to unilaterally deviate from this strategy profile. The strategies are mutually best responses and (U, L) is a Nash equilibrium.

There is an easy procedure to determine the set of Nash equilibria for games in matrix form. First, one successively goes through all the strategies of player 1 and marks the respective best response(s) of player 2. Then one repeats the whole procedure with the strategies of player 2 and marks player 1's best responses. If there are fields in which there are marks for both players, then the strategy profile associated with that field is a Nash equilibrium.

A Nash equilibrium is a prediction about the outcome of a game, but why should a game actually be played in such a way? One could argue that an important property of a Nash equilibrium is stability in the following sense: no player has an incentive to unilaterally deviate from the equilibrium strategy profile, because strategies are, by definition, best responses to each other. Players do not regret their choices of strategies once they find out what the other players are doing. This idea of consistency sounds plausible and rather innocuous. A potential problem is, however, that players make their choices simultaneously, that is without observing the strategies of all other players, and can commit to the strategies while the game is

Table 11.2 An example for a matrix game		L	R
	U	4,3	3,2
	D	2,1	1,4

Table 11.3 Dominant-strategy and Nash equilibria

	L	R
U	2,2	1,1
D	1,1	1,1

being played. Hence, each player has to be able to not only determine her own optimal strategy, but also the optimal strategies of the other players, and therefore to understand and solve the utility-maximization problems of these players. The concept of a Nash equilibrium requires both a large extent of implicit agreement between players that they are in fact seeking to find a Nash equilibrium, as well as strong cognitive abilities to be able to think through all the different strategic situations from the perspective of all the players. For instance, in the above example, player 2 needs to ponder which strategy player 1 will choose. If player 1 chooses 'D' instead of 'U', then her best response would not be 'L,' but 'R'. She needs to conjecture that player 1 will actually assume that a Nash equilibrium will be chosen and must then put herself into player 1's position. In the example, complexity is reduced, though, because player 1 has a dominant strategy. Because 'U' is always a best response, it is the best choice player 1 can make, independent of player 2's decision. Player 2, being aware of that, is able to predict that 1 will always choose this strategy, if she is rational. Hence, she will always choose 'L' herself. If a given player, i, has a dominant strategy, then the complexity of the game is significantly reduced, because it is easier for all the other players to make predictions.

One can conjecture that the predictive power of Nash equilibria is better in situations that are not very complex and if players are more experienced with the situation with which they are confronted.

As this subchapter has shown, the problem of cognitive overload can be reduced significantly, if the solution concept is not a Nash, but a 'dominant-strategy equilibrium.' In such an equilibrium, each player follows a dominant strategy and hence no player needs to conjecture about the strategy choices of all the other players, because her own optimal choice does not depend on the strategies of all the others.

▶ **Definition 11.4, Dominant-strategy equilibrium** A strategy profile, $s^{ds} = \{s_1^{ds}, \ldots, s_n^{ds}\}$, is called a *dominant-strategy equilibrium*, if the strategy of each player is a dominant strategy:

$$u_i(s_i^{ds}, s_{-i}) \geq u_i(s_i, s_{-i}) \text{ for all } s_i \in S_i, \text{ for all } s_{-i} \in S_{-i} \text{ and for all } i \in N.$$

Unfortunately, dominant-strategy equilibria exist only for a very limited class of games, such that it is rarely possible to predict the outcome of a game based on this equilibrium concept. Hence, using this concept instead of a Nash equilibrium does not really solve the problem.

A dominant strategy equilibrium is always a Nash equilibrium, but not *vice versa*. This property is exemplified by the game in Table 11.3.

This game has two Nash equilibria, (U, L) and (U, R), and each player has a dominant strategy, namely 'U' for player 1 and 'L' for 2. Therefore, (U, L) is also

Table 11.4 The game "Rock, Paper, Scissors" in matrix form

	R	P	S
R	0, 0	−1, 1	1, −1
P	1, −1	0, 0	−1, 1
S	−1, 1	1, −1	0, 0

a dominant-strategy equilibrium, while (U, R) is not. 'U' is only a best response if player 2 chooses 'R,' and similarly 'R' is only a best response if player 1 chooses 'U.'

Digression 31. Existence of a Nash Equilibrium

As we have seen when we have analyzed the game in Table 11.3, it is often not easy to predict the outcome of a game because there may be multiple equilibria. Another problem, which is at least as fundamental as the multiplicity, is the (non-)existence of Nash equilibria, a potential problem one already knows from the subchapter covering dominant strategies. Is it possible that a game has no Nash equilibrium? If so, then what would be a good prediction of the game's outcome?

An example for a game in which no Nash equilibrium exists is RPS. A matrix representation of the game can be found in Table 11.4. Whenever a player chooses a best response to the strategy of her opponent, the opponent must end up with a payoff that is smaller than the one that could be achieved by a different strategy, yielding her a utility of -1. Hence, there cannot be a profile of strategies that are mutually best responses and, thus, no Nash equilibrium exists.

A game that does not have an equilibrium is quite unsatisfactory, because this means one cannot make a prediction about the way people play it, which was why we started with game theory in the first place. Consequently, researchers started searching for a way out of this problem and found one in the idea of "mixed strategies." The idea is quite simple: put yourself in the position of a player in RPS. It is immediately clear that you want to avoid the other player knowing what you will do, because she could then exploit this knowledge, which would guarantee you a payoff of -1. Hence, how can you ensure that she does not know what you will do and is not able to predict it, either? One possibility is to delegate the strategy choice to a random generator that chooses each strategy with a given probability that you determine at the beginning. This is precisely the idea underlying mixed strategies. A *mixed strategy* is a probability distribution over the (as they will be called from now on) *pure* strategies at your disposal. If one allows players to choose probability distributions over pure strategies, then one increases the set of possible strategies, because each probability distribution over pure strategies also becomes a strategy – a mixed strategy. A Nash equilibrium, in which at least one player uses a mixed strategy, is called a *mixed strategy Nash equilibrium*.

However, what is the point of this exercise? In games like RPS, no Nash equilibrium exists in pure, but only in mixed strategies. In RPS, the equilibrium is easy to find: each player chooses a pure strategy with the probability of 1/3. For example, if player 1 chooses a pure strategy with that probability, then player 2 receives the following expected utility from each of her pure strategies:

$$
\begin{aligned}
u_2\left(R,(\tfrac{1}{3},\tfrac{1}{3},\tfrac{1}{3})\right) &= \tfrac{1}{3}\cdot 0 + \tfrac{1}{3}\cdot(-1) + \tfrac{1}{3}\cdot 1 = 0, \\
u_2\left(P,(\tfrac{1}{3},\tfrac{1}{3},\tfrac{1}{3})\right) &= \tfrac{1}{3}\cdot 1 + \tfrac{1}{3}\cdot 0 + \tfrac{1}{3}\cdot(-1) = 0, \\
u_2\left(S,(\tfrac{1}{3},\tfrac{1}{3},\tfrac{1}{3})\right) &= \tfrac{1}{3}\cdot(-1) + \tfrac{1}{3}\cdot 1 + \tfrac{1}{3}\cdot 0 = 0.
\end{aligned}
$$

Player 1's mixed strategy makes player 2 indifferent between all of her pure strategies and, thus, each of her pure strategies is a best response. This is, in turn, the precondition for her to be willing to randomize herself. If she randomizes herself with the same probabilities, then player 2 is also indifferent between all her pure strategies and each pure strategy, as well as the mixed strategy, is a best response. Therefore, it is a Nash equilibrium in mixed strategies, if both players randomize and choose each pure strategy with a probability of 1/3.

As the example shows, one can come up with a clear prediction of the game's outcome, if one allows for a more comprehensive concept of a strategy. It was one of John Nash's seminal contributions to show that such an equilibrium exists under very general conditions.

Result 11.1, Nash's theorem Every game with a finite number of players and a finite number of pure strategies has at least one Nash equilibrium in mixed strategies.

This result of Nash's theorem is of fundamental importance, because it guarantees that a prediction about the outcome of a game, based on the concept of a Nash equilibrium, is possible under very general conditions. I omit the proof of the theorem, because it involves advanced mathematical methods.

Another example of a game in which no Nash equilibrium exists in pure strategies is the penalty kick in soccer. The goalkeeper decides which part of the goal to defend, while the kicker simultaneously decides where to place the shot. If the goalkeeper conjectures the kicker's strategy correctly, then she successfully defends the shot; otherwise the kicker is successful. In order to be able to analyze this situation one can simplify and assume that each player has the pure strategies 'left,' 'middle' and 'right.' The game has a Nash equilibrium in mixed strategies, in which each player randomizes by choosing among the pure strategies with a probability of 1/3. Economists studied the behavior of goalkeepers and kickers based on data from the Italian and French professional soccer leagues. They found that the observed behavior was consistent with theoretical predictions.

11.4.1   Multiple Equilibria

This chapter has shown so far that some games, for example the one in Table 11.3, have multiple Nash equilibria. There are at least two problems caused by the multiplicity of equilibria. First, the predictive power of a theory that makes several predictions is limited and, second, it is only of limited use in supporting players with identifying optimal strategies. The problems are dramatic in the game represented by Table 11.3 because *any* strategy of a player can be rationalized, even if there are only two equilibria. The players have to, somehow, coordinate on one of the two equilibria in order to exclude some kinds of behavior as implausible. Without such a coordination, a formal analysis of the game is useless, from the point of view of the predictive power of the theory as well as from the point of view of giving advice how to play it.

One solution to this problem is to employ a stronger solution concept, for example an equilibrium in dominant strategies. The game in Table 11.3, for example, has two Nash equilibria, but only one equilibrium in dominant strategies. As argued before, not many games have equilibria in dominant strategies and, among them, there are some that have more than one.

Another possible solution to the problem of multiple equilibria is to hypothesize that players can coordinate on so-called "focal" strategies. The term *focal* was coined by Schelling (1960) and implies that some equilibria are, in a sense, more "salient" than others. However, the concept of focality is weak. It is not quite clear how to precisely define what makes an equilibrium focal and whether or not an equilibrium is focal depends on many things, such as the context of the respective game. In Schelling's own words (p. 57): "People can often concert their intentions or expectations with others if each knows that the other is trying to do the same. Most situations – perhaps every situation for people who are practiced at this kind of game – provide some clue for coordinating behavior, some focal point for each person's expectation of what the other expects him to expect to be expected to do. Finding the key, or rather finding *a* key – any key that is mutually recognized as the key becomes *the* key – may depend on imagination more than on logic; it may depend on analogy, precedent, accidental arrangement, symmetry, aesthetic or geometric configuration, casuistic reasoning, and who the parties are and what they know about each other." The idea of focal points is, therefore, not a full-fledged theoretical concept, but merely a heuristic one that helps determine how players behave in certain situations. Here is an illustrative example. Assume that you and another player have to pick one out of three numbers. If you pick

Table 11.5 Meeting in New York

	GCT	ESB	WS
GCT	3, 3	0, 0	0, 0
ESB	0, 0	1, 1	0, 0
WS	0, 0	0, 0	1, 1

identical numbers, then everybody wins CHF 10; otherwise, nobody gets anything. In that game, each pair of identical numbers is a Nash equilibrium and dominant-strategy equilibria do not exist. Now, assume the set of numbers you can pick from is 0.73285, 1 and 1.3857. In this situation, many people intuitively pick the integer 1. All pairs, $\{0.73285, 0.73285\}$, $\{1, 1\}$ and $\{1.3857, 1.3857\}$, are Nash equilibria, but only $\{1, 1\}$ is focal, although it is very difficult to theoretically identify why.

In some games with multiple equilibria, the equilibria can be ranked according to the payoffs or utilities that the players receive. If one equilibrium makes everyone better off than all the others, it is a strong candidate for a focal point.

▶ **Definition 11.5, Pareto dominance**   A Nash equilibrium is Pareto dominant, if each player's utility is strictly larger in it than in all other Nash equilibria.

An illustrative example for such a situation is depicted in Table 11.5. The basic story underlying this payoff matrix goes as follows. Two businessmen are planning to meet at noon in New York City, but have forgotten to fix a meeting point. The possible meeting points are the information desk at Grand Central Terminal (GCT), the main entrance to the Empire State Building (ESB), and the bull and bear statue at Wall Street (WS). If they do not meet, they get a utility of zero each. If they meet at ESB or WS, both of them get a utility of 1. However, because their favorite cafe is close to GCT, they get a utility of 3, if they manage to meet there.

The game has three Nash equilibria: all the strategy profiles where the businessmen go to the same place. However, since there are multiple equilibria, it is not possible to predict what the businessmen will end up doing on the basis of this solution concept alone. In addition, there are no equilibria in dominant strategies. However, the equilibrium (GCT, GCT) Pareto improves the others and, hence, might be focal. Using the idea of Pareto improvements as a means to select between equilibria is promising, because it can be assumed that people have a strong tendency to coordinate on the better ones.

Still, it has to be taken into account that, while the concept of Pareto dominance may often be helpful in predicting the outcome of a game, this is not always the case. First, it may be the case that equilibria cannot be ranked according to Pareto dominance, such that the concept is not applicable to these games. Second, there may be multiple Pareto dominant Nash equilibria. In these games it may be possible to reduce the number of plausible Nash equilibria, but the multiplicity problem cannot be overcome completely. If, for example, the utility of meeting at the ESB is also 3 for each player, then it is possible to exclude (WS, WS) as a "likely" equilibrium, but a prognosis about the game's outcome is still not possible; both the (GCT, GCT) and the (ESB, ESB) equilibria are Pareto dominant.

It is even possible for players to coordinate to play out a Pareto-dominated equilibrium, even though each of them would prefer a different outcome? Even the worst equilibrium (in utility terms) is an equilibrium and unilateral deviations are not beneficial. An example for such a situation is inefficient production standards, like the so-called QWERTY keyboard, which stems from the arrangement of letters on the (US) keyboard that begins with the sequence q,w,e,r,t,y. The arrangement of the letters on a keyboard was determined in the times of the mechanical typewriter. The purpose of its design was to maximize an *effective* typing speed. With mechanical typewriters, there is always the risk that the typebars will entangle, if one's typing is too fast. For that reason, the arrangement of letters on the QWERTY keyboard did not maximize the potential, but instead the effective typing speed. With the invention of the electric typewriter, the problem of entangled typebars was solved, but the then inefficient QWERTY arrangement remains in use until today. The standard is inefficient, but it is also an equilibrium. In that example, one of the reasons why it is hard to coordinate on another, more efficient equilibrium is that the expectations of the players are shaped by history. The new, more efficient equilibrium is counterfactual and it lives only in our imagination, whereas the other, less efficient equilibrium has been played out for years and decades. History can, therefore, be a more powerful focal mechanism than Pareto dominance is.

Another important example for multiple equilibria is public transportation. Suppose creating and maintaining a public transportation system has fixed as well as variable costs per user. In order to cover the fixed costs, users must pay taxes that equal the fixed costs divided by the number of users and the variable costs (for example as user fees). If the number of users is small, the costs per user are high, which implies that it is individually rational to rely on private transportation. If the number of users is high, costs per user are low and this can create a virtuous circle where people rely heavily on public transportation. Switzerland is an example for a country with a dense, reliable and affordable public transportation system, whereas most metropolitan areas in the USA heavily rely on private transportation.

Digression 32. The Economics of Social Media
The QWERTY keyboard may seem like an odd example for inefficient standards that is without much relevance for the functioning of the economy. The conclusion that coordination problems are only of secondary relevance would be premature, however, because the problem of multiple equilibria underlying the choice of inefficient production standards is at the heart of a lot of digital technologies. Take social media as an example. The attractivity of websites like Facebook or AirBnB depends on the number of users. The more users these websites have, the more attractive they are. This phenomenon is called a *network externality*. Network externalities can easily dominate quality differences between the different sites, like user friendliness, transparency or privacy. A platform that offers poorer quality may nonetheless survive (and even thrive), simply because it is used by a larger number of customers.

When one looks at these industries, one finds a typical pattern. In the early stages, there are usually several competing platforms, like Facebook, Friendster, MySpace or Xanga, and it is, *ex ante*, unclear which one will succeed. In the language of game theory, the game has multiple equilibria: one where the majority of users coordinate on Facebook and others where they coordinate on any one of the other platforms. Objective differences in the quality of the different platforms are a poor predictor for their future success. The number of users, however, is. The fastest-growing platforms are usually the ones that will outcompete the others and, once they dominate the market, it is very hard for new entrants to succeed, even if they offer much better quality. The large quantity of users protects the incumbent against market entries.

11.4.2   Collectively and Individually Rational Behavior

Another important topic that is, by now, better understood, because of game theoretic reasoning, is whether or not one should expect that self-interested behavior of individuals leads to outcomes that benefit a group's welfare. Game theory can play an important role in answering these questions by identifying mathematical structures that lead to certain equilibria. The structural characteristics of such a game, which lead to certain types of equilibria, can help social scientists to detect patterns that help them to interpret and grasp situations in the real world.

In order to illustrate this point, I will introduce one of the most famous games: the prisoner's dilemma. What is the historical background of this game? I discussed the First and Second Theorem of Welfare Economics in Chap. 5. According to these theorems, market equilibria are Pareto-efficient under certain conditions. The Coase Irrelevance Theorem has generalized these conditions and opened a perspective for a better understanding of the factors that explain differences between institutions: transaction costs. Way into the 20th century, many economists were convinced that the "invisible hand," as Adam Smith had coined it, is reality: if man follows his or her self-interest, then the interests of the rest of society are taken care of and there is no tension between individual self-interest and societal welfare.

This vision of a frictionless society can be illustrated by the following "invisible-hand game given in Table 11.6:" In this game, the two players have two strategies, M, F, each and both have a dominant strategy to play F. Hence, (F, F) is a unique Nash, as well as a dominant-strategy, equilibrium. It is, at the same time, Pareto-efficient.

Table 11.6 Invisible-hand game

	M	F
M	3, 3	5, 5
F	5, 5	10, 10

Table 11.7 Prisoner's
dilemma

	M	F
M	3, 3	10, 1
F	1, 10	7, 7

The "invisible-hand game" reveals no deeper truth about our social reality; in the end, it should not surprise one that it is possible to tinker with utilities such that a unique, Pareto-efficient equilibrium exists. The really important question is whether the game is meaningful to describe the social world.

Now that one has started to tinker with utilities, one will most likely end up with the game given in Table 11.7. Together with the invisible-hand game, this game became the most famous metaphor for the logic of social interactions. It was developed by the mathematicians Merrill Flood and Melvin Dresher. The name "prisoner's dilemma" is due to Albert W. Tucker, who adapted the game by Flood and Dresher, but framed it in a different context where two individuals have committed a crime and can either confess or not. (Both are better off collectively, if they do not confess, but each person is individually better off confessing. Hence, the name prisoner's dilemma.)

The game is a parable applicable to many economically relevant situations, for example the tragedy of the commons that was discussed in Chap. 6. Table 11.7 shows a prisoner's dilemma in matrix form. Two players can choose between two strategies, 'M' and 'F.' The central characteristic of the game is that it is optimal to choose 'M' for each player individually; that (M, M) is a dominant-strategy equilibrium. However, in a concerted effort, both players could increase their respective utility by choosing 'F.' Applied to the example of the tragedy of the commons, one can interpret the game as follows: two fishermen live on a lake, where they catch fish to make a living. While going out to catch fish, they have the choice between catching many fish ('M') or just a few ('F'). If both choose 'F,' both can sell only a smaller quantity, but at higher prices and the fishing grounds stay intact, which guarantees future income. One normalizes the utility associated with this strategy to 7. If both choose 'M', they can sell a lot, but prices are low and fish stocks dwindle due to overfishing. This leads to utilities of 3. If one fisherman chooses 'M,' while the other chooses 'F,' the fish stocks also dwindle, but to a lesser extent. The fisherman choosing 'M' sells a lot at moderate prices, while the other sells a small quantity. In this situation, the fisherman choosing 'M' gets a utility of 10, while the other receives only 1.

Because both players have the dominant strategy to choose 'M,' it seems clear, if one believes in invisible hands, that the equilibrium should have good welfare properties. But this is wrong. If both players could coordinate and play 'F,' then both would be better off. The decentralized decisions of the players are individually, but not collectively, rational.

11.4.3   Simple Games as Structural Metaphors

Coming back to the starting point of this chapter, the analysis of two-player two-   Φ
strategy games reveals a lot about the fundamental problems that can exist when
individual decisions are mutually dependent. These simple games illustrate the
problems societies are confronted with in a nutshell.

- There can be situations with a unique equilibrium that is also efficient. In such a
  situation, there is no tension between individual and collective rationality.
- There may be a unique equilibrium that is inefficient. In such a situation, there
  is tension between individual and collective rationality. Situations of this type
  are referred to as "cooperation problems," because individual incentives impede
  beneficial cooperation.
- Multiple equilibria may exist. Situations like these are called "coordination problems," because they represent the fundamental challenge to coordinate on an
  equilibrium.

The above classification of potential problems is useful, because it provides a
framework for interpreting problem structures in many different societal contexts.
Chapter 6 already analyzed the problem of overfishing (the tragedy of the commons)
and showed that it is inherently a cooperation problem. It can also be argued that
the social and economic causes of anthropogenic climate change are unresolved
cooperation problems.

Digression 33. Cooperation Problems and Externalities
This is a good point to hint at an important link between different concepts
that have been discussed in this book. I discussed the concept of externalities in Chap. 6. An externality exists, if the acts of an individual, A, have
an impact on the well-being of another individual, B, that A does not take
into consideration: it is a non-internalized interdependency. Looking at cooperation problems, like the prisoner's dilemma, one sees that it is exactly an
externality that is at the heart of the problem: the rational behavior of one individual makes the other individual worse-off, but the individuals do not find
a way to internalize this effect. Hence, cooperation problems are metaphors
for situations with mutual externalities, like anthropogenic climate change.

I also discussed the ontology of money in Chap. 3 and one can now interpret it
in the context of a simple game. Money has no intrinsic value and its value, as a
medium of exchange and storage of wealth, relies on a convention: an agreement
between people to accept money as a medium of exchange. If everybody complies,
the simplified exchange of goods has positive effects on the economy and this is also
an equilibrium. If a single individual stops accepting money, nothing bad happens
for the rest of society. However, if nobody accepts money as a medium of exchange,

it is rational for each individual to not accept money, either, and the economy has to rely on barter. The important fact is that both, a monetary and a barter economy, are equilibria. Due to the multiplicity of equilibria, one of the central challenges of an economy that relies on some abstract medium of exchange is to stabilize peoples' expectations, such that they believe in the convention and are willing to accept money. The stabilization of expectations is not always easy, as can be seen in times of economic crises when there a the danger of so-called "bank runs." A bank run is a situation in which many people lose trust in a bank's solvency and try to get back their savings. If enough people do that, then the belief becomes a self-fulfilling prophecy and the bank actually gets into trouble. Many phenomena on financial markets have a similar structure and are better understood once they are interpreted as coordination problems. Bank runs and financial crises are examples of why game theory is important in macroeconomics and finance and why it came to new fame during the global financial crises in recent years.

The three classes of problems described above are, in principle, prototypes for most of the problems that one will encounter during one's studies. If one keeps them in mind, it will be easier to understand the fundamental structure underlying the different theories.

Φ   **Digression 34. The Cold War as a Game**

Deterrence is the art of producing in the mind of the enemy the fear to attack. (Stanley Kubrick, Dr. Strangelove)

During the Cold War, the United States and the Soviet Union were in a nuclear stand-off. Thus, the RAND Corporation (a major US think tank) hired some of the world's top game theorists to study the situation. At the time, both nations had the same policy, "If one side launched a first strike, the other threatened to answer with a devastating counter-strike." This became known as Mutually Assured Destruction, or MAD, for short. Game theorists got worried about the rationality and, thereby, the credibility of MAD. The argument goes like this, "Suppose the USSR launches a first strike against the USA. At that point, the American President finds his country already destroyed. He doesn't bring it back to life by now blowing up the world, so he has no incentive to carry out his original threat to retaliate, which has now manifestly failed to achieve its point. Since the Russians can anticipate this, they should ignore the threat to retaliate and strike first. Of course, the Americans are in an exactly symmetric position, so they too should strike first. Each power will recognize this incentive on the part of the other, and so will anticipate an attack if they don't rush to preempt it. What we should therefore expect is a race between the two powers to be the first to attack." (Don Ross 2016)

This analysis led the RAND Corporation to recommend that the United States take actions designed to show their commitment to MAD. One strategy was to ensure that "second-strike capability" existed. A second strategy was to make leaders appear irrational. The CIA portrayed President Nixon as either insane or a drunk. The KGB, which appears to have come to the same conclusion as RAND, responded by fabricating medical records to show that General Secretary Brezhnev was senile.

Another strategy was to introduce uncertainty about the ability to stop a counterstrike, for example by building more nuclear missiles and storing them in numerous locations (which made it less likely that the President could stop all of them from being launched in the event of a Soviet attack). A third strategy was to make MAD credible by creating "doomsday machines": technologies that carry out a counterstrike automatically, without the ability of human beings to interfere. The USSR went so far as to create Perimeter, or Dead Head, which was the closest thing this world has ever seen to such a doomsday machine. It can automatically trigger the launch of intercontinental ballistic missiles, if a nuclear strike is detected by seismic, light, radioactivity and overpressure sensors.

It is commonplace to suggest that the strategic situation during the Cold War was a case of the prisoner's dilemma. However, it is far from obvious that the leaderships in either country in fact attached the necessary payoffs in their utility functions – preferring the destruction of the world to their own unique destruction – that would have been required for their situation to actually have been a prisoner's dilemma.

11.5 Extensive-Form Games

Up until this point, one has not been able to analyze situations where players choose their strategies sequentially instead of simultaneously. Many social phenomena cannot be adequately described as simultaneous-move games, because timing plays an important role.

If the order of play is important, then games are usually not depicted in matrix representation but with the help of a *game tree*. A game tree describes what actions any given player has at the different points in time and how these actions influence the further course of the game. Formally, a game tree is a directed graph with nodes as positions in a game, where the players have to make decisions and edges represent the possible decisions (moves). The nodes are also called *decision nodes* in game theory.

As an example for such a game tree, take Fig. 11.1. The game is a version of the so-called centipede game (it is called that, because the game tree looks a bit like a centipede, if there are enough decisions that the players have to make). There are two players, $i = 1, 2$, and three decision nodes, $T = \{1.1, 2, 1.2\}$. Player 1 has to

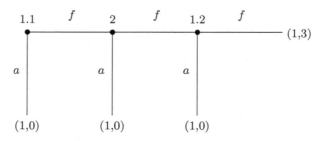

Fig. 11.1 Centipede game

make a decision at nodes 1.1 and 1.2, while player 2 only decides at node 2. Both players have a choice between the same actions at every node, $A_i^t = \{a, f\}, t \in T$.

The concept of a strategy is more complex than before. A strategy is a rule that determines an action for every *potential* node in the game. Because player 1 has to make a decision at two decision nodes, a strategy assigns an action to both, irrespective of whether both nodes are reached during the course of the game or not. This complete list of actions, one for each decision node where a player has to make decision, is called an *action profile*.

One has to specify a complete contingency plan for each player, because otherwise it would not be possible to solve the game. If a player contemplates her optimal strategy, she has to be able to figure out how the game ends, if she goes for this or for that strategy. This is only possible, if all players specify what they will do at each decision node.

The set of possible strategies of a player equals the set of possible action profiles. Player 1's strategy set is $S_1 = \{aa, af, fa, ff\}$ where, for example, $s_1 = af$ is interpreted as player 1 choosing a at decision node 1.1 and f at decision node 1.2. Because player 2 decides only once during the game, at node 2, her strategy set equals the set of actions she has at this node, $S_2 = A_2^2 = \{a, f\}$.

As in normal-form games, each strategy profile leads to an outcome, which is represented by the players' utilities. For instance, the strategy profile (af, f) implies that the game ends immediately and that players' utilities are $u_1(af, f) = 1$ and $u_2(af, f) = 0$.

Solution concepts are defined by means of an analogy to normal-form games and, hence, extensive-form games can basically be solved in the same way as games in normal-form once the strategies are defined. However, due to the more complex structure, there may be some problems related to the concept of a Nash equilibrium that did not exist before: Nash equilibria can be based on so-called "empty threats." In order to see what that means, take a look at the game in Fig. 11.2, the chainstore game (or market-entry game). Two firms, $i = 1, 2$, are potentially competing in a market. If firm 1 does not enter the market, NE, then the incumbent firm, 2, has a monopoly. If firm 1 enters, E, then firm 2 has two options: to start a price war, PW, or to accommodate, A. The game has two Nash equilibria in pure strategies: NE, PW and E, A. Because no player has a dominant strategy and no equilibrium

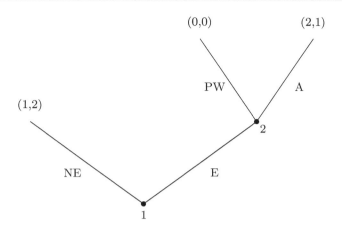

Fig. 11.2 Chainstore or market-entry game

is Pareto dominant, the concepts that were discussed thus far are of little help in determining the game's outcome.

However, one can use the sequential structure of the decisions to distinguish between the different equilibria. Because firm 1 makes her decision *before* firm 2, firm 2's choice '*PW*' is not credible. If firm 1 enters the market, firm 2's best response is '*A*.' The threat to start a price war if firm 1 enters is not credible, because it relies on the assumption that firm 1 does not enter. However, (NE, PW) is still a Nash equilibrium, since unilateral deviations are not beneficial for any firm.

A concept that will help one to identify such non-credible strategies is called *backward induction*. Intuitively, backward induction can be described as "thinking ahead and reasoning backward:" In a first step, one determines the individual "subgames" of a game, i.e. the parts of the game tree that can be interpreted and analyzed as independent games. For example, the chainstore game has two subgames: one starting with firm 2's decision node and another that is the whole game.

In a next step, one looks at the *terminal* subgames and determines the optimal actions chosen at these nodes. A subgame is a terminal subgame, if the game ends thereafter, no matter what the player who makes a decision does. In the chainstore game, the subgame that starts at firm 2's decision node is a terminal subgame, whereas the whole game is not because, if firm 1 enters the market, the game goes on and firm 2 makes a decision.

Once the optimal decisions at the terminal nodes have been determined, then the terminal subgame is replaced by the utilities the players get from the optimal play. For example, in the chainstore game, the terminal subgame starts at decision node 2 and firm 2's best action is to choose '*A*,' because this yields a higher utility than starting a price war '*PW*.' Hence, in this step, the last subgame is replaced by the utilities achieved through this action, (2,1). Replacing the terminal subgame with the utility vector makes the game tree "shorter" and there are new terminal subgames. This procedure needs to be repeated until the start of the game is reached.

In the chainstore example, this is the case after replacing node 2, when firm 1 faces the decision to enter the market, which gives her a utility of 2, or to stay out of the market, which gives her a utility of 1. Since entering the market gives her a greater utility, she will do exactly this and choose 'E.' Therefore, only one Nash equilibrium remains after backward induction, (E, A), and the other equilibrium, (NE, PW), which contained the empty threat to start a price war, is eliminated.

By solving the game from its end, one can reduce its complexity step by step. Players who need to determine their optimal choices at earlier nodes can rely on a continuation of the game that is always (at every decision node) optimal for each player.

Digression 35. Chess and the Existence of Backward-Induction Equilibria
One of the first formal game theoretic studies was Ernst Zermelo's analysis of the game of chess. Chess can be interpreted as an extensive form game between two players, but the game's complexity makes it impossible to write down the players' strategies, to draw a game tree, or to solve it (at least with today's means). However, Zermelo was able to show that there is an optimal, deterministic way to play chess. This result also illustrates why backward-induction equilibria must exist, if each player has a finite number of strategies.

Certain rules in chess guarantee that it cannot go on forever (see Article 5.2 of the official Fide chess rules) and, thus, every player has finitely many strategies. The conditions that Zermelo found to be necessary in his proof are hence met and it is, therefore, proven that either white has a winning strategy, or that black has a winning strategy, or that both can force at least a draw.

Until now, nobody has been able to find out whether white or black has a winning strategy or whether each player can force a draw. Therefore, of course, nobody knows the optimal strategy to play chess. Zermelo's result is, in this respect, a rather strange mathematical theorem: one knows that there is an optimal way to play chess, but one does not know what the optimal strategies are. Fortunately, one might say, because this is why the game of chess remains interesting.

Zermelo's theorem has important implication for other games, as well. First, it reveals that, under quite general conditions, a *pure* strategy equilibrium exists when players move sequentially. Furthermore, it shows that this equilibrium is not based on empty threats. These two points are of importance for the ability to predict the outcomes of extensive-form games.

11.6 Summary

One has seen that game theory is an analytical tool that helps one to analyze situations of strategic interdependence. This method has proven to be extremely versatile and has generated interesting insights far beyond the narrow field of economics, ranging from political science, law and business administration to evolutionary biology. A topic that I have not covered in this chapter is that the insights of game theory also paved the way for behavioral economics and neuroeconomics. Even in simple games, the required cognitive abilities for the players to find a Nash equilibrium are so high that it became apparent that rational-choice models of decision-making have poor predictive power in a number of situations. In addition, problems like the prisoner's dilemma spurred literatures on the cultural and genetic roots of cooperative behavior, which has been generating fascinating insights into the evolutionary and cultural forces that have shaped our brains and our perceptions of reality.

Digression 36. Games as Structural Metaphors: Further Examples   Φ
This chapter has already clarified that game theory is a method and that games with specific sequences of moves and payoff structures are problem structures, which are not tied to specific interpretations, but that can be used as metaphors for a wide array of social phenomena. This versatility is one of game theory's strengths, because it allows one to understand the strategic similarities between, apparently, very different social spheres. Here are some examples for social phenomena that have aspects of the chainstore game:

- **Military conflicts:** Situations that are very similar in their logic to the market-entry problem can be found in many military conflicts. Often, one party in a conflict threatens to attack another party, should that party continue with some provocative action. However, if there were an actual attack, both parties would be worse off.
- **Bailouts:** The state has an interest in ensuring that its major banks are managed in a way that makes situations of serious financial stress unlikely. However, if a major bank gets into financial trouble, the economic consequences for the rest of the country are so severe that the state bails it out. If banks anticipate this incentive, they know that they are at least partially insured against failure and so they have an incentive to invest in riskier strategies, which increases the likelihood that a bailout will become necessary. The major challenge for a state is, therefore, to make a no-bailout strategy credible. This is, of course, the exact situation that Switzerland, the USA and other European countries faced during the financial crises that started in 2007, and it also illustrates some of the EU's problems regarding institutional reforms in some of its member states.
- **Legalization of illegal immigrants:** Countries want to restrict and control illegal immigration. Therefore, it is in their best interest to signal a tough policy towards potential illegal immigrants in order to prevent them from attempting to migrate. It is in light of this background that the debate about the legalization of illegal immigrants in the USA can be understood. The Obama administration was largely in favor of legalizing this group of people. President Barack Obama said in a press conference on September 06

2014 that, although his "preference is to see Congress act," he intended to take unilateral action in order to give illegal aliens "some path" to "be legal," if Congress did not enact the sort of immigration legislation he wanted (at that time congress was being controlled by a Republican majority that was mostly against legalization). Advocates of the pro-legalization camp typically use two types of arguments to support their views: humanitarian and economic (illegal immigrants are, for example, an important part of the Californian agricultural industry). Opponents often argue that legalization sends the wrong signals, because it encourages immigration.

- **Touchiness can pay off:** Now, it is time to get to the really important stuff. Think of a typical situation in a partnership. You can stay home for the night with TV and crackers (your partnership has reached a mature stage) or you can go out with friends, but without your significant other. You think the latter alternative is much more fun, but only if your significant other does not create a scene the next morning. Your partner would be jealous if you go out without him or her, but he or she also shies away from making a scene. Thus, if you actually went he or she would give in and make the best out of the evening. However, he or she would profit from a reputation of being touchy.

The art and craft of a social scientist is to boil complex social phenomena down to their essential strategic structures. This is not always easy, as the discussion of the Cold War as a prisoner's dilemma game has shown, and a reconstruction of the above situations as chainstore games may be wrong or misleading in a given situation. Everyone is well aware that, if one has a hammer, everything looks like a nail and it is the same with game theory: if one has, for example, the prisoner's dilemma as a device for making sense of things, then suddenly everything looks like a cooperation problem.

Fascinating as these topics may be, it is now time to come back to the analysis of prototypical markets, which is why I had to cover game theory in the first place. Markets rarely fit to the ideals of perfect competition or monopoly and, next, I will apply the methods from game theory to creating a better understanding of the functioning of oligopoly markets. Usually, firms have some control over prices. However, that is limited by the existence of competitors. Thus, there are important strategic interdependencies that have to be taken into account, if one wants to make meaningful predictions about the functioning of these markets. Game theory is the analytical toolbox for achieving this.

References

Maugham, W. S. (1949). *A Writer's Notebook*. Arno Press.
von Neumann, J. (1928). *On the Theory of Parlor Games*. Princeton University Press.
Ross, D. (2016). Game Theory. In E. N. Zelta (Ed.), *Stanford Encyclopedia of Philosophy*, (Winter 2016 Edition). https://plato.stanford.edu/archives/win2016/entries/game-theory/.
Schelling, T. C. (1960). *The Strategy of Conflict*. Harvard University Press.
Zermelo, E., & Borel, E. (1913). *On an Application of Set Theory to the Theory of the Game of Chess*. Wiley Blackwell.

Further Reading

Binmore, K. (1994). *Game Theory and the Social Contract, Volume 1: Playing Fair*. MIT Press.
Binmore, K. (1998). *Game Theory and the Social Contract, Volume 2: Just Playing*. MIT Press.
Gibbons, R. D. (1992). *Game Theory for Applied Economists*. Princeton University Press.
Gintis, H. (2000). *Game Theory Evolving: A Problem-Centered Introduction to Modeling Strategic Behavior*. Princeton University Press.
Homer (1997). *The Odyseey*. Penguin Classics.
Kafka, F. (1976). *The Silence of the Sirens*. The Denver Quarterly.
Osborne, M. J. (2004). *An Introduction to Game Theory*. Oxford University Press.

Firm Behavior in Oligopolistic Markets   **12**

This chapter covers . . .

- how to apply techniques from game theory toward understanding firm behavior and equilibria in oligopolistic markets.
- the difference between and the significance of price and quantity competition on oligopolistic markets.
- how models of oligopolistic behavior can help one to better understand markets for oil, gas, etc.
- the logic of collusive behavior and the role of regulation in oligopolistic markets.
- how firms have to be organized that compete in such markets.

12.1 Introduction

A horse never runs so fast as when he has other horses to catch up and outpace. (Ovid (2002), Ars Amatoria)

Models of markets with perfect competition and monopolistic markets pinpoint extreme cases. They illustrate and help to understand how markets work. However, the stylized nature of these models makes it necessary for them to abstract from aspects of reality that may be relevant for understanding some markets. One such aspect – strategic interdependence of firm decisions – will be the topic of this chapter.

This chapter starts with a short summary of the central results from the theory of competitive and monopolistic markets:

- **Perfect competition:** There are many suppliers of an identical product and each seller assumes that she cannot influence the market price with her decisions. Under certain conditions, price-taking behavior leads to the "price-equals-marginal-costs" rule for the profit maximizing choice of output and, at the same time, to the Pareto efficiency of this type of market, because all potential gains from trade

© Springer International Publishing AG 2017
M. Kolmar, *Principles of Microeconomics*, Springer Texts in Business and Economics,
DOI 10.1007/978-3-319-57589-6_12

are exploited. Two additional conditions, however, have to be met to make this rule rational. On the one hand, producer surplus has to exceed the relevant fixed costs. On the other hand, the production technology has to induce increasing or constant long-run marginal costs. If competitive markets work, then market entry and exit will drive profits to zero, because positive (negative) profits encourage entry (force exits).

The managerial implications of these findings point towards the crucial importance of having an effective accounting system: marginal and average costs of production have to be precisely reflected in the relevant indicators. In addition, given that profits are approximately zero with constant returns to scale or, in the long run, the return on equity cannot exceed the return on debt, owners cannot expect larger profits from their investments than they would get in the capital market.

- **Monopoly:** Only one supplier of a product exists, which implies that customers see a relevant difference between this product and the closest substitute and that other firms cannot imitate it. Compared to competitive markets, a monopolist generates a higher producer surplus, such that it can sustain itself, even if fixed costs would drive competitive firms out of the market (if they are not too high). The efficiency of such a market depends on the monopolist's ability to discriminate prices. The closer the monopolist gets to the ideal of perfect price discrimination, the more efficient the market becomes. However, there is a tension between the efficiency of monopolistic markets and the distribution of rents between the firm and the consumers, because in the efficient solution, the monopolist is able to transform all rents into producer surplus.

  In order to implement the optimal policy, the firm needs more information than under perfect competition. In addition to an accounting system, it needs a market-research department that estimates price elasticities and helps to segment demand into different groups that are targeted individually.

These findings give some mileage in understanding firm behavior and the functioning of markets, but the important topic of the strategic interdependence of firms' decision making has been left out of the picture. Strategic interdependence does not play any role in a monopolistic market, by definition, and it does not play a role in competitive markets, because each single firm is too small to influence aggregates. It becomes important, however, if there is more than one firm that is sufficiently large to influence the market price such a decision made by one firm can influence the profit of another. This direct interdependency between firms' objectives follows the same logic as the one analyzed in Sect. 6 and can therefore, in principle, be analyzed with the same toolbox of property rights and transaction costs. A direct interdependency can occur if several firms sell homogenous goods, but also if they sell differentiated goods that are closely linked (which happens, if cross-price elasticities between the goods are non-zero).

The latter situation is, to some extent, always present for a monopolist, but it is usually left out of the analysis to avoid additional complexities. This chapter will also neglect the analysis of several monopolists whose profits are interdependent, because they sell similar products. Instead, it focusses on oligopoly markets in which few sellers supply a homogenous good. The assumption that the goods are perfect substitutes, from the consumers' point of view, simplifies the analysis and allows it to isolate the pure effect of strategic interdependence.

The central tool for understanding strategic interdependence is game theory and the definition of a game and Nash equilibrium will aid in the development of a solution concept. Firms have, in principle, two instruments to maximize profits, if they are selling a given product. Both instruments are, however, not independent, because they are linked by the market-demand function. This is why it is irrelevant, for the monopolist, whether he sets a price and lets quantities adjust passively or sets a quantity and lets the price adjust passively; both approaches lead to the same solution. This equivalence is lost in an oligopolistic market. As the following analysis will show, predictions for the functioning of an oligopoly market with price- and quantity-setting firms differ sharply. In order to understand the deeper reasons for this difference, one has to start by building models of price- and quantity-setting and then see what predictions they make.

The model of quantity setting is called the *Cournot model* and the model of price setting is called the *Bertrand model*, named after the French mathematicians Antoine Augustin Cournot, who developed his model as early as 1838, and Joseph Louis François Bertrand, who reworked the model by using prices in 1883. It is fascinating that Cournot's analysis anticipated a lot of concepts from economics and game theory, like supply and demand as functions of prices, the use of graphs to analyze supply and demand, reaction functions and the concept of a Nash equilibrium (limited to the oligopoly context).

Digression 37. The Stackelberg Model and the Value of Commitment   Φ

I can resist everything except temptation. (Oscar Wilde (1892))

There is a third model of oligopolistic decision making that goes back to Heinrich Freiherr von Stackelberg (1934). He returned to Cournot's original analysis, but assumed that two firms determine their quantities sequentially instead of simultaneously, as Cournot had assumed. This model will not be covered in this chapter, but I would like to focus attention on a figure of thought that emerged from this model and that proved to be of primary importance for economics and other social sciences: the idea of *commitment*.

It turns out that the firm that sets its quantity first (the "leader") has an advantage over the other firm (the "follower"), in comparison to the Cournot model. However, if this were the case, both firms would like to be the leader and the factors that determine leadership are not obvious. Both firms would do whatever they could to be able to choose their strategy first. What is necessary is the existence of some mechanism or device to be available to one firm, but not to the other, which allows the firm to make its leadership position credible. Such a mechanism is called a *commitment device*.

The appreciation of the economic role and value of such a device offers an important new perspective on a number of social phenomena. One can interpret them as reactions to commitment problems. According to Dubner and Levitt (2005), a commitment device is "a means with which to lock yourself into a course of action that you might not otherwise choose but that produces a desired result."

The ultimate commitment device can be found in Homer's Odyssey, where Ulysses puts wax in his men's ears so that they could not hear and had them tie him to the mast so that he could not jump into the sea, to make sure that he does not fall prey to the song of the sirens. (Franz Kafka (1931) sees this as "[p]roof that inadequate, even childish measures, may serve to rescue one from peril.")

Commitment problems exist on the individual as well as on the social level. Fitness goals are a good example of an individual commitment problem. Most people would like to exercise a little more, drink less alcohol or eat healthier food. However, if it is time for a run, a friend asks if one is ready for a second glass of wine, or one has the choice between chocolate cake and broccoli, one can resist everything other than the temptation to give in. What would be needed in these situations is a device that forces one to stick to one's resolutions. Some argue that emotions, like shame and embarrassment, can be interpreted as such a device: assume that one publicly announces a fitness goal ("I will run the Berlin Marathon next year"). If one makes such a public announcement and fails to stick to one's goals, one's friends will ridicule one and one will feel ashamed, which helps one's future self persevere. These emotions make deviations from one's plans costly (in this case, in a purely psychological sense), which is the most important property of a credible commitment device: if one wants to stick to a savings plan, sign a long-term contract that is costly to cancel; if one wants to prepare for an exam, lock oneself into a room without internet access and give the key to a friend, who will be away for the weekend; and so on.

The prisoner's dilemma is the main example of a social commitment problem: both players would profit from a device that makes the cooperative strategy credible. If the dilemma is used as a metaphor for social interactions in general (a mainstream view since Thomas Hobbes claimed that life before organized, civil societies was solitary, poor, nasty, brutish, and short), then the state can be interpreted as one big attempt to make cooperation credible. This idea can refer to institutions like the rule of law, property rights, and their enforcement by means of material sanctions and punishments, but it can also refer to culture in general, where credibility stems from "softer" sanctions like feelings of guilt and shame.

Commitment problems have also been shown to be at the heart of phenomena like inflation and taxation. The phenomenon is also known as the *time inconsistency* of decision making. For the case of monetary policy, politicians have an interest in promising low inflation for the future, in order to control the expectations of the people. However, once tomorrow comes, increasing inflation can have positive, short-run effects, like increasing employment. Hence, the announcement of a low-inflation goal may not be credible, if the government cannot commit to it, leading the economy into a high-inflation equilibrium. Independent central banks with high degrees of discretion in monetary policies are widely seen as a commitment device that can solve this problem. If the central banker's objective is a zero-inflation policy, then taking away discretion from politicians can, in the end, help them in achieving their goals. The same is true for taxation. If a government wants to encourage investment, then it should announce very low rates of capital taxes but, once the investments are sunk and the factories are built, the corporations are locked in and it is rational for the politician to increase taxes again. If this incentive is anticipated, then firms will not invest in the first place. One of the reasons why Switzerland is considered an attractive place for investments is because it managed to establish a reputation for not falling prey to this incentive. A lack of such a reputation can be a serious impediment to economic development.

The above mentioned firms are "locked in" with their investments. This *lock-in effect* is a widely used business practice that helps firms to make profits. Software standards are a good example. In order to be able to use software, one usually has to make large investments of time and effort. These investments lock one into a standard because, *ex-post*, after one has made the investments, the opportunity costs of switching to another standard (called *switching costs*) are higher than *ex-ante*, before one committed to it. This asymmetry in opportunity costs can be exploited by firms for setting higher prices for software updates, and so on.

Evolutionary biologists have used commitment problems to explain the evolution of moral sentiments, by arguing that the evolution of emotions that make cooperation rational (not in a material, but in a psychological sense) have a positive effect for the survival of groups.

A problem with any credible commitment device is that they reduce flexibility. If the future can be perfectly foreseen, then commitment incurs no additional costs but, the more uncertain the future becomes, the more risky it is to constrain one's choices. What would have happened to the epic poem "Odyssey" if Ulysses, tied to the mast, had drowned because of an unforeseen storm that hit his ship before he passed the island of the sirens? He would not be remembered for his brilliance and cunning intelligence, but but for his ability to drown himself in an attempt to control his virility. That is not exactly the type of story that would be remembered forever.

12.2   Cournot Duopoly Model

In the Cournot duopoly model, it is assumed that two profit-maximizing firms, U_1 and U_2, simultaneously plan the quantities y_1 and y_2 of a homogenous good that they want to sell in a given period of time. Quantities are chosen from the set of positive real numbers (including zero). They produce with a technology that, for given factor prices, leads to linear or overproportionately increasing cost functions, $C_1(y_1)$ and $C_2(y_2)$. Furthermore, there is an inverse demand function, $P(y_1 + y_2)$, that gives the market price for market supply, $y_1 + y_2$, (customers see both goods as perfect substitutes). In order to keep the analysis simple, assume that all firms are completely informed about all the cost functions and the inverse demand function, and that all of this is common knowledge.

If the profits of the firms are denoted by π_1 and π_2 and can be written as:

$$\pi_1(y_1, y_2) = P(y_1 + y_2) \cdot y_1 - C_1(y_1), \quad \pi_2(y_1, y_2) = P(y_1 + y_2) \cdot y_2 - C_2(y_2).$$

From the managers' points of view, the problem is that profits depend not only on the firm's own strategy, but also on the strategy chosen by the other firm, because the market price is a function of total quantity. In order to solve this problem, assume that manager 1 (2) expects that the other firm will supply a quantity y_2^e (y_1^e). The managers determine the optimal quantities given these expectations. The first-order conditions for the profit-maximizing strategies are:

$$\frac{\partial \pi_1(y_1, y_2^e)}{\partial y_1} = \underbrace{\frac{\partial P(y_1 + y_2^e)}{\partial y_1} \cdot y_1 + P(y_1 + y_2^e) \cdot 1}_{= MR_1(y_1, y_2^e)} - \frac{\partial C_1(y_1)}{\partial y_1} = 0$$

for firm 1 and

$$\frac{\partial \pi_2(y_1^e, y_2)}{\partial y_2} = \underbrace{\frac{\partial P(y_1^e + y_2)}{\partial y_2} \cdot y_2 + P(y_1^e + y_2) \cdot 1}_{= MR_2(y_1^e, y_2)} - \frac{\partial C_2(y_2)}{\partial y_2} = 0$$

for firm 2.

Both conditions have a simple economic interpretation: for an expected production level of the competitor, a firm chooses its quantity such that the marginal revenue of the last unit produced equals the unit's marginal costs. This condition corresponds to the condition of a non-price discriminating monopolist with the exception that marginal revenues depend on the expectations of the other firm's production decision.

If one solves the first-order conditions for the respective decision variables, y_1 and y_2, one gets two functions $Y_1(y_2^e)$ and $Y_2(y_1^e)$, which determine the optimal quantity for one firm for a given expected supply of the other firm. These are the so-called *reaction functions* of the two firms.

Points on the reaction functions imply that firms behave optimally for any given expectation of the other firm's strategy. However, plans do not have to be mutually consistent. There can be situations where both firms start with expectations, y_2^e and y_1^e, choose their strategies optimally, but end up with quantities that deviate from the expectations of the other firm, $Y_1(y_2^e) \neq y_1^e$ or $Y_2(y_1^e) \neq y_2^e$. In order to guarantee consistency, one has to require that expectations and actual behavior coincide, $Y_2(Y_1(y_2^e)) = y_2^e \wedge Y_1(Y_2(y_1^e)) = y_1^e$: The best response of firm 2 to the best response of firm 1, at an expected quantity of y_2^e, is equal to the expected quantity y_2^e, and the best response of firm 1 to the best response of firm 2, at an expected quantity of y_1^e, is equal to the expected quantity y_1^e.

This is another way to say that one is looking for a Nash equilibrium in the game. Formally, a Nash equilibrium of a Cournot duopoly model is completely characterized by $Y_2(Y_1(y_2^e)) = y_2^e \wedge Y_1(Y_2(y_1^e)) = y_1^e$.

The general characterization of the Nash equilibrium does not contain anything interesting from an economic point of view, because it is just a formal way to say that firms follow their objectives rationally and that their behavior is mutually consistent. In order to gain more economic understanding, this chapter will proceed by assuming that the demand function is linear and that the cost functions are identical and linear, $p(y_1 + y_2) = a - b \cdot (y_1 + y_2)$, $C_1(y_1) = c \cdot y_1$, $C_2(y_2) = c \cdot y_2$ with $a > c > 0$ und $b > 0$. These functional specifications are called the *linear model*. A lot of the understanding that one can get from this model carry over to more general models with nonlinear functions for either demand or cost, or both. The model is illustrated in Fig. 12.1. In the figure, $y_1 + y_2$ is plotted along the abscissa and demand, as well as the marginal-cost functions, are plotted along the ordinate. The marginal-cost function intercepts the ordinate at c; the demand function interrupts at a and has a slope of $-b$.

From a mathematical point of view, there are a lot of different ways to determine the equilibrium. This subchapter will cover a long and rather complicated way for the purpose of exercise, by first computing the profit functions (in order to have a lean notation, one can skip the explicit mention of expected values):

$$\pi_1(y_1, y_2) = (a - b \cdot (y_1 + y_2)) \cdot y_1 - c \cdot y_1, \quad \pi_2(y_1, y_2) = (a - b \cdot (y_1 + y_2)) \cdot y_2 - c \cdot y_2.$$

They can be simplified to:

$$\pi_1(y_1, y_2) = (a - c - b \cdot y_2) \cdot y_1 - b \cdot y_1^2, \quad \pi_2(y_1, y_2) = (a - c - b \cdot y_1) \cdot y_2 - b \cdot y_2^2.$$

The next step is to determine the first-order conditions:

$$\frac{\partial \pi_1(y_1, y_2)}{\partial y_1} = (a - c - b \cdot y_2) - 2 \cdot b \cdot y_1 = 0$$

and

$$\frac{\partial \pi_2(y_1, y_2)}{\partial y_2} = (a - c - b \cdot y_1) - 2 \cdot b \cdot y_2 = 0.$$

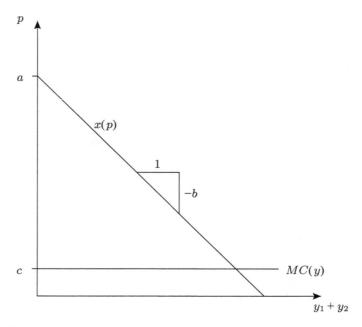

Fig. 12.1 The linear model

Firm 1's first-order condition is depicted in the left panel of Fig. 12.2. Marginal costs are constant at c. Marginal revenues intersect the ordinate at $a - b \cdot y_2$ and have a slope of $-2 \cdot b$. They are falling with the supply of the other firm. A comparison with the monopoly case is illustrative: the marginal revenues of a non-price-discriminating monopolist have the same slope, $-2 \cdot b$, but they intersect the ordinate at a. One can, therefore, think of a Cournot duopolist i as a monopolist with a "curtailed" demand function $\tilde{a} = a - b \cdot y_j$. If one solves both conditions for the respective decision parameters, one ends up with the reaction functions:

$$Y_1(y_2) = \begin{cases} (a - c - b \cdot y_2)/(2 \cdot b), & \text{if } y_2 \leq (a - c)/b \\ 0 & \text{if } y_2 > (a - c)/b \end{cases},$$

$$Y_2(y_1) = \begin{cases} (a - c - b \cdot y_1)/(2 \cdot b), & \text{if } y_1 \leq (a - c)/b \\ 0 & \text{if } y_1 > (a - c)/b \end{cases}.$$

They are illustrated in Fig. 12.2 where y_1 is plotted along the abscissa and y_2 along the ordinate. Given that firm 1's reaction function has y_2 and firm 2's reaction function has y_1 as explanatory variables, one has to look from the abscissa and ordinate simultaneously to understand the figure. The "flatter" graph is firm 2's reaction function, which has the traditional orientation. The "steeper" graph is firm 1's reaction function, which is symmetric to firm 2's, but with the opposite orientation.

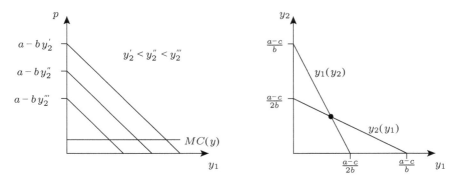

Fig. 12.2 Marginal revenues and marginal costs of the oligopolists (*left*) and reaction functions (*right*)

The figure reveals the following. (1) $Y_1(0) = Y_2(0) = (a - c)/(2 \cdot b)$: thus, if the other firm produces nothing, the remaining duopolist behaves like a monopolist. (2) The profit-maximizing quantity of a firm falls with an increase in the quantity of the competitor. (3) The Nash equilibrium is given by the interception of both reaction functions. Only at that point are the actual and the expected supplies of the firms consistent with each other.

Formally, one can find the Nash equilibrium by inserting the reaction functions into one another, $Y_1(y_2) = (a - c - b \cdot y_2)/(2 \cdot b)$ and $Y_2(y_1) = (a - c - b \cdot y_1)/(2 \cdot b)$. A solution to these equations is given by

$$y_1^* = \frac{a - c}{3 \cdot b}, \quad y_2^* = \frac{a - c}{3 \cdot b}$$

for the individual equilibrium supplies and

$$y^{CN} = y_1^* + y_2^* = \frac{2(a - c)}{3 \cdot b}$$

for the equilibrium market supply.

Compare this solution to the monopoly solution, $y^M = \frac{a-c}{2 \cdot b}$, and to the solution under perfect competition, $y^{PC} = \frac{a-c}{b}$. (One can obtain this number from the "price equals marginal costs"-rule, which implies for the linear model that $a - b \cdot y = c$. Solving for y gives the result.) It follows that $y^M < y^{CN} < y^{PC}$, which reveals a lot about the effects of competition: comparing a monopoly with a duopoly and this, in turn, with perfect competition reveals that competition reduces inefficiency. However, with only two firms, the competitive forces are not strong enough to enforce the solution under perfect competition. Accordingly, the equilibrium price in a duopolistic market lies between the price in a monopolistic market and the price under perfect competition, because the demand function is monotonically decreasing. The different prices can be determined as markups on marginal costs:

$$p^{PC} = c < p^{CN} = c + \frac{a-c}{3} < p^M = c + \frac{a-c}{2}.$$

These markups play an important role as "rules of thumb" in the management literature, because they allow it to quickly assess the profitability of a market. They depend on the elasticity of demand, which is, in and of itself, a function based on the customers' tastes and incomes (as reflected in a and b), as well as the competitiveness of the market, expressed by the number of firms. The markup under perfect competition is zero and it is smaller in a Cournot duopoly than in a monopoly.

12.3　The Linear Cournot Model with n Firms

The above analysis suggests that the Cournot model builds a bridge between the model of non-price-discriminating monopolies and the model of perfect competition. In order to define this insight more precisely, it makes sense to analyze the equilibrium of an oligopolistic market with an arbitrary number of firms, to see how the number of competitors influences the outcome. The following paragraphs will determine the Nash equilibrium for the linear model with n firms in the market. For this purpose, some additional notation is needed. Denote the supply of any firm i by y_i and the supply sum of all firms, except for firm i, by y_{-i}. Then, firm i's profit equation is:

$$\pi_i(y_i, y_{-i}) = (a - b \cdot (y_i + y_{-i})) \cdot y_i - c \cdot y_i, \quad i = 1, \ldots, n.$$

Given the quantity supplied by all other firms, firm i's profit-maximizing supply can be determined by the first-order condition:

$$\frac{\partial \pi_i(y_i, y_{-i})}{\partial y_i} = (a - c - b \cdot y_{-i}) - 2 \cdot b \cdot y_i = 0, \quad i = 1, \ldots, n.$$

In general, there are n first-order conditions and n unknown variables $y_1, \ldots, y_n$. If one assumes that identical firms behave identically in equilibrium, i.e. for any two firms i and j $y_i = y_j$, one can replace y_i with y and y_{-i} with $(n - 1) \cdot y$. This substitution reduces the system of equations to one:

$$(a - c - b \cdot (n - 1) \cdot y) - 2 \cdot b \cdot y = 0.$$

If one solves this equation for y, one obtains the Nash equilibrium quantity of a representative firm as $y^* = (a - c)/((n + 1) \cdot b)$. Market supply $n \cdot y^*$ is then given by $n/(n + 1) \cdot (a - c)/b$. To understand this result, compare it to the one under perfect competition, which was determined as $(a - c)/b$. Then carry out the comparative-static analysis with respect to n by treating n as a continuous variable (which it is not, but the assumption facilitates the analysis):

$$\frac{\partial y^*}{\partial n} < 0, \quad \frac{\partial (n \cdot y^*)}{\partial n} > 0.$$

Two implications follow: first, individual supply is falling due to the number of firms; the more competitors there are, the less each single firm produces. Second, market supply is increasing due to the number of competitors. Even though each single firm produces less, if more competitors are on the market, this effect is over-compensated by the sheer number of firms. Now let the number of firms become very large, such that one obtains $\lim_{n \to \infty} n/(n+1) \cdot (a-c)/b = (a-c)/b$ in the limit: the market tends towards the equilibrium under perfect competition, if the number of firms gets very large. The other extreme is a monopolistic market, $(n = 1)$. In this case, the optimal quantity is $(a-c)/(2 \cdot b)$: the result from the monopoly model.

Alternatively, one can look at the markup the firms can charge. The equilibrium price is given by $p^* = a - b \cdot n \cdot y^*$, which is equal to $p^* = c + (a-c)/(n+1)$. It follows that the markup is decreasing in the number of firms and converges to zero, if n becomes very large. Hence, the Cournot model provides a theoretical foundation for the idea that competition drives a market towards efficiency: the more competitors there are, the smaller the individual firm's influence is on the outcome of the market. If the number of firms becomes arbitrarily large, then the influence of a firm completely disappears and it behaves as a price taker and sells according to the efficient "price-equals-marginal-costs" rule.

12.4 The Bertrand Duopoly Model

In order to see how price- instead of quantity-setting influences the behavior in such a market, assume that the duopolists choose the prices p_1 and p_2 instead of quantities. All other assumptions from the previous model persist and prices are assumed to be positive, real numbers (including zero). The only exception is that one directly assumes constant and identical marginal costs for both firms, $C_1(y_1) = c \cdot y_1, C_2(y_2) = c \cdot y_2$. Price competition with more general cost functions is very difficult to analyze formally and the fundamental ideas of price competition are contained in the simplified model.

The firms' profits are analogous to the previous model, but with the exception that, this time, prices are the strategic variables. Customers are confronted with two prices and they will choose their preferred firm and their optimal demand accordingly. Hence, $x_1(p_1, p_2)$ and $x_2(p_1, p_2)$ are the demand functions relevant for the two firms, for any given pair of prices p_1 and p_2. The profit functions become:

$$\pi_1(p_1, p_2) = p_1 \cdot x_1(p_1, p_2) - c \cdot x_1(p_1, p_2),$$
$$\pi_2(p_1, p_2) = p_2 \cdot x_2(p_1, p_2) - c \cdot x_2(p_1, p_2).$$

Both firms set prices simultaneously and independently. In order to be able to do so, they have to form expectations about the other firm's price p_1^e, p_2^e. A Bertrand-Nash equilibrium is a pair of prices, p_1^* and p_2^*, such that both firms maximize their profits given price expectations for the other firm, and these expectations are correct, $p_1^e = p_1^*, p_2^e = p_2^*$.

The maximization problems are non-standard, because the profit functions are not continuous in prices. Both goods are perfect substitutes from the point of view of the customers, so they will always buy the cheaper one. Assume that one firm charges a price that is a little bit higher than the price of the competitor. In that case, no one will buy from this firm. If the firm lowers the price just a little bit to undercut its competitor's price, then all customers will change their minds and now buy from this firm instead. The firm can meet this demand, because it can produce with constant marginal costs and without any capacity constraint. Hence, demand is non-continuous at this point.

An example is two neighboring bakeries that are on the way to work for a number of people. If one bakery sets a higher price for a croissant than the other bakery, then no one will buy there (one abstains from queuing or transaction costs of queuing). Hence, demand as a function of both prices can be written as follows. Let $X(p_i), i = 1, 2$ be the market demand function:

$$x_1(p_1, p_2) = \begin{cases} X(p_1), & p_1 < p_2 \\ 0.5 \cdot X(p_1), & p_1 = p_2 \\ 0 & p_1 > p_2 \end{cases},$$

$$x_2(p_1, p_2) = \begin{cases} X(p_2), & p_1 > p_2 \\ 0.5 \cdot X(p_2), & p_1 = p_2 \\ 0 & p_1 < p_2 \end{cases},$$

using the convention that consumers will be split up equally between the two firms, if prices are identical.

The non-continuity of the profit functions implies that one cannot characterize the best-response functions using partial derivatives of the profit functions. The non-continuity occurs at $p_1 = p_2$ because, at this point, demand switches from one firm to the other. To characterize best responses, the following paragraphs will focus on firm 1. A similar argument holds for firm 2, because of the symmetry of the problem.

If the purpose is to characterize just one equilibrium, then the task is simple: start with the conjecture that both firms offer a price that equals marginal costs, $p_1 = p_2 = c$. Market demand splits equally between the firms for this pair of strategies and both firms make zero profits. If a firm sets a higher price, it loses the demand and still makes zero profits. If it sets a lower price, it wins over all the customers, but sells at a price that is lower than its marginal costs, so it incurs losses. In other words, it cannot improve its profits by deviating to another price, which is the definition of a Nash equilibrium. Therefore, $p_1^* = p_2^* = c$ is a Bertrand-Nash equilibrium.

It is slightly more complex to prove that the equilibrium is unique. In order to show uniqueness, start with the scenario in which at least one firm sets a price below marginal costs. This price leads to losses for at least one firm (the one with the lower price). This firm can avoid these losses by increasing its price above that of its competitor. (If both firms set equal prices, the same logic applies.) Now, assume that at least one firm sets a price that is strictly larger than its marginal costs. If the other firm sets a price below marginal costs, then one is back at the case analyzed above. Thus, assume that the other firm sets a price above or equal to its marginal costs. If they are equal to marginal costs, both firms make a profit of zero, because one of them has no customers and the other is selling at marginal costs. The firm that is selling at marginal costs can increase its profits by increasing its price a little bit, making sure that it is above marginal costs, but below the price of the competitor. If the price is larger than marginal costs, but smaller than the competitor's price, it wins the whole market and also makes a profit. However, it is not rational for the competitor to stick to the higher price. He can increase his profits by slightly undercutting the other price, making sure that it is still above marginal costs. In this case, he wins over the market, which increases profits from zero to something strictly positive. Last, but not least, one has to focus on situations in which both firms set equal prices above marginal costs. In this case, they share the market equally, making positive profits. Denote the prices by $p > c$. Formally, this leads to $\pi_1(p, p) = 0.5 \cdot X(p) \cdot (p - c) > 0$. What happens if firm 1 deviates to a price $p_1 = p - \epsilon$, where ϵ is a small positive number, $\epsilon > 0, \epsilon \to 0$? Given that all customers buy from firm 1 now, profits become $\pi_1(p - \epsilon, p) = X(p - \epsilon) \cdot (p - \epsilon - c)$. Given that the firm wins half of the market by this change, there exists an ϵ that is small enough such that profits go up.

To summarize, the above line of reasoning has shown that the equilibrium is, in fact, unique. The model of Bertrand price competition has a stark implication: price competition drives prices all the way down to marginal costs. This result is remarkable: even with only two firms, the market behaves as if it were perfectly competitive. This result has an important implication for competition policy: the number of firms in a market is, in general, a poor indicator for the functioning of the market. No conclusive evidence about the intensity of competition can be drawn from the number of firms alone. Further information about the type of competition is necessary.

This result has been derived under very specific assumptions, especially regarding the absence of capacity constraints and identical marginal costs. In order to figure out how robust the results are, one must start with an analysis of the consequences of different marginal costs, $c_1 < c_2$. In this case, setting prices equal to marginal costs leads to different prices and only the low-cost firm 1 is able to sell its products. However, it no longer has an incentive to stick to a price that equals marginal costs, because it can still serve the whole market at higher prices, as long as it sets a price below firm 2's marginal costs (which define the lower limit for the price of this other firm). The exact strategy of firm 1 depends on the difference between both firms' marginal costs. Let p_1^M be the price that firm 1 would set, if it had a monopoly.

- If $c_2 > p_1^M$ then firm 1 is able to set the monopoly price without being threatened by firm 2. Due to a sufficiently large cost differential, firm 1 has a *de facto* monopoly, even though another firm exists that could enter the market. Firm 1 is protected against market entries, due to its cost leadership.

- If $c_2 < p_1^M$ then firm 1 cannot enforce the monopoly price, because it would encourage market entry by firm 2. This case is not only interesting because of its economic implications, but also because it shows a tension between economic intuition and mathematical modelling, where one has to ask which source is more trustworthy: one's intuition or the results from the theoretical model. Here is the problem: intuitively one would expect that the low-cost firm would set the highest price it can that is still lower than the marginal costs of the competitor, i.e. $p_1 = c_2 - \epsilon$ with $\epsilon > 0, \epsilon \to 0$. Such a price keeps firm 2 out of the market and is, at the same time, as close to the monopoly price as possible. Such a strategy does not exist from a mathematical point of view, however, because the set $p_1 < c_2$ is an open set (the boundary $p_1 = c_2$ does not belong to it). Hence, for each price, $p_1 = c_2 - \epsilon$, there exists a larger price, $\tilde{p}_1 < c_2 - 0.5\epsilon$, that leads to higher profits, which follows from the denseness of real numbers. The implication of the denseness of real numbers is that firm 1 has no optimal strategy, which in turn implies that there is no Nash equilibrium. This result is highly unsatisfactory, because intuition tells one that this is highly unlikely; that this problem is merely an artefact of an abstract property of real numbers.

  One way to bring intuition in line with the mathematical model is to impose a certain "granularity" on the set of admissible prices. If one assumes that prices are elements of a finely structured set of possible prices (the smallest change in prices could, for example, be 1/10 of a Rappen), then an equilibrium exists where firm 1 chooses the highest price lower than the marginal costs of the second firm (provided that it is higher than its own marginal costs).

  If the granularity of prices solves the problem, one may ask why this assumption was not used right from the beginning. The reason is twofold. First of all, the necessary notation would be more complex. Second of all, discrete price changes have unintended side effects of their own. For example, in the case of identical marginal costs, one would get the potential for multiple equilibria or positive profits in the equilibrium. These problems illustrate the role mathematics plays in economics: there is no deeper truth behind the mathematical formalism used in most theories. Mathematics helps one to understand the logical structure of arguments: it does no more nor less.

12.5 Conclusion and Extensions

The Cournot and the Bertrand models lead to radically different predictions about the functioning of oligopolistic markets. The natural question then becomes which model is more adequate to describe oligopolistic behavior. Unfortunately, the answer to this question is not that simple. The Cournot and Bertrand models are only the tip of the iceberg of models of oligopolistic behavior that have been developed over the years and that focus on different aspects of firm strategies in such a market environment. Firms can, for example, also compete in the positioning of their products, technological innovations, marketing or reputation. It depends on the specific industry, maybe on the exact period of time, as well as on other factors that are hard to predict, whether a market is more adequately described by quantity or by price competition. While both models are useful, a meta theory that explains and clarifies the conditions under which each model is more adequate is still missing.

In a nutshell, it can be argued that the Bertrand model is useful for the analysis of price wars. It shows that the results of the model of perfect competition may also hold in markets with few firms. This has important methodological consequences, because it implies that the much easier model of perfect competition can also be used to analyze industries with few competitors, as long as there is evidence that they engage in price competition.

The Cournot model is useful for the analysis of firms' behavior in less competitive situations. It builds a bridge between the monopoly model and the model of perfect competition, because it predicts a continuous adjustment from the monopolistic to the perfectly competitive equilibrium as the number of firms increases.

Economists have tried to develop a "unified" approach to the Cournot/Bertrand problem. An interesting one is to disentangle the problem of an oligopolist into two stages. The idea is to assume that a firm's production capacity has to be planned at a relatively early stage (stage 1) and that the firm is then committed to produce within the chosen capacity constraint. The production decision (stage 2) takes place under conditions of price competition. Interestingly, such a two-stage game is able to predict Bertrand-type price competition in periods of low demand and overcapacity (capacity constraints are not binding). At the same time, the market transforms into Cournot competition, if capacity constraints are binding. Given that firms try to avoid overcapacities (they are costly), Cournot competition can therefore be regarded as the normal case, if demand is relatively predictable. However, if demand fluctuates widely over time, there will be periods of Bertrand competition again and again.

Independently of whether one is confronted with price or quantity competition, firms have a strong incentive for coordinated or collusive behavior. The reason is that the joint industry profits are maximal, if the firms coordinate on the monopolistic solution and share the profits equally. To see this assume, on the contrary, that industry profits would be maximized in the oligopolistic equilibrium. If this were the case, the monopolist could imitate the oligopolists and choose the Cournot or Bertrand solution instead. The fact that a profit-maximizing monopolist prefers another solution shows that he must be better off. Thus, it is in the interest of the oligopolists to collude and constrain their production in an attempt to move closer to the monopolistic outcome, which creates a tension between profits and efficiency of the market. Different strategies are possible to achieve this goal.

- Firms can try to make explicit price-fixing agreements. However, this is illegal in most countries, exactly because it would make the market less efficient. Hence, firms have developed more subtle strategies to coordinate their outputs.
- One way of reducing competition is through a merger or an acquisition (M&A). These measures usually have to be approved by the national or supranational competition authorities. However, even if M&As are not an option, in practice it

is sometimes possible to gain control over some other firm's strategies by complicated cross-ownership or holding structures.

- It is also possible to reach implicit agreements on prices or quantities that fly below the radar of the competition authorities. These agreements are relatively easy to achieve, because of the limited number of firms that all operate in the same industry but, at the same time, difficult to enforce. However, enforcement is crucial, because every single firm has an incentive to break the agreement and sell a little more at a lower price. The reason is that the monopoly solution is not a Nash equilibrium, so every single firm can profit from unilaterally deviating from a non-equilibrium strategy. Coordination in an oligopolistic market has the structure of a prisoner's dilemma. A way out of this dilemma opens, if firms compete repeatedly. If firms compete not only today, but also in the future, then trust can build and they can, in principle, punish deviations from cooperative behavior over time. The exact conditions under which cooperation can be stabilized, by repeated interactions, are complicated to characterize, but an important factor is how forward-looking firms are. If they focus heavily on the present, then future gains and losses are of only secondary importance, which makes the enforcement of cooperative behavior difficult.

Digression 38. The Prisoner's Dilemma and Frames of Reference
From the point of view of the competing firms, Cournot and Bertrand equilibria have the character of the prisoner's dilemma: both firms could be better off by coordinating on the monopoly solution, but individual rationality leads them to a different outcome.

At this point, one could argue that, as with the prisoners in their interrogation rooms, this solution is no dilemma at all, because the general public profits from the inability of the firms or prisoners to cooperate. The prisoners are guilty and end up in jail and the outcome of oligopolistic competition is closer to the Pareto optimum than the monopolistic one is.

What this discussion shows is that the perception of a problem depends on the frame of reference. Oligopolistic competition is a cooperation problem from the point of view of the firms, but not from the point of view of society. On the contrary, society can make use of the dilemma structure between the firms to make markets more efficient.

Thus, the existence of a cooperation problem does not automatically imply that society should do something about it. It depends on the frame of reference (the most adequate one from a normative perspective), whether a cooperation problem is perceived as a vice or as a virtue.

Empirical industry studies usually identify many factors that influence market behavior, but that change rather frequently, which makes it very difficult to empirically identify and control, *ceteris paribus*, experiments to test the theory. One way out of this dilemma is to test the theories in the lab by means of market experiments.

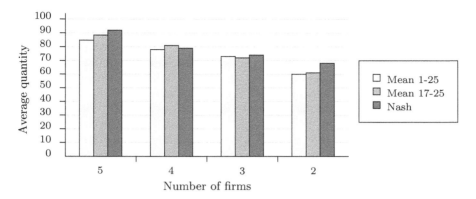

Fig. 12.3 Experimental results about collusion and the intensity of competition

The advantage of this approach is that the researcher can control a lot of the relevant factors by the design of the experiment. However, the validity of experiments is limited, because participants are aware that they are not in real markets, but in the lab. There is an extensive debate about the so called *external validity* of experiments that this chapter will not cover. Instead, this subchapter will briefly summarize the main findings from the literature on experimental oligopoly theory.

In experiments about Cournot quantity competition studies find a lot of support for the predictions of the model, if the experiment is run for a single round and subjects are anonymous and cannot communicate with each other. Repeated interaction and the possibility to communicate reduces the intensity of competition and collusive behavior becomes more likely. However, collusion is fragile and depends on the number of firms (players) in the experiment. In a duopoly, collusive behavior can be frequently observed, but it breaks down quickly, if the number of players increases. With four firms (players), the intensity of competition is generally higher in experiments than is predicted by the theory and the solution converges very quickly to the competitive equilibrium. Figure 12.3, which is taken from Huck, S., Normann, H.T., Oechssler, J., 2004, illustrates this result.

The number of firms, or players, is denoted on the abscissa, and average quantities are displayed along the ordinate. The experiment ran for 25 rounds. The left bar shows the average production levels over all rounds. The middle bar depicts the average production levels over the last nine rounds. The right bar gives the theoretical prediction of the Cournot model. One finds the pattern described above: with only two firms, production is lower than theoretically predicted, which points towards some form of implicit collusion. The incentive for collusive behavior is, however, more limited towards the end of the experiment, which supports the idea that cooperative behavior is supported by repeated interactions because, the closer one gets to the end, the less credible future punishments become. A second result of the experiment is that total production increases in the number of firms.

The Bertrand model has also been experimentally tested. Dufwenberg and Gneezy (2000) found that the experimental findings are in line with the theoretical predictions, if the number of firms is three or larger. With only two firms and repeated interactions (in which pairs of players are randomly matched in each round to make sure that the above-mentioned punishment mechanisms cannot work), the authors found that prices are consistent with the theoretical predictions during the first rounds, but increase later on in the experiment. Apparently, the realization of the bitter truth that the "price-equals-marginal-costs" rule leads to zero profits made the players creative in findings ways to implicitly collude. This seems to be possible in duopolistic situations, even if players are randomly matched but, as soon as the number of players exceeds two, it apparently becomes too complicated to figure out collusive strategies.

Digression 39. The Three Cs of Economics
Chapter 11 explains that games can be interpreted as structural metaphors that allow one to gain insight into the logic of individual decision making and collective outcomes. It asserts that society has to overcome two types of challenges, if it wants to alleviate scarcity, cooperation problems and coordination problems. At the beginning of this chapter, the argument is brought that a third type of problem exists: commitment. Commitment problems lie at the heart of the solution to cooperation, as do coordination problems. To see why, take a prisoner's dilemma as an example. In this cooperation problem, players would like to mutually coordinate on the the cooperative strategy, but individual rationality makes cooperation not credible. Hence, what is missing is a commitment device that allows them to overcome the credibility problem. Coordination problems have a different logic, but commitment mechanisms play a crucial role, as well. If all players could publicly commit to a specific strategy, the equilibrium-selection problem would be solved.

Hence, coordination, cooperation and commitment problems define the structural landscape of economics. This is why they can be called the three Cs of economics.

Such a structural approach to economics has two main advantages:

- First, the simplicity of the the three Cs approach gives one a frame of reference for the interpretation and understanding of societal problems. Is it a coordination problem or a cooperation problem? What kind of commitment device might help to overcome it? Additionally, if there is no problem, then what kind of commitment mechanism is in the background that helps in stabilizing the efficient outcome?

  Here are two examples that illustrate this approach: Chapter 10 demonstrated that externalities can be interpreted as unresolved cooperation problems. Hence, the next step is to think about commitment mechanisms that help in internalizing them. On the other hand, previous chapters have argued that a complete set of competitive markets leads to efficiency under certain assumptions. The commitment device in the background is a system of perfectly enforced property rights. But is this the end of the story? Who enforces the property rights and is it in the interest of this person to do so? Does one have to dig deeper to identify commitment mechanisms for law enforcers? etc.

- Second, the three Cs are a tool for future studies. When one begins to study more elaborate and advanced economic theories, it is easy to lose track of the basic story underlying the theory. Yet most, if not all, theories are variations of coordination or cooperation problems, plus some more or less elaborate ideas on commitment. Approaching these theories with a three-Cs perspective helps one to make sense of them. It also helps one to scrutinize the basic ideas of these theories. Is the problem at hand adequately described as a coordination or cooperation problem? Are the institutions the theory focuses upon convincing, in the sense that they are credible commitment devices and, if not, why?

Part III of this book gave an introduction into the functioning of different prototypical markets. The following table summarizes the main findings from those chapters.

Overview of market structures (long run)	Sellers	Buyers	Price	Profits	Efficiency
Perfect competition	Many (homogeneous goods)	Many	$p = MC$, in the long run $p = \min AC$	$\pi = 0$	Efficient
Bertrand oligopoly	Few (same cost structure)	Many	$p_B = MC$	$\pi_B = 0$	Efficient
Cournot oligopoly	Few	Many	$p_C > MC$	$\pi_C > 0$	Inefficient
Monopoly (no price discrimination)	One	Many	$p_M > p_C > MC$	$\pi_M > \pi_C > 0$	Inefficient
Monopoly (1st degree price discrimination)	One	Many	$p_M^j = $ individual j's willingness to pay	$\pi_M = $ maximum sum of CS and PS $> \pi_M > \pi_C > 0$	Efficient
Monopolistic competition	Many (heterogeneous goods)	Many	$p = MC + \mu = AC$, $\mu = $ mark-up	$\pi = 0$	Inefficient

References

Dubner, S. J., & Levitt, S. (2005). *Freakonomics*. William Morrow.
Dufwenberg, M., & Gneezy, U. (2000). Measuring Beliefs in an Experimental Lost Wallet Game. *Journal of Games and Economic Behavior*, 30(2), 163–182.
Huck, S., Normann, H., & Oechssler, J. (2004). Two Are Few and Four Are Many: Number Effects in Experimental Oligopolies. *Journal of Economic Behavior and Organization*, *53*(4), 435–446.
Ovid (2002). *Ars Amatoria (The Art of Love)*. Modern Library.
von Stackelberg, H. (1934). *Marktform und Gleichgewicht*. Wien: Springer.
Wilde, O. (1892). *Lady Windermere's Fan*. Bloomsbury Publishing.

Further Reading
Belleflamme, P., & Peitz, M. (2015). *Industrial Organization: Markets and Strategies*. Cambridge University Press.
Tirole, J. (1988). *The Theory of Industrial Organization*. MIT Press.

Part V
Appendix

A Case Study

13

This chapter covers ...

- how to apply the theoretical insights from the last chapters in order to gain a better understanding of a specific market or industry.
- how legal, technological and economic aspects of an industry work hand in hand in determining the functioning of markets.
- how to interweave empirical facts with economic theory to thereby build a case.
- some facts about the Europan aviation industry.
- how all these facts influenced Swiss Air in the years before its grounding.

13.1 The Grounding of Swissair as a Case Study for the Use of Economic Theory

> Swissair's collapse this week stranded thousands of passengers world-wide; saw its planes blocked in London; and left fliers holding potentially worthless tickets. A widespread feeling in Switzerland is that the airline that was the national pride was finished off by the banks that embody its national character of reliable, no-nonsense business. Yet its undoing may have been something very un-Swiss: bad management. [...] The plunge in air traffic after the Sept. 11 terrorist attacks in the U.S. pushed Swissair over the edge, but it has been flirting with collapse for months. [...] The nightmare began as a grand plan for growth. Like scores of other companies from this small, land-locked country, Swissair grew into a global player. 'An inflexible regulatory environment and some poor investments' crippled Swissair, says Damien Horth, an airline analyst at ABN-Amro in London. 'Poor management by Swissair in terms of its acquisitions and not controlling its associates well' proved fatal. (The Wall Street Journal, October 02, 2001)

We have covered a lot of ground in the last chapters and one should, by now, have a decent understanding of the functioning of prototypical markets, how they contribute to welfare and their weaknesses. We have also devoted a lot of pages putting the theories into perspective and applying them toward getting a better understanding of the societal phenomena that are characteristic for today's societies. The case

© Springer International Publishing AG 2017
M. Kolmar, *Principles of Microeconomics*, Springer Texts in Business and Economics,
DOI 10.1007/978-3-319-57589-6_13

studies have been relatively short, however, and have been tailored to specific theories, or even to specific aspects of a theory. What is still missing is a case study of sufficient complexity that it allows one to bring different theories together and to discuss the adequacy of the different theories for gaining an understanding of complex economic and social issues.

This chapter is an attempt at filling this gap and at illustrating how economic theories can be used to analyze and to better understand developments in markets and industries. The case that I am analyzing is the spectacular grounding of Swissair, a former Swiss airline. In order to be able to do so, one has to combine insights from different chapters. As one will see, a narrow economic focus is not sufficient to get a grip on this case. Rather, one has to embed economic analysis into the integrative approach that is fostered in this book, in order to gain a better understanding of the different factors that contributed to the insolvency of this once proud airline. One will see that legal, political, managerial and cultural aspects played important roles in this case. However, this analysis will also mention the limitations of the theoretical framework that has been developed in this book. Some aspects of the aviation industry require more elaborate market models, and I will briefly show how one can use the theories presented in this book to address these issues and to develop the theories in the directions necessary to understand these more complex aspects.

The market structure of an industry reflects the technology of production, the size of the market, as well as the legal and regulatory framework in which the firms operate. Changes in any of these factors can trigger deep, structural changes that impact on the number of competitors in a market, as well as the way competition works. Some of the reasons for the grounding of Swissair cannot be understood, if one does not take these factors into consideration. In the following pages, I will briefly summarize the most important theoretical insights from the preceding chapters that are relevant for a better understanding of the airline industry and apply them to the Swissair case. They provide one with a toolkit that allows one to better understand some of the key factors that determine the functioning of the airline industry. If one is still familiar with them, one can skip this subchapter. However, there are some additional properties of the industry that have to be taken into consideration for a comprehensive understanding, which require more advanced theories and therefore have to be left out of consideration. The main purpose of the following case study is, therefore, twofold:

1. It should help one to see how economic theories can be used in order to better understand real-world phenomena, how to select the most adequate theories, and how to use one's insights to gain an understanding of the case. The fact that one has to conduct a thought experiment under "laboratory conditions," which leave out some important aspects of the problem, does not compromise this approach, but instead creates a relatively accessible foundation. Hence, one is not aiming to present a "full-scaled" report on the case, but rather a version that allows one to apply and restrict one's attention to the theories that one has learned throughout the previous chapters.

2. It illustrates how economics, law and management can and should work hand in hand to better understand the logic of social phenomena. In the end, good political and managerial decisions become more likely, if they are built on such an integrative approach.

13.2   Some Facts About the Aviation Industry in Europe

On March 31, 2002, after 71 years of service, Swissair ceased operation. This was the official endpoint of an economic downturn that led a once major international airline into bankruptcy. The airline prospered well into the 1980s, when it was one of the five major airlines in Europe. It was known as the "Flying Bank," due to its financial stability, and it was considered a national icon in Switzerland. How is it possible that a "Flying Bank" can turn into a money burner within 20 years? Which factors contributed to the demise of this airline?

A major event like the grounding of an airline can never be traced back to only a few causal factors. Reality is messy and one should shy away from oversimplifications. An economic analysis of the European aviation industry sheds at least some light on the case and makes some of the aspects that contributed to the grounding more transparent. However, an economic analysis of the case only gets one so far. In the end, the interplay between legal, technological and institutional factors created an environment in which managers had to act and define strategies for their firms. This environment may have been relatively hostile towards an airline like Swissair, but there is no direct causal chain from the changing economic logic of the industry to the demise of Swissair.

After World War 2, air traffic increased rapidly and many airlines profited from the political regulation of the markets that created national, *de-facto*, monopolies. During the 1960ies and 1970ies, Swissair was considered one of the best airlines of the world and made huge profits. Things began to change in the 1980ies when the European Community started a process of liberalization of the community air transport market, to which the member states committed themselves in 1986, and that also became relevant for Switzerland. In order to create a single market for air transport, the EU liberalized its air transport sector in three stages, called "packages." This process culminated in the third package, adopted in 1993 and extended in 1997. It introduced the freedom for any airline of a member state to provide services within the EU and the freedom to provide "cabotage," the right for an airline of one member state to operate a route within another member state. This single market was extended to Norway, Iceland and Switzerland in the following years.

This process of liberalization gradually changed the market structure from a system of regionally partitioned monopolies into a system of interregional competition, leading to a period of "cutthroat" competition in a market with too many, too small airlines, i.e. a form of competition where it is clear that some firms will be forced to leave the market. The following analysis will show that it was clear from the onset that this change in the political regulation of the industry would eventually lead to a consolidation and concentration of airlines. Different airlines began from different starting positions in this process of predatory competition. The "Flying Bank" Swissair had a head start, because of its large asset holdings and huge liquidity. However, the fact that Switzerland rejected taking part in the European Economic Area in 1992 was a huge disservice to Swissair, because the emerging common airline market was, for that reason, not a level playing field. Here are a few examples of the obstacles that confronted them: Swissair planes were not allowed to take up passengers during intermediate landings in EEA countries and Swissair was not allowed to sell tickets for sections within EEA member countries.

13.3   Applying Economic Theory to Understand the Aviation Industry

Now, one can lay down some principles that govern optimal firm behavior in a monopolized aviation industry. One should focus one's attention on two different technological characteristics that are of major importance: the technology-induced cost structure of an airline and the bundling problem that results from the network structure of the product. These factors influence the pricing strategies of the airlines and, thereby, its profits. The network of flights offered by an airline determines its portfolio of different products, which implies that all airlines offer a different product portfolio, even if there may be some routes for which they compete directly.

Digression 40. Additional Aspects
This analysis gives only a very broad concept about airline pricing. There are at least three additional aspects in reality that complicate pricing, but that also make pricing in this industry intellectually fascinating. (1) There is no spot market for flights, so demand for a specific flight drops more or less stochastically over time before the flight. This implies that there is no single price on such a "dynamic" market, but a time-dependent price function. Prices may vary over time, depending on load factors, and so on. (2) Each flight has a given capacity, which implies that the marginal costs of an additional passenger are very small before and very large after the capacity threshold is reached (one would have to change the aircraft). Hence, any cost-plus pricing rule would discriminate prices at this point. The resulting problem is known as *peak-load pricing* in the literature. (3) An airline offers a network of different connections, which implies that there are complementary, as well as substitutive, edges in each network and airlines also compete with respect to their network structures.

In order to be able to analyze the effect of liberalization on the industry, remember that the model of oligopolistic quantity setting (Cournot competition) includes the case of perfect competition (and thereby also Bertrand Competition) as a special case, with a perfectly elastic demand function. Hence, one can restrict one's attention to a short repetition of this model in order to be able to distill the main messages for the Swissair case.

13.3.1 Costs

Chapter 7 explained that a firm's total costs, $C(y)$, are the sum of fixed costs, FC, and variable costs, $VC(y)$, the last of which depend on the quantity produced, y. In the airline context, y may be, for example, the number of passengers transported from A to B or the frequency of flights offered. Hence, one can describe the total costs as:

$$C(y) = VC(y) + FC.$$

One important characteristic of the aviation industry is the structure of the airlines' cost functions: fixed costs are a significant share of total costs because the logistic infrastructure is, at least in the short run, largely independent of the occupancy rate. Fixed costs do not influence the pricing policy of a firm, but are relevant for profits and for determining whether a firm stays in a market or has to exit it.

Fixed costs are, to a large degree, capacity costs: that is, the depreciation and financing costs of the aircraft fleet and its maintenance, the costs of the supporting infrastructure, as well as the costs of landing rights and the handling of passengers at airports (contracts are usually longer term).

An airline's variable costs, for a given flight, encompass gasoline, onboard services like free drinks and meals (if they exist), and so on. If one breaks down costs to a single passenger, not even gasoline costs are variable. Hence, depending on the level of aggregation, variable costs are relatively unimportant in this industry. This leads to the following observation: average total costs of a firm are:

$$AC(y) = \frac{C(y)}{y} = \frac{VC(y)}{y} + \frac{FC}{y}.$$

Average costs are decreasing over a given range when y is small and, depending on the structure of the variable costs, they might even be decreasing for all y. In order to see this, note that $VC(0) = 0$ by definition. Because fixed costs are large, average costs decrease over a significant range of y and decrease over the whole range (for all y), if marginal costs are constant or decreasing. Assume, for simplicity, that variable costs are linear,

$$VC(y) = c \cdot y, \qquad c > 0,$$

which implies that marginal costs are constant and equal to c. Hence, the average-costs function is:

$$AC(y) = c + \frac{FC}{y}.$$

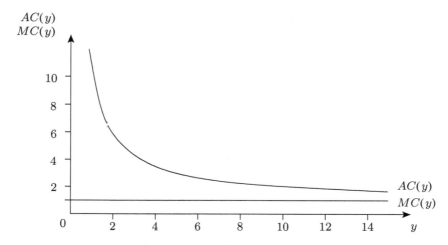

$AC(y)$
$MC(y)$

10

8

6

4

2

0   2   4   6   8   10   12   14

$AC(y)$
$MC(y)$

y

Fig. 13.1 Average and marginal costs

Average costs decrease strictly for all y and converge to c as y grows large, because $AC'(y) = -FC/(y^2) < 0$ and $\lim AC(y)|_{y \to \infty} = c$. In Fig. 13.1, one can see this relationship in the special case of $FC = 10$ and $c = 1$.

What are the implications of this type of cost structure for the functioning of the industry? First, a single firm can operate a given network more efficiently than two competing firms can that share the market:

$$C(y) = FC - c \cdot y < 2 \cdot C(y/2) = 2 \left(FC + \frac{1}{2} \cdot y \right) = 2 \cdot FC + c \cdot y.$$

There are size effects, because an increase in output decreases average costs. Because of this property, industries like the aviation industry have a tendency for concentration, because fixed costs limit the number of competitors that can profitably operate in the market. However, it indicates that the number of firms, for which this is the case, is rather limited. Hence, starting from a situation with a large number of protected airlines, it is very likely the case that competition leads to concentration; some airlines will not be able to survive the process of market liberalization, as soon as they start to compete on some segments of the networks. Airlines were forced to play musical chairs when the European Community decided to liberalize the market. However, to the extent that average costs are downward sloping, the process of market concentration is potentially efficiency-enhancing, because it reduces the total costs of production.

13.3.2   The Linear Cournot Model with n Firms

In the following subchapter, assume that the effect of an increase in competition can be captured by the Cournot model of oligopolistic competition. Taken literally, the model could only be applied to flights for which airlines directly compete (e.g. Zurich – Frankfurt), because the products have to be perfect substitutes. The assumption of Cournot competition is, however, innocuous insofar as that the qualitative results do not depend on this specific market model. The assumption of monopolistic or Bertrand competition would lead to similar conclusions. Also, more elaborate pricing strategies, or more complicated models of network competition, with imperfect substitutability, would also leave the qualitative results unchanged. If the qualitative results are robust in this sense one can – remember the epistemic status of positive theories that was discussed in Chap. 1 – go for a simple model.

In order to be able to understand the effects of competition, one can determine the Nash equilibrium for a linear Cournot model, with n airlines that compete in (some segment of) the market. I follow the notation from Chap. 12. (If one is still familiar with the n-firm model from Chap. 12, then one can skip the derivation of the Nash equilibrium and jump immediately to the conclusions.)

Assume, for simplicity's sake, that the demand for the services of a given airline is linear and has the following form:

$$p = a - b \cdot Y,$$

where Y is market supply, $a > c$ denotes the maximal willingness to pay in the market and b quantifies how price-sensitive the market is. Denote by y_i, the supply of a single airline i, and by y_{-i}, the supply sum of all airlines except i. Then, airline i's profit equation is:

$$\pi_i(y_i, y_{-i}) = (a - b \cdot (y_i + y_{-i})) \cdot y_i - c \cdot y_i - FC, \quad i = 1, \ldots, n.$$

Given the quantity supplied by all other firms, firm i's profit-maximizing supply can be determined by the first-order condition, which establishes the well-known "marginal-revenues-equals-marginal-costs" rule:

$$\frac{\partial \pi_i(y_i, y_{-i})}{\partial y_i} = (a - c - b \cdot y_{-i}) - 2 \cdot b \cdot y_i = 0, \quad i = 1, \ldots, n.$$

A single airline, i, behaves like a monopolist on a "curtailed" market with a market-demand function of $p = a' - b \cdot y_i$, where $a' = a - b \cdot y_{-i}$. Figure 13.2 depicts this situation for a curtailed demand function, $p = 11 - y$, and cost function, $C(y) = y + 10$. In this situation, the airline's optimal output is $y^M = 5$, with a corresponding price of $p^M = 6$. Per-unit profit would be $p^M - AC(y^M) = 6 - 3 = 3 > 0$. (Remember that this result need not be a Nash equilibrium: declaring it as such would require specifying all the parameters to make sure that all the other airlines are on their reaction functions, as well.)

This formulation reveals that the relevant demand function for a single airline, $p = a' - b \cdot y_i$, depends on the total quantity supplied by all other airlines, which captures the effect of competition in this model: an increase in y_{-i} shifts the curtailed demand function inwards; an increase in the total supply of the competitors

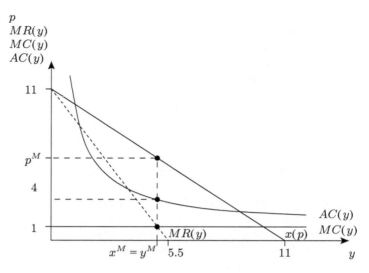

Fig. 13.2 The airline's optimal policy for a given curtailed demand function

has the same consequences as a reduction in the demand for aviation services does. In the example displayed in the figure, the price is above average costs, which implies that the airline makes positive profits. If competition has the effect of shifting the curtailed demand function inwards, it is easy to see that there is a point at which price equals average costs, such that the airline can no longer operate profitably. This is the point at which the airline is forced out of the market, if it cannot cut costs or make its services more attractive to the customer.

If one assumes that all other airlines supply the same quantity, $y_{-i} = (n - 1)y$, it becomes apparent that this downward shift in the curtailed demand function may be a result of an increase in the quantity supplied by the competitors, holding the number of competitors constant, or the result of new airlines competing in the market. The model, therefore, makes very sharp predictions about the effect of market liberalization on a single airline: as soon as new airlines start to compete with the formerly monopolistic network of, for example, Swissair, this increase in competition "steals" part of the demand from Swissair, which eventually reduces profits to zero. How long it takes before an airline starts making losses depends on its fixed and variable costs.

In order to derive more detailed results, one has to solve for the Nash equilibrium. In general, the first-order conditions specify a system of n equations and n unknowns, $y_1, \ldots, y_n$. If one assumes that identical firms behave identical in equilibrium so that, for any two firms i and j, $y_i = y_j$ holds, one can replace y_i with y and y_{-i} by $(n - 1) \cdot y$. One ends up with one equation and one unknown variable:

$$(a - c - b \cdot (n - 1) \cdot y) - 2 \cdot b \cdot y = 0.$$

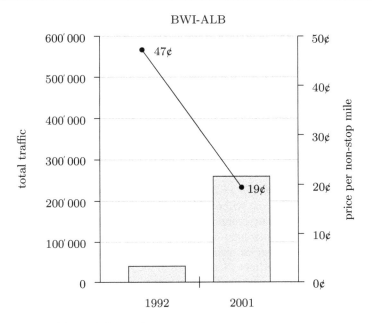

Fig. 13.3 Prices and demand after market entry

If one solves this equation for y, one obtains the equilibrium quantity of a firm as $y^* = (a-c)/((n+1) \cdot b)$ and market supply $n \cdot y^*$ is $n/(n+1) \cdot (a-c)/b$. One can also derive the market price, as well as the airline's profits:

$$p^* = \frac{a+nc}{n+1},\tag{13.1}$$

$$\pi^* = \frac{(a-c)^2}{(n+1)^2 b} - FC.\tag{13.2}$$

These findings allow a more detailed analysis of the effect of competition (which one can interpret as an increase in the number of competitors n in the market). First, looking at equilibrium prices, one finds that $\partial p^*/\partial n < 0$: an increase in competition brings prices down and, given that demand and prices are inversely related, leads to an increase in total demand.

Reliable data for this effect exists for the US-market. Figure 13.3 demonstrates the potential empirical magnitude of this effect for a single route, the Baltimore (BWI) to Albany (ALB) market for the 1992-2001 period. The entry of Southwest Airlines into the market decreased the average fare by 61%, causing passenger demand to increase by 641%.

The downward pressure on prices, which is caused by competition is, in and of itself, no problem; on the contrary, it can be argued that it makes the market more efficient, because it brings the equilibrium closer to a Pareto-efficient allocation. However, a look at profits reveals that competitive pressure can lead to deeper structural changes in the market. The first term in the profit condition is revenues minus variable costs. It defines the gross margin that an airline can use to cover its fixed costs. This gross margin decreases as the number of competitors increases, which implies that a number of competitors exists, due to which airlines start to make losses. Denote this number by $\bar{n}$. It can be determined from (13.3) by setting profits equal to zero:

$$\pi^* = \frac{(a - c)^2}{(\bar{n} + 1)^2 \cdot b} - FC = 0 \Leftrightarrow \bar{n} = \frac{a - c}{\sqrt{b \cdot FC}} - 1. \tag{13.3}$$

If, for example, $b = 1, c = 1, a = 2001$ and $FC = 1.000.000$, then the maximum number of firms that can exist without taking losses is $\bar{n} = 2$. Of course, given that market liberalization heats up competition and competition brings down profits, the maximum number of airlines that can survive in a regulated, quasi-monopolized market exceeds the number of airlines that can survive on a liberalized market. Hence, if the number of airlines that operate in the regulated market, $\hat{n}$, exceeds $\bar{n}$, then cutthroat competition sets in. This was exactly the situation European airlines were confronted with at the beginning of market liberalization.

If cutthroat competition is the effect of market liberalization, one may ask why airlines were willing to enter new, formerly protected markets. The answer to this question follows the logic of Cournot competition and the market-entry game. Assume that an airline does not offer a direct flight from A to B before market liberalization and that it considers opening this route. The incumbent had a monopoly before liberalization and will suffer from market entry. However, even if total market profits fall after market entry, because of competition, it is still profitable to enter, as long as profits are positive. The fact that the incumbent is worse off is irrelevant for the entry decision (one is facing a cooperation problem).

It could also be argued that the picture painted above is incomplete, because an airline like Swissair could compensate the loss in profits in its formerly protected markets (routes) by entering other markets (routes); competition is not a one-way street. In order to see why this logic is flawed, one has to remember that monopoly profits exceed the sum of oligopoly profits in a market. Hence, if two former monopolists on markets A and B start competing with each other on both markets, total profits are reduced. Therefore, the additional profits from entering new markets cannot compensate the loss in profits in one's former monopoly market. The only case where it is, theoretically, possible that an airline can increase its total profits is a situation with asymmetric competition, where the airline enters more new markets that there are competitors entering its formerly monopolized market. (In this example, this effect would trivially occur, if the airline operating market B decides not to enter market A, but the airline operating market A enters market B.) As this chapter already suggested, Swissair was not able to compete on a par with other European airlines because it did not belong to the EEA. Hence, it was much more difficult for Swissair to compensate for losses in the home market by entering new ones than it was for its competitors from the EEA.

The managers of the airlines could have known that market competition would eventually lead to market concentration by either insolvencies of some carriers, mergers and acquisitions, or strategic alliances. What was not clear from the outset, however, was whether Swissair would be among those airlines that survived this process, despite its handicap.

13.3.3 Extension: Network Choice, Acquisitions, and Strategic Alliances

The effect of market liberalization was, of course, that airlines started to operate flights on routes that had previously been monopolized. Any new route has an effect on the network structure of an airline and this network structure is such an important factor for an understanding of the functioning of the market that one has to devote some time and energy to this fact so that one can fully appreciate it. In order to do so, one can use the models in the toolbox gained from previous chapters as heuristics so one can develop an understanding of more complex technological and market structures.

Its network structure is an important element for the success of an airline. An airline's network is the collection of routes or connections that it offers. This network structure is important for at least two reasons. First, it is an important determinant of the airline's total costs. Depending on the size and structure of the network, costs may differ and it is, therefore, of great importance to develop and structure the network efficiently. Second, the size and structure of the network influences demand, because it influences the potential customer's willingness to pay.

The second argument can be illustrated by means of the following example. Assume an airline offers a flight between two cities, A and B, and considers extending its network by offering a new connection between cities B and C. The first effect is, of course, that this new flight creates demand from those passengers who want to travel from B to C. However, there is a second, indirect effect, because the new connection creates additional demand by those passengers who want to travel from A to C, who can now be served by this airline. Hence, from the airline's point of view, the value of network $A - B - C$ exceeds the sum of values of sub-networks $A - B$ and $B - C$, which is a simple form of a positive network effect. The fact that network effects imply that the total is more than the sum of its parts has implications for the optimal network structure, from the point of view of the demand side: the network should be extended by including additional routes, if the additional revenues from an additional route (including network externalities) exceed the additional costs (airport charges, direct operating costs of the flight, etc.).

The first, cost-saving argument can be illustrated by the same three-cities model with cities A, B, and C. Assume that, for given prices m, passengers want to travel from any city to either of the others (A to B, B to A, A to C, C to A, B to C, and C to B). An airline has, basically, two options at its disposal. The first is to offer the full set of connections (FCN, fully connected network): that is, to offer the capacity to accommodate demand for each route. The second is to use one of the

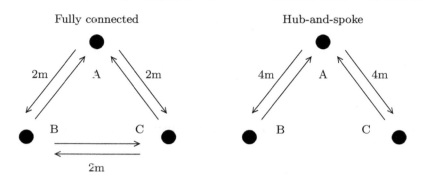

Fig. 13.4 Different network structures, fully connected and hub-and-spoke

cities, say A, as a hub that is connected to both other cities, but to not connect the others directly with each other. Passengers who want to fly from B to C, or *vice versa*, need to fly via A. This structure is called a hub-and-spoke (HSN) network. Figure 13.4 shows the two network types and the corresponding traffic flows on each route.

In order to illustrate how network structures influence costs, I introduce a simplified model that illustrates how network structure influences an airline's costs. For simplicity, the study only encompasses a single airline, but the general message remains valid in a competitive environment. The tree different routes are denoted by AB, BC and CA. Demand is $2 \cdot m$ for each route, m for each direction. The airline's total costs are a function of the number of routes it operates and of the number of passengers on each route. There are fixed costs for operating a route, F, and variable costs on each route, i, are a function of the total number of passengers, k, transported on that route times marginal costs, c, so:

$$VC_i(k_i) = c \cdot k_i^\alpha.$$

where $\alpha \geq 1$ is a parameter that determines whether marginal cost increase ($\alpha > 1$) or are constant ($\alpha = 1$). Note that k is not necessarily equal to $2 \cdot m$ because, if the HS-network is chosen, nobody flies on BC, but all the BC passengers have to take a detour *via* A, and analogously for all passengers who want to go from C to B. In that case, $k_{AB} = k_{CA} = 4 \cdot m$. If the airline offers a fully connected network, $k_i = 2 \cdot m$, then total costs are

$$TC^{FCN} = 3 \cdot c \cdot (2m)^\alpha + 3 \cdot FC = 3 \cdot c \cdot 2^\alpha \cdot m^\alpha + 3 \cdot FC.$$

If the airline decides to operate a HS network instead, then total costs are:

$$TC^{HSN} = 2 \cdot c \cdot (4 \cdot m)^\alpha + 2 \cdot FC = 2 \cdot c \cdot 4^\alpha \cdot m^\alpha + 2 \cdot FC.$$

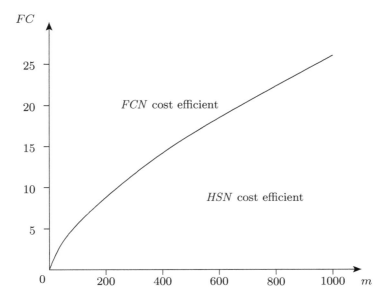

Fig. 13.5 The function describes the locus on which the total cost of operating the different networks are equal, given $\alpha = 3/2$ and $c = 1$. Below the graph is the area in which a HSN type would be cost efficient; above is the area in which a FCN would be cost-efficient

HS is more cost efficient for the airline than FC, if and only if:

$$TC^{HSN} < TC^{FCN} \Leftrightarrow m < \left(\frac{FC}{c \cdot (2 \cdot 4^\alpha - 3 \cdot 2^\alpha)} \right)^{\frac{1}{\alpha}} .$$

In the case of constant marginal costs, $\alpha = 1$, the above condition simplifies to

$$m < FC/(2c).$$

Hence, the relationship between the demand for a given route and the fixed-to-variable-costs ratio is crucial for the optimality of a network structure. Large fixed costs make it, *ceteris paribus*, more likely that a HSN is more cost efficient. This relation is depicted in Fig. 13.5.

Cost efficiency, however, is not the only factor an airline has to take into consideration when it optimizes its network for a given set of possible connections. From the point of view of a customer, it may make a difference whether she flies directly from Zurich to Copenhagen or whether she has to change planes in Frankfurt. Usually, the willingness to pay is lower in the latter case. The optimal network structure, therefore, reflects an optimal compromise between cost efficiency and the willingness to pay.

At the point in time at which the European market started to liberalize, experience from the liberalization of the US market (which went through a similar process more than a decade earlier) already suggested that a HSN is superior to a FCN and most European airlines adopted a HSN structure (Air France with a hub in Paris, Lufthansa with a hub in Frankfurt, KLM with a hub in Amsterdam,...). Swissair, however, resisted that trend and maintained three, large-scale airports in Zurich, Basel and Geneva. In fact, Swissair operated a network that was neither purely FCN nor HSN but, given the size of the Swiss market, it was closer to a FCN. While this decentralized structure did not really make the available network more attractive for travellers, because the geographic distance between the three cities is negligible by international standards and the local market is too small to justify such a network, it was (and is) costly to maintain. The additional costs were estimated to be in the high double-digit million Swiss Francs per year.

The decision to maintain three, comparably large, international airports in a small country like Switzerland was, to a great extent, political. Especially keeping the airport in Geneva was a political decision to manifest the equality between the French and German speaking parts of the country. However, political decisions that do not follow the logic of markets have their price and, in the case of Swissair, that price was substantial. The cost-inefficient network structure further reduced the airline's profits over a period of time when increased competition was already driving fares down and cutting back profits.

The analysis up until this point has shown that the effects of liberalization of the airline market are rather complex. First, opening a new route in a formerly protected market has the competitive effects analyzed before. Airlines will start opening new routes, if they can increase their profits by doing so, even if the overall effect is that industry profits go down. For this isolated effect, strategic alliances, mergers and acquisitions are forms of collusive behavior that can increase the airline, as well as the industry profits by coordinating the strategies of the airlines. Hence, the analysis predicts that there is a strong tendency to move in the direction of a more concentrated industry, but the overall welfare effects of concentration are unclear, because concentration allows the airlines to move closer in the direction of a monopolistic solution. Second, opening new routes may have positive network externalities, as well as cost-saving effects, if the extended network can be organized in a more cost efficient way. There are two ways to achieve such a goal: either by an extension of an airline's own network (internal growth), by the formation of strategic alliances, or by mergers and acquisitions (external growth). Both strategic options have their own advantages and disadvantages and it would be beyond the purpose of this case study to analyze them in detail.

13.4   How About Swissair?

The analysis of the last sections has shown that market liberalization was likely to erode profits by intensifying competition and that decreasing average costs could easily make airlines unprofitable. Internal, as well as external, growth strategies

Table 13.1 Main political, legal and economic factors that contributed to Swissair's problems

	Cause	Effect
Political factors	Inefficient network structure within Switzerland	Inefficient cost structure
Economic factors	Importance of fixed costs	Limits the number of competitors
	Increased competition on routes	Pressure on prices and quantities
Legal factors	Non-membership in EEA	Inefficient cost and route structure

(alliance formation, mergers and acquisitions) were key to managing profits during a period of time when the whole industry was likely to consolidate. It is, therefore, no surprise that the world's first and largest global alliance, Star Alliance, was founded in 1997.

European market liberalization was more of a challenge than an opportunity for Swissair, because the vote against the ratification of the EEA Treaty in 1992 implied that Switzerland had to renegotiate the restrictive, bilateral air service agreements with every single EU member-state. Additionally, equal access for Switzerland-based airlines to the EU market was granted only in combination with the Agreement on Free Movement of Persons, which was not fully in force before 2004. Table 13.1 gives an overview of the main factors that contributed to Swissair's demise.

Given the above arguments and, despite of the impediments that resulted from the non-membership in EEA, it seems straightforward that Swissair had its own growth strategy, the Hunter strategy, with the objective to reach a 20 percent market share in Europe. Its aim was to increase its market share by the acquisition of smaller airlines instead of by entering into alliance agreements. (In 1989, however, Swissair was the first European airline to seal a partnership agreement with the overseas carrier, Delta Airlines. Part of the arrangement was a mutual, 5% equity swap. One year later, a similar deal was made with Singapore Airlines.) These airlines created the so-called "Qualifier Group." Table 13.2 gives an overview of the acquisitions, as of 2000.

As can be seen, the Hunter strategy exclusively targeted airlines from smaller European countries like Belgium, Austria, Finland, Hungary, Poland, Portugal and Ireland and, thereby, bypassed the more important and mature markets in Italy, Germany and France.

The idea was to funnel traffic from the accessed markets through Zurich and Brussels to establish two principal hubs within Europe. One key problem with this strategy was, however, a lack of network externalities and cost synergies, because the different networks did not fit together well. On top of that, the strategy diluted the company's valuable brand, because of the lower quality standards of the acquired carriers (effectively only carriers that had been shunned by the other alliances). The dilution of the brand further undermined Swissair's ability to extract premium fares from its passengers.

Table 13.2 Holdings of Swissair as of 2000

Airline	Equity stakes
Air Europe	49.0
Volare Air	49.0
Air Littoral	49.0
Austrian Airlines	10.0
AOM France	49.5
Balair/CTA Leisure	100.0
Crossair	70.5
Cargolux	33.7
LOT Polish	37.6
LTU Group	49.9
Portugalia	42.0
South African Airways	20.0
Sabena	49.5
Ukraine International Airlines	5.6
TAP Air Portugal	34.0

13.5 Concluding Remarks

Economic analysis is not like a crime novel where, in the end, the detective manages to perfectly solve the case and to identify the culprit. In economics, there usually is no single culprit and the best an economist can hope for is to identify some of the more important contributing factors, which are related to the industry, regulatory and, ultimately, market context in which Swissair was embedded. Management failure may have played another important role, upon which economists can only speculate without further information. Nevertheless, as the above analysis has shown, management happens within a political, legal and regulatory, as well as economic context that, together, created a huge handicap for Swissair.

1. Increasing competition drove profits down and created a situation of cutthroat competition in which some carriers could not survive.
2. The decision not to ratify the EEA Treaty made it difficult for Swissair to get a foothold in the profitable EEA markets.
3. Growth strategies had to take this strategic disadvantage into consideration and Swissair had to acquire what was left on the market. The resulting network structure was far from optimal. Market share alone was not a good objective.

The purpose of this case study was not to develop a detailed analysis of the economics of aviation industry in general, or the insolvency of Swissair in particular, but to illustrate how different economic theories, combined with empirical facts about politics and law, can be used to better understand certain aspects of reality.

Mathematical Appendix

14

This chapter covers ...

- an introduction into functions with several variables.
- an introduction into linear equations.
- the concept of elasticities.

> Do not worry about your difficulties in Mathematics. I can assure you mine are still greater. (Albert Einstein)

> If I were again beginning my studies, I would follow the advice of Plato and start with mathematics. (Galileo Galilei)

> (1) Use mathematics as shorthand language, rather than as an engine of inquiry. (2) Keep to them till you have done. (3) Translate into English. (4) Then illustrate by examples that are important in real life. (5) Burn the mathematics. (6) If you can't succeed in 4, burn 3. This I do often. (Alfred Marshall)

The purpose of scientific theories is to develop hypotheses about causal relationships and to test them empirically. This is why the mathematical concept of a *function* is very important in both the natural and the social sciences. A function is a mapping from a set of explanatory variables onto a set of explained variables. One should know simple functions from high school: in order to define a function, it is usually assumed that a variable x, which is an element of some set X, and a variable y, which is an element of some set Y, exist and that y is related to x by some mapping $f : X \rightarrow Y$. Such a function is the easiest representation of a causal mechanism. If one states that $y = f(x)$, one means that some "state" y is caused by some "state" x and the function $f(.)$ represents this causal relationship between x and y. One calls x the *explanatory* and y the *explained* variable, because y is caused or "explained" by x via the function $f(.)$. Look at the following example: an individual demand function $x(p)$ assumes a relationship between a market price p and a quantity x that the consumer is willing to buy at this price. This is a causal relationship that is represented by the function $x(.)$ and for which the price, p, is the explanatory and the quantity, x, is the explained variable.

© Springer International Publishing AG 2017   319
M. Kolmar, *Principles of Microeconomics*, Springer Texts in Business and Economics,
DOI 10.1007/978-3-319-57589-6_14

The simple, one-explanatory–one-explained-variable function is convenient, but often too simplistic to appropriately cope with economic phenomena. In social systems there are usually several factors that causally determine some outcome. In the case of individual or market demand for some good, i, for example, it is not only the price of this good, p_i, that determines demand, but also the prices for other goods, as well as the income of the individual. With n goods, one would, therefore, have prices, $p_1, \ldots, p_i, \ldots, p_n$, and income, b, that explain demand, x_i, and one has to denote this by means of a demand function that depends on all these variables, $x_i(p_1, \ldots, p_i, \ldots, p_n, b)$. Otherwise, one would not be able to fully understand the causal mechanisms at work.

There are two important fields of application for functions that represent causal mechanisms. First, it might be important to understand how the change in one explanatory variable changes the explained variable because, in empirical tests, it is often possible to measure changes in some variables, but not their absolute values. In order to describe those changes one can use the concept of the partial derivative of a function. The next subchapter introduces and works with partial derivatives.

Second, there are important cases in which a causal system is described by several functions. In markets, for example, both supply, $y(p)$, and demand, $x(p)$, are of importance. Supply and demand are mappings from explanatory to explained variables. In such situations, it is a standard problem to analyze whether it is possible to find values of the explanatory variables that are consistent with some constraints on the explained variables. In the case of supply and demand, such a constraint is the condition that supply equals demand, $x(p) = y(p)$ (equilibrium). If one asks if a price exists such that supply equals demand, one asks, from a mathematical point of view, if a value p exists such that $x(p) - y(p) = 0$. In other words, one is looking for the root of the equation $x(p) - y(p)$. This will be done in the subchapter after the next.

Functions are rather abstract and complicated tools. In order to avoid complications, assume throughout this book that the domain, as well as the codomain, of all functions are the set of real numbers and that all functions are continuous and have no "kinks." Why this is important, as well as more general properties of functions, will be discussed in math class.

14.1 Functions with Several Explanatory Variables

This subchapter now leaves the demand and supply context behind to talk about functions more generally. Most people are familiar with the $y = f(x)$ notation of functions. (y no longer stands for supply, but for an arbitrary explained variable and x no longer stands for demand, but for an arbitrary explanatory variable, from now on.) For a function with only one explanatory variable, it is possible to use a very lean notation in order to be able to describe a change in the explained variable that is caused by a (small, infinitesimal) change in the explanatory variable: $f'(x)$. For example, the derivative of $f(x) = x^2$ is denoted as $f'(x) = 2 \cdot x$. There is nothing wrong with this notation, but it is not sufficiently precise, if one faces a problem with several explanatory variables. Assume that there are two explanatory variables x_1 and x_2, and denote by $y = f(x_1, x_2)$ the causal relationship. If one denotes derivatives as $f'(x_1, x_2)$ one cannot distinguish between changes in x_1 or x_2. One, therefore, has to introduce a way to denote derivatives that solves this problem. In principle, there are several ways to do so. For example, one could use the notation $f^1(x_1, x_2)$, $f^2(x_1, x_2)$ for the derivatives with respect to x_1 and x_2. However, this is not the usual convention.

Let $x_1, \ldots, x_n$ be the explanatory variables. One is interested in the changes of the function f evaluated at some point $a_1, \ldots, a_n$, which is caused by some infinitesimal change in x_i, holding all other explanatory variables constant (comparative statics). The most common notation for these so-called *partial derivatives* is given by:

$$\frac{\partial f(a_1, \ldots, a_n)}{\partial x_i}, i = 1, \ldots, n.$$

The notation $f(a_1, \ldots, a_n)$ reminds one that one is looking for the derivative of the function at a specific point $(a_1, \ldots, a_n)$. The "∂"-sign is pronounced as "del" and is reminiscent of the definition of partial derivatives by means of the difference coefficient,

$$\frac{\partial f(a_1, \ldots, a_n)}{\partial x_i} = \lim_{dx_i \to 0} \frac{\overbrace{f(a_1, \ldots, a_i + dx_i, \ldots, a_n) - f(a_1, \ldots, a_n)}^{=df(a_1, \ldots, a_n)}}{dx_i},$$

$i = 1, \ldots, n.$

The notation "d" represents a discrete change in x_i and $f(.)$, respectively, and ∂ indicates the limit of this change, if dx_i becomes arbitrarily small (converges to zero).

In order to be able to work with partial derivatives, one has to generalize the rules of differentiation. Here are the most important ones:

Additive functions Let $f(x_1, \ldots, x_n) = g(x_1, \ldots, x_n) + h(x_1, \ldots, x_n)$; then

$$\frac{\partial f(a_1, \ldots, a_n)}{\partial x_i} = \frac{\partial g(a_1, \ldots, a_n)}{\partial x_i} + \frac{\partial h(a_1, \ldots, a_n)}{\partial x_i},$$

$i = 1, \ldots, n.$

Product rule Let $f(x_1, \ldots, x_n) = g(x_1, \ldots, x_n) \cdot h(x_1, \ldots, x_n)$; then

$$\frac{\partial f(a_1, \ldots, a_n)}{\partial x_i} = \frac{\partial g(a_1, \ldots, a_n)}{\partial x_i} \cdot h(a_1, \ldots, a_n) + g(a_1, \ldots, a_n) \cdot \frac{\partial h(a_1, \ldots, a_n)}{\partial x_i},$$

$i = 1, \ldots, n.$

Quotient rule Let $f(x_1, \ldots, x_n) = g(x_1, \ldots, x_n)/h(x_1, \ldots, x_n)$; then

$$\frac{\partial f(a_1, \ldots, a_n)}{\partial x_i}$$

$$= \frac{\dfrac{\partial g(a_1, \ldots, a_n)}{\partial x_i} \cdot h(a_1, \ldots, a_n) - g(a_1, \ldots, a_n) \cdot \dfrac{\partial h(a_1, \ldots, a_n)}{\partial x_i}}{(h(a_1, \ldots, a_n))^2},$$

$i = 1, \ldots, n.$

Chain rule For a number of scientific problems, the causal chain between explanatory and explained variables is more complex, because the effect of some explanatory on the explained variable is mediated by some "intermediate" variable. For example, it could be that some variable, x_i, has an influence on the intermediary variable z, $z = g(x_i)$, and z has an influence on y, $y = \tilde{f}(x_1, \ldots x_{i-1}, z, x_{i+1}, x_n)$. (For simplicity, assume that there is no direct effect of x_i on y, which will generalize the analysis in the next section. One calls this function $\tilde{f}(.)$, because it is a function of z and one has to be able to distinguish it from $f(.)$, which is a function of x_i. One can denote this structure as $y = f(x_1, \ldots x_{i-1}, x_i, x_{i+1}, x_n) = \tilde{f}(x_1, \ldots x_{i-1}, g(x_i), x_{i+1}, x_n)$.

The individual demand function can be used as an example. One has assumed that individual demand is a function of prices and income, b. If one further assumes that income is, itself, determined by some other factors, like qualification, then one gets a chain of causal effects: qualification determines income and income determines demand.

In a situation like this, one gets the following rule for the differentiation of $f(.)$ with respect to x_i:

$$\frac{\partial f(a_1, \ldots, a_n)}{\partial x_i} = \frac{\partial \tilde{f}(a_1, \ldots, a_n)}{\partial z} \cdot \frac{\partial g(a_i)}{\partial x_i}.$$

The above expression is intuitive: x_i has an influence on z. This effect is captured by the second term of the product. The induced change in z, in turn, influences y. This is captured by the first term.

If x_i has an additional direct effect on y, one gets a function $y = \tilde{f}(x_1, \ldots x_{i-1}, x_i, x_{i+1}, x_n, z)$. The derivative with respect to x_i must, therefore, also include this direct effect:

$$\frac{\partial f(a_1, \ldots, a_n)}{\partial x_i} = \frac{\partial \tilde{f}(a_1, \ldots, a_n, z)}{\partial x_i} + \frac{\partial \tilde{f}(a_1, \ldots, a_n, z)}{\partial z} \cdot \frac{\partial g(a_i)}{\partial x_i}.$$

A frequent application of partial derivatives is to estimate the effect of a discrete change, or simultaneous changes, in the explanatory variables on the explained variable (for example, because only discrete changes can be measured empirically). This can be done by means of the total differential.

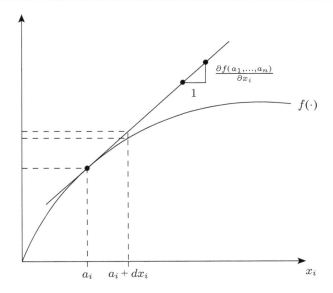

Fig. 14.1 Linear approximation of a function at a point a_i

Total differential Take $f(x_1, \ldots, x_n)$ and consider a simultaneous change in the explanatory variables dx_i. Then, the total effect is given as:

$$df(a_1, \ldots, a_n) = \frac{\partial f(a_1, \ldots, a_n)}{\partial x_1} dx_1 + \ldots + \frac{\partial f(a_1, \ldots, a_n)}{\partial x_n} dx_n.$$

In order to understand this expression, assume that all changes are zero except for x_i. Then, the total differential simplifies to:

$$df(a_1, \ldots, a_n) = \frac{\partial f(a_1, \ldots, a_n)}{\partial x_i} dx_i.$$

The right-hand side is a *linear* function of x_i, because the partial derivative is evaluated at a given point $a_1, \ldots, a_n$. However, this means that one can estimate the effect of an explanatory variable on y by means of a linear approximation, which is sometimes also called the *linear form*. Figure 14.1 illustrates this method.

Graphically speaking, the slope of the tangent line is equal to the partial derivative of the function at a given point. As can be seen, for discrete changes in x_i there is a gap between the true effect on y and the effect that is measured by the linear approximation: the linear approximation overestimates the true effect, in this example. However, if dx_i becomes very small, the "error" becomes arbitrarily small and vanishes in the limit for an infinitesimal change in x_i. One of the reasons why linear approximations are popular is that linear systems can be analyzed by means of linear algebra, which is powerful and simplifies the analysis considerably.

The following will reveal how the above rules can be used to determine derivatives of specific functions.

Example 1 Let $f(x_1, x_2) = x_1^2 + x_2$; then

$$\frac{\partial f(x_1, x_2)}{\partial x_1} = 2 \cdot x_1, \quad \frac{\partial f(x_1, x_2)}{\partial x_2} = 1.$$

The additive structure of the function implies that the different variables do not influence each other. As a consequence, the partial derivatives are independent of the other variable.

Example 2 Let $f(x_1, x_2) = x_1^2 \cdot x_2$; then

$$\frac{\partial f(x_1, x_2)}{\partial x_1} = 2 \cdot x_1 \cdot x_2, \quad \frac{\partial f(x_1, x_2)}{\partial x_2} = x_1^2.$$

The multiplicative structure implies that the partial derivatives, with respect to one variable, also depend on the other variable. However, in order to determine the derivative, the other variable can be treated as a number, because it is, in fact, a number, given that the partial derivative is an exercise in comparative statics (which means that all other variables are treated as constants).

Example 3 Let $f(x_1, x_2) = x_1^2 / x_2$; then

$$\frac{\partial f(x_1, x_2)}{\partial x_1} = \frac{2 \cdot x_1 \cdot x_2 - x_1^2 \cdot 0}{(x_2)^2} = \frac{2 \cdot x_1}{x_2},$$

because of the quotient rule.

All other rules that one has learned in school remain applicable to this generalized problem. If, for example, the problem is to determine the derivative of $f(x) = 10 \cdot \ln[x]$, with respect to x, it follows that $f'(x) = 10/x$. One can use this function to generalize the rules in the direction of functions with more than one variable. In order to do so, recognize that the above function has multiple variables already, because it is a function of x as well as 10, $f(x, 10)$, because 10 influences the result. Now, assume that one is not only interested in the partial derivative of this function at 10, but also at $9, 11, \ldots$. In this case, one can either determine the derivative for each case separately, or one can replace the specific number 10 by a dummy variable. If one redefines x by x_1 and calls the dummy variable x_2, one ends up with a new function $f(x_1, x_2) = x_2 \cdot \ln[x_1]$. However, now one has crossed the border from standard to *multivariate analysis*. The partial derivative of this function, with respect to x_1, can now be determined:

$$\frac{\partial f(x_1, x_2)}{\partial x_1} = \frac{x_2}{x_1}.$$

It is time for a quick plausibility check: if $x_2 = 10$, one gets $10/x_1$, which is reassuring. What one sees, from this example, is that one treats all the explanatory variables that stay constant in the same way as one has always treated numbers and the reason is that they are, in fact, numbers. The only difference is that they are written in an abstract way. In addition, one can, of course, also analyze the effect of a change of x_2 on y:

$$\frac{\partial f(x_1, x_2)}{\partial x_2} = \ln[x_1].$$

14.2   Solution to Systems of Equations

Economists are interested in equilibria, because they tell them something about the logical consistency of the assumptions of a model. As already stated, an equilibrium exists, if there is a price such that supply equals demand. Supply and demand, however, are both functions, which implies that the previous chapters have implicitly talked about a property of mathematical objects (functions). If $x(p)$ and $y(p)$ are the market-demand and market-supply functions, an equilibrium is a price p^*, such that $x(p^*) = y(p^*)$. One can, alternatively, rearrange this condition to get $x(p^*) - y(p^*) = 0$: excess demand has to be equal to zero. If one looks at the problem from this perspective, one can see that the economic problem of the existence of an equilibrium is equivalent to the mathematical problem of the existence of a root of a function, the excess demand function $ED(p) := x(p) - y(p)$.

Most students will have touched the problem of the existence of roots in high school: a function has a maximum or minimum, if its first derivative is zero. The intermediate-value theorem is useful, in this respect, because it specifies sufficient conditions that guarantee the existence of a root of a function $ED(p)$: $ED(.)$'s domain has to be closed, $ED(.)$ has to be continuous and at least two prices, p and p', exist, such that $ED(p) < 0 < ED(p')$.

In order to be able to analyze problems like the one above, one needs a little knowledge about how to solve functions. The above problem is very simple, because it only has one equation in one explanatory variable: $ED(p) = 0$. In a number of more realistic situations, the problem is more complex, however. Assume, for example, that there is not one, but two markets, with goods 1 and 2, and one wants to know if prices exist that equilibrate both markets simultaneously. The mathematical problem becomes:

$$ED_1(p_1, p_2) = 0 \wedge ED_2(p_1, p_2) = 0,$$

with $ED_1(.)$, $ED_2(.)$ being the excess-demand functions for both markets, which are functions of both prices, p_1 and p_2. The mathematical problem is to find a solution to a system of two equations and two unknowns.

In reality, there are many more goods and services that are simultaneously traded in markets, such that one has to specify n markets with excess-demand functions and an equilibrium exists, if the system of n equations in n unknowns has a solution. This is a rather involved problem, which is why I restrict my attention to, at most, two equations and two unknowns and one also restricts one's attention to linear

functions, most of the time in this book. Here, I denote the explanatory variables
by x_1, x_2, the explained variables by y_1, y_2 and the causal mechanisms by $y_1 = f_1(x_1, x_2)$, $y_2 = f_2(x_1, x_2)$.

Assume that one has to identify a pair of explanatory variables, x_1^* and x_2^*, that
set both functions equal to zero, $f_1(x_1^*, x_2^*) = 0 \wedge f_2(x_1^*, x_2^*) = 0$. As can be
conjectured from the intermediate-value theorem, it is not guaranteed that such a
solution exists for general functions. However, if both equations are linear, one can
use methods from *linear algebra* to identify the solution. Let

$$f_1(x_1, x_2) = a_1 + b_1 \cdot x_1 + c_1 \cdot x_2, \quad f_2(x_1, x_2) = a_2 + b_2 \cdot x_1 + c_2 \cdot x_2$$

be a linear system of equation with $a_1, b_1, c_1, a_2, b_2, c_2$ as the exogenous parameters
of the equations. (a_1, a_2) are the intercepts and the other parameters measure the
respective slopes. The problem of finding a zero is then given as:

$$a_1 + b_1 \cdot x_1^* + c_1 \cdot x_2^* = 0 \wedge a_2 + b_2 \cdot x_1^* + c_2 \cdot x_2^* = 0.$$

This problem has a unique solution, if the two equations are not parallel:

$$x_1^* = \frac{a_1 \cdot c_2 - a_2 \cdot c_1}{b_2 \cdot c_1 - b_1 \cdot c_2}, \quad x_2^* = \frac{a_1 \cdot b_2 - a_2 \cdot b_1}{b_2 \cdot c_1 - b_1 \cdot c_2}.$$

These formulas give one the general solution to the problem. In order to make sure
that the denominator does not become zero, one has to, in addition, assume that
$b_2 \cdot c_1 - b_1 \cdot c_2 = 0$ is excluded. If one inserts specific numbers, one can see what
the general solution implies.

One can calculate the above solution with a little effort by, for example, solving
the first equation for x_1, which yields $x_1 = -a_1/b_1 - c_1/b_1 x_2$. This equation
is an intermediate step that can be used to eliminate x_1 in the second equation,
$a_2 + b_2 \cdot (-a_1/b_1 - c_1/b_1 x_2) + c_2 \cdot x_2 = 0$. Now, one is left with only one equation
with one unknown variable that can be solved for x_2.

This approach comes to an end, if one is confronted with a problem with more
than two variables and unknowns. In such a case, one can use techniques from
matrix algebra to characterize a solution.

Another problem may exist, if the equations are not linear. It would be far beyond
the scope of this textbook to dig deeper into the solution of systems of nonlinear
equations.

14.3 Elasticities

The measurement and comparison of changes is very important in economics and
market research. So-called *elasticities* are a bread-and-butter concept with which
everyone should be familiar. This subchapter will introduce the problems to which
elasticities provide an answer and introduce the concept formally.

Assume one wants to know how demand $x(p)$ reacts to price changes. To be more specific, I will analyze the demand for bread and will assume that the demand function is linear, $x(p) = 100 - p$. Additionally, the price is in Swiss Francs and the quantity is in kilos.

An obvious candidate for the measurement of the effect of price changes is the partial derivative of the demand function:

$$\frac{dx}{dp} = x'(p) = -1.$$

This finding has a very straightforward interpretation: an increase in the price of bread by one Swiss Franc reduces the demand by one kilo.

This is a perfectly reasonable and informative statement and one could leave it at that. However, it has one disadvantage that limits its usefulness in practice: the instrument depends on the units in which one measures the dependent, as well as the independent, variable. Why is this a problem? Assume that one measures bread in grams instead of kilos. In this case, the demand function would be $x(p) = 100,000 - 1,000 \cdot p$ and the partial derivative becomes:

$$\frac{dx}{dp} = x'(p) = -1,000.$$

This is, again, a perfectly reasonable number: an increase in the price of bread by one Swiss Franc reduces the demand by 1,000 grams. However, without knowing the units of measurement, one cannot compare the two numbers and, at first glance, one could conclude that they are referring to completely different markets.

The same thing happens if one measures the price in Rappen instead of Franks. The demand function becomes $x(p) = 100 - 0.01 \cdot p$, and the first derivative is

$$\frac{dx}{dp} = x'(p) = -0.01:$$

an increase in the price of bread by 1 Rappen reduces the demand for bread by 0.01 kilos (or 10 grams).

This dependence on the units of measurement also limits the usefulness of the instrument, because it makes it difficult to compare changes between countries that use different currencies. However, it is a potentially interesting question to ask if Swiss customers react more or less strongly to price changes than, for example, the French customers. Nevertheless, even within a country, it may be interesting to understand if the demand for bread reacts more or less strongly to price changes than does the demand for smartphones and it is very hard to make the units of measurement for these two products commensurable.

This is why economists use a measure that is independent of the units of measurement. The basic idea is to focus on relative instead of absolute changes. The absolute change in demand is given by dx and the relative change can be constructed by dividing the absolute change by some reference level x^r:

$$\text{relative change in demand} = \frac{\text{absolute change in demand}}{\text{reference level of demand}} = \frac{dx}{x^r} = \frac{x - x^r}{x^r}.$$

The same can be done for price changes. Let dp be the price change and p^r the reference price, one gets:

$$\text{relative change in price} = \frac{\text{absolute change in price}}{\text{reference level of price}} = \frac{dp}{p^r} = \frac{p - p^r}{p^r}.$$

The relative changes are independent of the units of measurement, because they cancel out: if the numerator is measured in, for example, kilos or Swiss Francs, the denominator is measured in kilos or Swiss Francs, as well. Relative changes can be transformed into percentage changes, by multiplying them by 100.

Now that the units of measurement have been eliminated, one can come back to the initial question of how to measure changes in demand that are caused by changes in prices. An *elasticity* relates the relative change of one variable (demand) to the relative change in another variable (price):

$$\text{price elasticity of demand} = \frac{\text{relative change in demand}}{\text{relative change in price}}$$

or, more formally:

$$\epsilon_p^x = \frac{dx/x}{dp/p} = \frac{dx}{dp}\frac{p}{x}.$$

This elasticity is called the *price elasticity of demand* and it measures the percentage change in demand that is caused by a 1% change in the price.

If one allows for infinitesimal changes in prices, one can use partial derivatives to characterize elasticities:

$$\epsilon_p^x = \frac{dx/x}{dp/p} = \frac{dx}{dp}\frac{p}{x} = \frac{\partial x}{\partial p}\frac{p}{x}.$$

The elasticity one gets for infinitesimal changes is also called *point elasticity*.

This determines one important elasticity, but the concept can also be used to determine changes in demand that are caused by changes in other explanatory variables, as well: for example, income levels or prices of other goods. Definitions 14.1–14.3 cover the most commonly used elasticities of demand. The following notation is used: the demand for good i is a function of the price of good i, p_i, as well as of the prices of other goods j, p_j, as well as income b.

▶ **Definition 14.1, Price elasticity of demand** The price elasticity of demand measures the percentage change in the demand for good i that is caused by a 1% change in the price of good i:

$$\epsilon_{p_i}^{x_i} = \frac{dx_i/x_i}{dp_i/p_i} = \frac{dx_i}{dp_i}\frac{p_i}{x_i} = \frac{\partial x_i}{\partial p_i}\frac{p_i}{x_i}.$$

▶ **Definition 14.2, Cross-price elasticity of demand** The cross-price elasticity of demand measures the percentage change in the demand for good i that is caused by a 1% change in the price of good j:

$$\epsilon_{p_j}^{x_i} = \frac{dx_i/x_i}{dp_j/p_j} = \frac{dx_i}{dp_j}\frac{p_j}{x_i} = \frac{\partial x_i}{\partial p_j}\frac{p_j}{x_i}.$$

▶ **Definition 14.3, Income elasticity of demand** The income elasticity of demand measures the percentage change in the demand for good i that is caused by a 1% change in income:

$$\epsilon_{b}^{x_i} = \frac{dx_i/x_i}{db/b} = \frac{dx_i}{db}\frac{b}{x_i} = \frac{\partial x_i}{\partial b}\frac{b}{x_i}.$$

The same type of question can also be asked for changes in supply. I will focus on the most commonly used elasticities in the following definitions. Assume that supply y_i is a function of the price of the good p_i and of wages w and interest rates r.

▶ **Definition 14.4, Price elasticity of supply** The price elasticity of supply measures the percentage change in the supply of good i that is caused by a 1% change in its price:

$$\epsilon_{p_i}^{y_i} = \frac{dy_i/y_i}{dp_i/p_i} = \frac{dy_i}{dp_i}\frac{p_i}{y_i} = \frac{\partial y_i}{\partial p_i}\frac{p_i}{y_i}.$$

▶ **Definition 14.5, Wage elasticity of supply** The wage elasticity of supply measures the percentage change in the supply of good i that is caused by a 1% change in the wage level:

$$\epsilon_{w}^{y_i} = \frac{dy_i/y_i}{dw/w} = \frac{dy_i}{dw}\frac{w}{y_i} = \frac{\partial y_i}{\partial w}\frac{w}{y_i}.$$

▶ **Definition 14.6, Interest elasticity of supply** The interest elasticity of supply measures the percentage change in the supply of good i that is caused by a 1% change in the interest rate:

$$\epsilon_{r}^{y_i} = \frac{dy_i/y_i}{dr/r} = \frac{dy_i}{dr}\frac{r}{y_i} = \frac{\partial y_i}{\partial r}\frac{r}{y_i}.$$

Elasticities can be positive or negative. Economists usually use the convention to talk about elasticities in absolute values (i.e. the modulus of the function), unless this is misleading. This convention allows them to use the following qualitative categories (expressed in absolute terms):

▶ **Definition 14.7, Elastic reaction**   A variable reacts elastically to a change in some other variable, if the elasticity is larger than 1.

▶ **Definition 14.8, Inelastic reaction**   A variable reacts inelastically to a change in some other variable, if the elasticity is smaller than 1.

▶ **Definition 14.9, Isoelastic reaction**   A variable reacts isoelastically to a change in some other variable, if the elasticity is equal to 1.

Note that these properties are local measures. A function can be elastic at one point and inelastic at some other point.

Index

A

Absolute advantage, 27
Action profile, 274
Additive functions, 321
Advertising, 114, 223, 231
Affectual forecasting, 185
Aggregate level, 7
Alienation, 37
Allocation, 86
Altruism, 8
Anchoring, 95, 237
Anthropocentric, 110, 121
Antibiotics, 121
Arbitrage, 80, 235, 247
Asymmetric information, 117, 238
Autonomous vehicles, 80
Autonomy, 4, 38, 85
Average costs, 194, 210, 249, 307
Average costs, decreasing, 210, 225, 307
Average costs, minimum, 210, 218
Average fixed costs, 195
Average variable costs, 195

B

Backward induction, 275
Backward-induction equilibria, 276
Bailouts, 277
Bank run, 272
Barter economy, 47
Basic research, 139
Behavioral biases, 95, 237
Bertrand model, 283, 291
Bertrand(-Nash) equilibrium, 291, 296
Best response, 258, 287
Binary relation, 146
Biodiversity, 121
Bonus payments, 202, 248

Brand, 51, 222, 317
Brand image, 223
Brand name, 223
Break-even point, 184
Bretton-Woods system, 48
Bridge tolls, 138
Budget constraint, 159
Budget line, 159
Budget set, 159
Bundling, 243, 306
Buyer, 48, 55, 92, 117, 227

C

Capacity, 64, 112, 137, 292, 295, 306
Capital, 64, 103, 186, 218
Capital, human, 65
Capital, physical, 65
Capital, social, 65
Capital, symbolic, 65
Cash flow, 78
Centralized (government) planning, 40, 101, 115
Ceteris paribus, 14, 36, 64, 119, 152, 224, 231, 296, 315
Chain rule, 322
Cheapest cost avoider principle, 110
Chess, 276
Choice set, 40, 146
Circularity, 12
Civil law, 47
Class action, 107
Classical economics, 21
Climate change, 57, 118, 271
Club goods, 132, 137, 225
Coase Irrelevance Theorem, 115, 269
Cod, 134
Coercive power, 94

© Springer International Publishing AG 2017
M. Kolmar, *Principles of Microeconomics*, Springer Texts in Business and Economics,
DOI 10.1007/978-3-319-57589-6

Cold war as a game, 272
Collective bargaining, 51
Collude, 295
Collusion, 297
Collusive behavior, 295, 316
Commitment device, 283, 298
Commitment problems, 284, 298
Common goods, 132
Common knowledge, 261, 286
Commons problem, 135
Comparative advantage, 26, 73
Comparative statics, 60, 165, 321
Compensated demand, 180
Complements, (perfect), 61, 93, 152, 173
Completeness, 8, 147
Completeness of markets, 115
Concave, 32, 205
Congestion, 118
Consequentialism, 21, 84
Constrained optimization problem, 164
Consumer surplus, 88, 138, 238
Consumer-choice problem, 164
Consumption bundle, 31, 144, 150
Continuity, 72, 148
Contract law, 45, 122, 140, 219
Contract theory, 101, 189
Convex, 33, 150
Convexity, 148
Cooperation problems, 271, 298
Coordination problems, 271, 298
Corporate social responsibility, 124, 218
Cost equation, 188, 218
Cost function, 188, 205, 227, 286, 307
Cost-minimization problem, 193
Cost-plus pricing, 229, 306
Costs, 21, 183, 202, 228, 307
Cournot model, 283, 286, 290, 308
Cournot(-Nash) equilibrium, 289, 296
Critical rationalism, 17, 60, 165
Cross-price elasticity of demand, 223, 282,
     329
Cutthroat competition, 305

D
Data collection, 114
Decision nodes, 273
Decision theory, 7, 56, 143, 145
Declarations, 40
Deep ecology, 110, 123
Demand, 48, 56, 87, 107, 158, 223, 287, 320
Demand function, individual, 58, 170, 319
Demand function, inverse, 89, 228, 286
Demand function, market, 59, 89, 247, 292

Demand schedule, individual, 58
Demand schedule, market, 58
Deontology, 84
Descriptive statements, 19
Differentiated products, 223, 250
Division of labor, 20, 33, 83
Dogmatism, 12
Dominant strategy, 259
Dominant-strategy equilibrium, 263
Doomsday machines, 273
Double coincidence of want, 48
Duhem-Quine problem, 18
Dynamic pricing, 237

E
Effective altruism, 5
Efficiency in consumption, 86
Efficiency in production, 86, 189
Efficient, 86, 100, 189, 218, 236, 256, 291,
     299, 311
Ego depletion, 96
Elastic reaction, 330
Elasticities, 223, 246, 326
Emissions, 105
Empirically testable, 165
Endogenous variable, 57, 164
Engel curve, 170
Epistemology, 17, 41
Equilibrium, 69, 90, 111, 217, 232, 287, 311,
     325
Equilibrium concept, 260
European aviation industry, 305
Evolutionary biology, 9, 129, 215
Excess-demand function, 72, 325
Excludable, 132
Exclusion, 132
Exclusion costs, 130
Exclusive control over some necessary
     resource, 224
Existence of an equilibrium, 71, 173, 178, 325
Exogenous variable, 57
(Expected) utility-maximization hypothesis,
     258
Explained variables, 57, 163, 191, 319
Explanatory variables, 57, 163, 191, 288, 319
Exploitation, 35, 133
External effects, 104
External validity, 297
Externality, 104, 189, 202, 219, 243, 271, 298

F
Fact-value dichotomy, 20
Fairness, 8, 83, 110, 136

Falsification, 17
Financial literacy, 96
Financial markets, 78, 272
Financial reporting, 187
Fireworks, 139
Firm, 39, 55, 101, 143, 189, 202
Firm-as-production-function, 189
First theorem of welfare economics, 90, 100, 219
First-degree price discrimination, 226, 232
Fixed costs, 194, 209, 249, 268, 282, 307
Fixed costs, contractual, 194, 214
Fixed costs, technological, 194, 209, 210, 253
Flying Bank, 305
Focal points, 266
Frames of reference, 296
Freedom, 5, 36, 45, 95, 305
Fully connected network, 313
Future generations, 116

G
Gains from trade, 32, 51, 73, 90, 106, 209, 230
Gains from trade, material, 32
Gains from trade, subjective, 32
Game, 215, 256, 283
Game theory, 7, 255, 283
Game tree, 273
Games as structural metaphors, 271, 277, 298
General equilibrium, 36, 70
General-equilibrium theory, 72
Giffen goods, 60
Giffen paradox, 178
Globalization, 33, 125, 220
Good state, 85

H
Happiness paradox, 127
Head tax, 93
Hedonic pleasure, 156
Hedonic treadmill, 128
Hedonic well-being, 156
Historical school of economics, 22
Hold-up problem, 36
Homo Oeconomicus, 7, 149
Homogeneous of degree zero, 166
Homothetic utility function, 168
Hub-and-spoke network, 314

I
Imperfect foresight, 118
Imputed interest, 187
Incentive mechanism, 84, 202
Income effect, 181
Income elasticity of demand, 329

Income-consumption path, 170
Incomplete markets, 105
Incompleteness of contracts, 117
Indifference curve, 150
Indifference relation, 146
Individual level, 7
Inefficient, 86, 104, 220, 231, 268, 299
Inelastic reaction, 330
Inferior good, 61, 177
Infinite regress, 12
Input, 63, 103, 188, 207
Institutional economics, 22
Institutions, 3, 39, 73, 83, 100, 269
Instrumentalism, 15
Interaction level, 7
Interdependencies, 26, 80, 104, 189, 271
Interdependencies between generations, 116
Interdependent, 26, 79, 102
Interest elasticity of supply, 329
Intergenerational justice, 116
Internalization, 104
Internalizes, 102
Inverse function, 88, 191, 228
Invisible-hand game, 269
Isoelastic reaction, 330

L
Landsgemeinde, 120
Law and economics, 47, 100, 111
"Law" of supply and demand, 71
Legal system, 46, 107, 139
Legitimate use of force, 46
Lex Mercatoria, 46
Lexicographic preferences, 157
Liability law, 122
Libertarian political philosophy, 100
Lighthouse services, 114
Linear algebra, 323
London Congestion Charge, 119

M
Macroeconomics, 10, 272
Management accounting, 187
Management of common goods, 136
Mandatory standards, 122
Marginal costs, 195, 203, 221, 228, 286, 306
Marginal costs, constant, 208, 229, 282, 288, 291, 307
Marginal costs, decreasing, 208, 225
Marginal costs, increasing, 207, 282
Marginal product, 190, 207
Marginal rate of substitution, 151
Marginal revenues, 203, 221, 228, 286

"Marginal revenues = marginal costs" rule, 203, 221, 228, 286
Marginalist revolution, 21
Market design, 256
Market failure, 104, 231
Market for lemons, 117
Market integration, 34, 248
Markets, 22, 37, 39, 45
Markets as production technologies, 79
Markup, 229, 252, 289, 299
Marshallian demand functions, 165
Maximization, 8, 187, 292
Maximization, (expected) utility, 8, 162, 258
Maximization, preferences, 8, 160
Maximization, profit, 114, 201, 202, 218, 225, 228, 258, 283
Maximum sustainable yield, 133
Meat market, 123
Mental accounting, 215
Microeconomics, 10, 143
Micro-foundation of macroeconomics, 10
Military conflicts, 3, 277
Minimum state, 100
Mixed strategy, 264
Mixed strategy Nash equilibrium, 265
Model, 13, 49
Money, 4, 36, 41, 47, 271
Money as means of storing value, 48
Money as medium of exchange, 47
Money as unit of accounting, 48
Money illusion, 166
"Money-pump" argument, 147
Monopolistic competition, 51, 249
Monopoly, 49, 196, 222, 274, 282, 288, 312
Monotonicity, 148
Motives of action, 150
Multiple equilibria, 72, 178, 264, 266, 294
Multivariate analysis, 324
Münchhausen Trilemma, 12

N
Narrative fallacy, 185
Nash equilibrium, 260, 264, 274
Nash's theorem, 265
Natural monopolies, 139, 225
Natural rights, 84
Naturalistic fallacy, 19
Negative externality in consumption, 113
Negative externality in production, 111
Neoclassical economics, 13, 20, 166
Network choice, 313
Network externality, 225, 268
New institutional economics, 20, 22

Night-watchman state, 100, 117, 132
No distortion at the top, 241
Non-excludable, 132
Non-identity problem of intergenerational justice, 116
Non-rival in consumption, 130
Non-verifiable contracts, 117
Normal goods, 60
Normal-form game, 259
Normative theory, 18, 84, 87

O
Objective function, 143, 203
Ockham's razor, 14
Oligopoly, 49, 281
Ontology, 41
Opportunity cost, 5, 17, 28, 32, 111, 151, 160, 183, 187
Ordinal concept, 154
Ordinal preference ordering, 156
Ordinary goods, 60
Output, 52, 63, 188

P
Paradigm, 13, 21
Parallel imports, 248
Pareto dominance, 267
Pareto efficiency, 86, 90, 97, 113, 121, 157
Pareto improvement, 86, 267
Pareto principle, 205
Partial derivative, 321
Partial equilibrium, 69, 70
Patents, 52, 224
Perfect (first-degree) price discrimination, 226, 232
Perfect competition, 49, 55, 201
Physiological wants, 4
Player, 256, 257
Point elasticity, 328
Point of saturation, 152
Pollination services, 111
Polluter-pays principle, 109
Polypoly, 49
Position good, 126
Positive externality in consumption, 113
Positive externality in production, 111
Positive theory, 18, 72, 215
Preference ordering, 8, 148, 150, 153, 154, 156, 160, 168
Preferences, 8, 27, 56, 146, 165
Prescriptive statements, 19
Price competition, 291
Price effect, 180, 228

Price elasticity of demand, 231, 329
Price elasticity of supply, 329
Price fencing, 243
Price taker, 49
Price wars, 295
Price-consumption path, 170
"Price-equals-marginal-costs" rule, 203, 281, 298
Principal-agent theory, 189
Principle of minimal harm, 109
Prisoner's dilemma, 269, 284, 296
Private goods, 132
Procrastination, 147
Producer surplus, 90, 96, 210
Product rule, 321
Production function, 189
Production-possibility frontier, 16, 30, 32, 189
Property rights, 40, 45, 100, 104, 110, 125, 136
Proximate causes, 9
Public goods, 125, 132, 139
Public monopolies, 224
Public space, 120

Q

Quantity effect, 228
Quasi-concave, 168
Quotient rule, 322
QWERTY keyboard, 268

R

Rational, 7, 8, 13, 21, 96, 115, 149, 187, 215, 269, 277, 298
Rationality, bounded, 9, 237
Rationality, instrumental, 8, 149
Rationality, value-based, 8, 149, 150
Reach, 130
Reaction function, 259, 286
"Realistic" assumptions, 9, 15
Redistributive policies, 93
Reflective equilibrium, 84
Relative scarcity, 92
Representative democracy, 106
Resale (markets), 235, 247
Reservation price, 90
Residual control rights, 46
Revealed preferences, 8, 94, 99, 227
Revealed-preference approach, 8
Rights-based approach to markets, 47
Rival in consumption, 130

S

Satisficing, 187

Scarcity, 4, 9, 16, 26, 39, 86, 100, 102, 163, 298
Scarcity, objective, 4
Scarcity, subjective, 4
Scientific theory, 10, 12, 15, 17, 189, 319
Second theorem of welfare economics, 91, 236, 269
Second-degree price discrimination, 226, 238
Self-fulfilling, 78, 272
Selfishness, 8, 85, 128, 150
Self-selection constraint, 240
Seller, 45, 48, 55, 90, 117, 299
Sequential integration, 35, 79
Set of alternatives, 143, 146
Shared fantasies, 40
Short versus long run, 209
Simultaneous integration, 34
Single European Market, 247
Social comparisons, 126
Social convention, 41, 48
Social media, 225, 268
Social science, 11, 26, 283, 319
Specialization and trade, 27, 33, 36, 39, 73
Speech acts, 40
Stability of an equilibrium, 72
Status, 40, 60, 126, 245
Strategic interdependence, 50, 256, 281
Strategy, 257, 274, 286
Strategy profile, 257
Strict convexity, 149, 160, 163, 176, 178
Strict preference relation, 146
Structural landscape of economics, 298
Subgames, 275
Substitutes, (perfect), 61, 100, 152, 171, 222, 231, 286, 292
Sunk costs, 195, 214
Sunk-cost fallacy, 215
Supply, 48, 49, 56, 63, 69, 71, 143, 216, 286
Supply function, individual, 63, 213
Supply function, market, 63, 89
Supply schedule, 65
Sustainability, 121, 219
Swissair, 303

T

Taming the passions, 36
Tax, 3, 19, 60, 80, 93, 122, 125, 219, 284
Technology leader, 224
Technology of production, 52, 64, 189, 304
Terminal subgames, 275
Theory of the firm, 101, 190
Third-degree price discrimination, 226, 245
Three Cs of Economics, 298

Trademarks, 224
Transaction costs, 22, 102, 106, 115, 130, 132, 269
Transfer prices, 104, 115
Transitivity, 8, 147
"Truncated" demand function/"truncated" supply curve, 108, 230
Tuition fees, 137

U
Ultimate causes, 9
"Unified" approach to the Cournot/Bertrand problem, 295
Unintended consequences, 76
Uniqueness of an equilibrium, 72, 178
United Nations Convention on the Law of the Sea, 133
Utility function, 154, 168
Utility-maximization hypothesis, 258

V
Vaccination, 113

Value judgment, 12, 18, 150
Value judgment, basic, 19
Value judgment, nonbasic, 19
Variable costs, 195, 209
Verification, 17
Virtue ethics, 84

W
Wage elasticity of supply, 329
Weak preference relation, 146
Welfarism, 21, 84, 116, 123
Wellbeing, 84, 87
Williamson puzzle, 101
Willingness to pay, 87, 90, 94, 107, 226, 227, 238, 309
Willingness to sell, 90, 107, 143

Y
Yield management, 237

Z
Zero-profit equilibrium, 250